READING AND UNDERSTANDING MORE MULTIVARIATE STATISTICS

READING AND UNDERSTANDING MORE MULTIVARIATE STATISTICS

EDITED BY
LAURENCE G. GRIMM
AND PAUL R. YARNOLD

AMERICAN PSYCHOLOGICAL ASSOCIATION
WASHINGTON, DC

Copyright © 2000 by the American Psychological Association. All rights reserved. Except as permitted under the United States Copyright Act of 1976, no part of this publication may be reproduced or distributed in any form or by any means, or stored in a database or retrieval system, without the prior written permission of the publisher.

First Printing—June 2002

Published by
American Psychological Association
750 First Street, NE
Washington, DC 20002

Copies may be ordered from
APA Order Department
P.O. Box 92984
Washington, DC 20090-2984

In the U.K., Europe, Africa, and the Middle East, copies may be ordered from
American Psychological Association
3 Henrietta Street
Covent Garden, London
WC2E 8LU England

Typeset in Futura and New Baskerville by EPS Group Inc., Easton, MD

Printer: Goodway Graphics, Springfield, VA
Cover Designer: Kathleen Sims Graphic Design, Washington, DC
Technical/Production Editor: Amy J. Clarke

The opinions and statements published are the responsibility of the authors, and such opinions and statements do not necessarily represent the policies of the American Psychological Association.

Library of Congress Cataloging-in-Publication Data
Reading and understanding MORE multivariate statistics / edited by
Laurence G. Grimm and Paul R. Yarnold.
p. cm.
Includes bibliographical references and index.
ISBN 1-55798-698-3 (softcover : acid-free paper)
1. Multivariate analysis. 2. Psychometrics. I. Grimm, Laurence G.
II. Yarnold, Paul R.
QA2780.R32 2000
519.5′35—dc21 00-035556

British Library Cataloguing-in-Publication Data
A CIP record is available from the British Library.

Printed in the United States of America

Contents

Contributors

William C. Black, Louisiana State University, Baton Rouge
Fred B. Bryant, Loyola University of Chicago, IL
Laurence G. Grimm, University of Illinois at Chicago
Joseph F. Hair, Jr., Louisiana State University, Baton Rouge
David H. Henard, Texas A & M University, College Station
Laura Klem, University of Michigan, Ann Arbor
Michael J Strube, Washington University, St. Louis, MO
Bruce Thompson, Texas A & M University, College Station, and Baylor College of Medicine, Waco
Kevin P. Weinfurt, Duke University, Durham, NC
Raymond E. Wright, SPSS, Inc., Chicago, IL
Paul R. Yarnold, Northwestern University Medical School and University of Illinois at Chicago

Preface

Since the 1995 release of *Reading and Understanding Multivariate Statistics*, we have been gratified by the reception that it has enjoyed from audiences in psychology, business, education, and medicine. Because we received so many correspondences that asked "What about a chapter on multivariate technique X?" and "Might you consider tackling some foundational and cutting-edge topics in measurement?," we decided to edit another volume, *Reading and Understanding MORE Multivariate Statistics*. Like its predecessor, this volume does not portend to teach readers how to actually perform the statistical procedures discussed. Instead, the book accomplishes the same aims as the first volume: to present the fundamental conceptual aspects of common multivariate techniques and to acquaint the reader with the assumptions, statistical notations, and research contexts within which the analyses are often used. When we originally conceived the previous volume, we envisioned our audience primarily as consumers of research. We wanted to enable them to intelligently read and interpret the methodology and results sections of research articles. We were delighted to learn that instructors of graduate courses in statistics use the book as a supplement to standard statistics texts, as the groundwork before teaching students how to perform the actual analyses. We were also pleased to learn that active researchers use the book as an entree to an in-depth study of a multivariate analysis. Following the axiom "if it ain't broke, don't fix it," we endeavored to preserve the defining features of the preceding volume.

Each chapter is written to allow the reader to transfer what has been learned from the chapter to an understanding of published work that uses the analysis under discussion. Consequently, when encountering an article that uses one of the multivariate analyses included in this book, the reader will be able to understand why the author used the particular analysis, the meaning of the statistical symbols associated with the analysis, the assumptions of the analysis, the interpretation of summary tables and diagrams, and the substantive conclusions logically derived from the analysis.

The chapters are written for an audience with no formal exposure to multivariate statistics; a grasp of univariate statistics is essential, however. Concepts and symbols are presented with minimal reliance on formulas. The authors provide example applications of each statistical analysis, the underlying assumptions and mechanics of the analysis, and a discussion of an interesting working example. Some topics are more complex than others and, consequently, require greater persistence by the reader. With careful attention, the reader should be able to study the chapters and achieve an understanding of many multivariate procedures and approaches to measurement, without getting lost in the minutia of mathematical formulas and derivations that, mercifully, are not presented. Each chapter includes a glossary of terms and symbols that provides a quick and easy way to access succinct definitions of fundamental concepts discussed in the text.

In our preceding volume, the chapter authors explain the following statistical techniques in language accessible to the readers: multiple regression and correlation, path analysis, principle-components analysis and exploratory and confirmatory factor analysis, multidimensional scaling, analysis of cross-classified data, logistic regression, multivariate analysis of variance, discriminant analysis, and meta-analysis. In both volumes, we selected the statistical methods that our readers would most likely encounter in mainstream scientific journals. In this volume, the methods include cluster analysis, repeated measures analysis, and structural equation modeling. In both volumes, we also chose some cutting-edge methods that we believe will become popular in the future. In this volume, survival analysis is included because there is an increasing interest in the longer term outcomes of interventions in psychotherapy, health psychology, and medicine. Canonical correlation analysis appears in this volume because we believe that it is a much-neglected analysis, is more complex than many other multivariate analyses and thus in need of a relatively simple explication, and is bound to become more popular as researchers broaden their concept of the laboratory to include community settings and the wealth of measures used therein. Q-technique factor analysis is offered, in part, as a corrective. Often researchers are interested in types of people, yet psychological research, using factor analysis, tends to emphasize variables rather than people.

In a break with the tradition of the previous volume, we decided to include three chapters on measurement that do not strictly fall under the rubric of multivariate statistics. There are several reasons for including chapters on measurement. We have received numerous requests from readers to apply the same straightforward, understandable treatment of measurement topics that we apply to multivariate statistics. We believe one reason for such requests stems from the fact that measurement, as an independent topic, has been usurped by the standard, first-year graduate sequence of advanced univariate statistics, followed by an introduction to multivariate statistics. Furthermore, there are several technical reasons to discuss measurement issues in a book that addresses advanced statistics. For example, measurement is a fundamental component of confirmatory factor analysis and structural equation modeling, both of which are encountered frequently in a broad spectrum of academic disciplines. In addition, measurement must be a consideration when choosing a statistical model: The first issue to consider when choosing a model concerns the etiology of the measurement aspects underlying the data. Articles in journals that discuss measurement issues use multivariable statistics, and that is what the measurement chapters in this book are about: how to read and understand statistics used in that context. Of course, general topics applicable to all statistical analyses, such as basic levels of measurement (see chapter 1), reliability, validity, and generalizability theory (included in this volume) are also applicable to the topics addressed in our previous volume. We make no effort to provide a comprehensive treatment of measurement. Instead, we include two chapters, one on reliability and one on validity, that reflect cornerstones of behavioral research. A chapter on item response theory represents cutting-edge analytic technology that is just beginning to emerge in the fields of psychology and, particularly, education.

We encourage readers to send us comments on the book and to suggest topics that would be of interest in future volumes (Larry G. Grimm, lgrimm@uic.edu; Paul R. Yarnold, p-yarnold@nwu.edu).

Acknowledgements

Among the APA staff who have been most helpful to us in bringing our efforts to light for the second time around, Mary Lynn Skutley in Acquisitions and Development deserves our appreciation. Her supportive email messages throughout the project kept us motivated and focused, without engendering guilt when we missed deadlines. Peggy Schlegel, our development editor, was a treasure. For both volumes, Peggy displayed a level of knowledge of complex statistics beyond anything we would have ever hoped for.

We were also fortunate to have two external reviewers, Andy Conway and one who wished to remain anonymous. They provided substantive suggestions that were indispensable in the revision process.

Finally, our greatest gratitude goes to our contributing authors. It is unusual to single out one particular author, but we would be remiss in not mentioning Bruce Thompson. His contributions to this book have been substantial, and he saved the project by writing two additional chapters at the 11th hour. Thank you, Bruce.

READING AND UNDERSTANDING MORE MULTIVARIATE STATISTICS

Introduction to Multivariate Statistics

Laurence G. Grimm and Paul R. Yarnold

Multivariate statistics is a broad and somewhat ambiguous term that is applied to a collection of analytical techniques that provide for the simultaneous analysis of multiple measures. Researchers have been using multivariate and multivariable designs for decades, the former referring to multiple dependent variables and the latter to multiple independent variables with one dependent variable. Most multivariate techniques were developed in the first half of the 20th century, but they awaited the advent of computers, especially microcomputers and canned software packages, before they became accessible to most researchers. Until recently, the data from many multivariate designs were analyzed piecemeal with univariate analyses. For example, a between-groups design, with three dependent variables, led to three univariate *F* tests.

The increasing popularity of multivariate analyses has advantages and disadvantages. On the plus side, we do not live in a two-variable world, and most interesting relations are among sets of variables. The ability of multivariate analyses to handle the complexity of numerous variables must be tempered, however, by the thoughtful consideration of which variables to include in the study. Because software packages can easily handle extremely large data sets, one must resist the temptation to haphazardly throw in a multitude of measures to see what pops out as important. Ill-conceived research questions, unreliable measures, inadequate sample sizes, and low power are problems that cannot be circumvented by the most sophisticated analysis. Furthermore, because multivariate analyses realistically can only be conducted with computer programs, there is a tendency to let the designers of the program do the thinking. In addition, the ease of "point-and-click" software interfaces risks removing the researcher from the immediacy of the data. There is nothing inherent in software or multivariate analyses that automatically dictates this mindlessness, but there is an irony here. Multivariate analyses, and their necessary accompanying software, allow researchers to ask complex questions, but the ease of obtaining computer outputs can give rise to an illusion of scientific discovery because, after all, "the computer arrived at an answer." The admonition "garbage in, garbage out" is more true now than ever. Finally, just because multivariate analyses can answer complicated questions does not mean that we should always ask complicated questions. As concluded by Wilkinson and the APA Task Force on Statistical Inference (1999),

> although complex designs and state-of-the-art methods are sometimes necessary to address research questions effectively, simpler classical approaches often can provide elegant and sufficient answers to important questions. Do not choose an analytic method to impress your readers or to deflect criticism. If the assumptions and strength of a simpler method are reasonable for your data and research problem, use it. (p. 596)

Purpose of This Book

Reading and Understanding MORE Multivariate Statistics acquaints readers with an array of multivariate analyses and measurement topics in a way that helps them understand the research context, assumptions, notations, and interpretations associated with each. No attempt is made to teach readers how to perform the analysis under discussion. Rather, the authors present fundamental conceptual aspects of multivariate techniques and measurement topics, and explain these procedures in simple, intuitive terms, using as few equations as possible. The authors have also provided suggested readings for those who want to achieve a deeper understanding of the topic at hand. A glossary of terms and symbols is included at the end of each chapter for easy reference. The only assumption that the authors make is that readers have a grounding in univariate statistics.

For graduate students, this book is a useful companion to the many standard texts that teach the students to perform multivariate analyses in statistics courses as preparation for doing their own research. Some researchers may already use a particular multivariate analysis in their work (e.g., discriminant analysis) but may have little knowledge of others (e.g., cluster analysis).

The book is equally helpful to those who consume research involving multivariate statistics: clinicians, program consultants, educators, administrators, and a range of others who need to use behavioral and social sciences research results in their work. Many times, these individuals are tempted to skip over the Results section of research articles that use multivariate statistics, thereby missing important implications that the results may have in their area of interest.

As with our first volume, *Reading and Understanding Multivariate Statistics,* we selected topics based on the analyses that contemporary readers would most likely encounter in scientific journals, such as repeated measures analysis, cluster analysis, and structural equation modeling. We also chose topics that we predict will become increasingly popular in the future, such as survival analysis and Q-technique factor analysis. In addition, because measurement must be a consideration in choosing a statistical model, we have included chapters that cover reliability, validity, and generalizability theory. An additional chapter on item response theory is included because it represents cutting-edge analytic technology that is just beginning to emerge in many fields.

What Precisely Is Multivariate Analysis?

It is impossible to provide a precise definition of *multivariate analysis* because the term is not used consensually. Multivariate analysis takes on a different meaning depending on the research context and the scale of measurement used to define variables. Strictly speaking, however, multivariate statistics provide a simultaneous analysis of multiple independent and dependent variables (Tabachnick & Fidell, 1989). More generally, designs are considered multivariate if they involve two or more dependent measures.

Consider a comparison between one-way analysis of variance (ANOVA) and multivariate analysis of variance (MANOVA). Suppose a researcher has a design involving one independent variable and three dependent variables. Using the univariate ANOVA approach to test the difference between the group means, three F tests would be performed, one test for each dependent variable. The MANOVA procedure applied to the same data would yield one omnibus test statistic, not three. The simultaneous nature of the MANOVA refers to the fact that the three dependent variables are combined in such a way that a new variable, or a "linear composite" (see below), is established. If the resulting test statistic is significant, it means that the two groups differ with respect to the composite variable. The significant F value may be due to a difference among the means of any, or all, of the dependent variables. It may also be due to the combination of the dependent variables, however. Thus, it should not be surprising if, after a significant multivariate F statistic is obtained, none of the subsequent univariate F tests are significant. The concept of a linear composite is ubiquitous in multivariate analyses, and it is to that topic that we now turn.

Composite Variable

A *composite variable,* also called a *synthetic variable* or *variate,* is an empirically derived linear combination of two or more variables, with each variable assigned a weight by the multivariate technique. For example, imagine that on the basis of a statistical analysis a researcher creates a new variable, "predicted success in graduate school," that is defined as 0.71 times one's score on the GRE verbal examination plus 0.5 times one's score on the GRE quantitative examination. The new variable, predicted success in graduate school, is an example of a composite variable. The variables are chosen by the investigator, and the composite variable is determined by the analysis. In many ways, the variate is the very essence of a multivariate analysis because it is arrived at by simultaneously analyzing a set of observations. For example, in multiple regression, there is one variate, and it is the multiple correlation; in discriminate analysis, the variate is a discriminate function, and there are typically $k - 1$ variates (k = number of groups); in canonical correlation there are at least two variates, one for the set of dependent variables and one for the set of independent variables. The way in which a synthetic variable is interpreted depends on the particular analysis and, ultimately, on one's understanding of the single variables that are "synthesized." The variate is interpreted in its own right, albeit the degree to which each variable contributes to the composite is usually of interest.

Latent Variable

A *latent variable,* also called a *latent,* or *hypothetical construct, factor* (in the context of factor analysis) or *underlying dimension,* cannot be measured directly and is operationalized by one or more other measured variables, sometimes referred to as *(multiple) indicators* or *manifest variables.*

For example, the theoretical concept "anxiety" (the latent variable) might be operationalized in terms of a statistical function applied to one's scores on measures, including physiological responses, self-reports of apprehension, and behavioral observations. A latent variable can be anticipated prior to an analysis or explained after a statistical analysis. Not all multivariate analyses involve measurement of latent variables (e.g., cluster analysis), but for some it is the entire point of the analysis. For example, the numerous variations of factor analysis all share the common goal of reducing a larger set of variables to a smaller set, with each factor representing an underlying construct. In structural equation modeling (SEM), (confirmatory) factor analysis is embedded in the technique and comprises the measurement model aspect of SEM—that aspect of SEM responsible for tapping latent constructs, as opposed to the aspect of SEM that addresses the relation among constructs. In multidimensional scaling (MDS), preferences or similarity judgments are used to infer underlying dimensions. For instance, one might infer that respondents use the dimensions of cost and attractiveness in offering preferences for certain automobiles.

Scales of Measurement

Measurement is the assignment of numbers to objects or events according to predetermined rules. Because different assignment rules exist, the same number can have different meanings, depending on the rules used to assign the numbers. Most statistics books in the behavioral sciences use a scale typology provided by Stevens (1946), but there are other typologies (Velleman & Wilkinson, 1993). All of the contributors in this book use Stevens's terms, and we present a brief, somewhat oversimplified, summary of these concepts.

All measures can be categorized as either *metric* (quantitative) or *nonmetric* (qualitative). Under the nonmetric rubric are the nominal and ordinal scales. The assignment of numbers using a nominal scale is a labeling activity. When using a *nominal scale,* one cannot interpret the numbers as anything other than the names of things. If men are assigned the number 1 and women are assigned the number 2—a process called *dummy coding*—it does not mean that men are twice as likely as women. When two categories are used, the nominal measure is called *dichotomous*. When there are two or more levels of a nominal variable, researchers often use the term *categorical variable.* Common examples of categorical measures are ethnicity, religion, political affiliation, behavioral outcomes such as responded–did not respond, and membership in one or another fixed condition of an experiment. It may seem odd to include categorization as a form or level of measurement, but as Coombs (1953) stated, "this level of measurement is so primitive that it is not always characterized as measurement, *but it is a necessary condition for all higher levels of measurement*" (p. 473, emphasis in original).

Note that the difference between categories is one of kind rather than degree; this is a fundamental characteristic of the nominal scale of measurement. In addition, the members of a category are viewed as similar with respect to the criterion or criteria used in the categorization schema. Whether the investigator has used a good method of categorization can only be judged within a practical or theoretical context.

An *ordinal scale* shares a feature of the nominal scale in that observations are categorized; however, ordinal numbers have a particular relation to one another. Larger numbers represent a greater quantity of something than smaller numbers. The ordinal scale represents a rank ordering of some attribute. Even though ordinal scales are quantitative, there is only a limited sense in which quantity is implied. Rankings reflect more or less of something, but not how much more or less of something. One commonly used example of such a scale is known as a *categorical ordinal* measure, in which all of the (usually small number of) possible response categories are given a reference label. For instance, patients might be asked to rate their satisfaction with their psychotherapy sessions. The researcher uses a 5-point ordinal categorical scale, on which 1 = *completely dissatisfied,* 2 = *somewhat dissatisfied,* 3 = *indifference,* 4 = *somewhat satisfied,* and 5 = *completely satisfied.* Here, the magnitude of the difference between codes indicating indifference and complete satisfaction (i.e., $5 - 3 = 2$ units) would be considered as representing twice the difference between indifference and marginal satisfaction (i.e., $4 - 3 = 1$ unit) were the scale to be considered interval: a position that seems difficult to defend.

A third level of measurement is the interval scale. An *interval scale* possesses the qualities of the nominal scale in that different numbers represent different things and is also like the ordinal scale in that different numbers reflect more or less of something. In addition, however, the interval scale has the property that numerically equal distances on the scale represent equal distances on the dimension underlying the scale. The Fahrenheit scale is an interval scale. The difference between the temperature of 80°F and 85°F is the same amount of heat (measured in units of mercury) as the difference between 90°F and 95°F.

The distinction between an ordinal and an interval scale is not always easily made, especially in the behavioral sciences. For instance, is the interval between an IQ of 100 and 105 the same as the interval between an IQ of 45 and 50? The numerical distance between the scores is the same, but it is not the numerical distance between numbers that defines the difference between ordinal and interval scales. It is the underlying dimension that the scale is tapping that is important. Is the difference in amount of intelligence between an IQ score of 100 and 105 the same as the difference in the amount of intelligence between IQ scores of 45 and 50? Technically speaking, the type of scale is not a feature of the data. Calling a scale interval instead of ordinal is a statement about one's belief about the characteristics of the phenomenon that is being measured.

A *ratio scale* possesses all of the properties of an interval scale, with the addition of a meaningful absolute zero point. Data collected using an interval or ratio scale are referred to as metric data. The Fahrenheit scale is not a ratio scale because zero is not the complete absence of heat. A measure of length is a ratio scale because there is an absolute zero point (i.e., a point of no length). With an absolute zero point, any number on a ratio scale can be used to establish its standing relative to any other number on the scale; in addition, a given number also represents an absolute amount of something. One characteristic of a ratio scale is the fact that a number that is mathematically twice as large as another number represents twice as much of a thing that is being measured. IQ, for example, does not possess this characteristic because someone with an IQ of 100 is not twice as smart as someone with an IQ of 50. However, because the underlying dimension of height has an absolute zero point, a person with a height of 80 inches is twice as tall as someone with a height of 40 inches. Time is another variable that has an absolute zero point. Therefore, one participant's reaction time of 0.50 seconds is twice as fast as another participant's reaction time of 0.25 seconds. The number of errors, or problems solved, are two additional examples of ratio scales.

In the behavioral sciences, the kinds of variables that lend themselves to ratio scales are less common than variables measured on an ordinal or interval scale. No measures of achievement, aptitude, personality traits, or psychopathology have a meaningful absolute zero point.

S. S. Stevens (1946, 1951) offered a typology of measurement scales that has functioned as a dominating schema for the selection of which analyses can be performed on what scale of measurement. It did not take long for authors of statistics textbooks to adopt this typology (e.g., Siegel, 1956), and today it is common to see textbooks provide decision trees that begin with "what scale of measurement?" and branch to the "allowable" analytic technique. It also did not take long for the debate to begin (Lord, 1953). It is beyond the scope of this chapter to do justice to the complex mathematical and philosophical issues that have arisen around measurement issues; the interested reader can receive a brief orientation to the topic by reading Velleman and Wilkinson's (1993) *Nominal, Ordinal, Interval, and Ratio Typologies Are Misleading.* A central issue in the debate is succinctly summarized by Pedhazur and Schmelkin (1991): "The major source of the controversy regarding measurement and statistics in sociobehavioral research is whether most of the measures used are on an ordinal or an interval level" (p. 28). In the strong form of Stevens's position, only nonparametric analyses can be used with ordinal data. Many statisticians, however, would agree with Labovitz (1970; see also Zimmerman & Zumbo, 1993; and Wilson, 1971, for a contrary view):

> Empirical evidence supports the treatment of ordinal variables *as if* they conform to interval scales. . . . Although some small error may accompany the treatment of ordinal variables as interval, this is offset by the use of more powerful, more sensitive, better developed, and more clearly interpretable statistics with known sampling error. (p. 515, emphasis in original)

Statisticians will continue to debate the appropriate correspondence between levels of measurement and methods of data analysis, but all would agree that more attention needs to be paid by behavioral scientists to increasing the validity, reliability, and sensitivity of our measurements.

Chapters in This Book

The following chapters in this text discuss measurement topics and a range of frequently used statistical procedures. Why are there different types of multivariate analyses? Because each multivariate procedure is appropriate for a different type of research question; each analysis has its own objective purpose. In addition, however, in deciding which multivariate procedure is appropriate for the analysis of one's data, it is important to consider the measurement scale of each of the variables in the research design.

The remainder of this chapter provides an admittedly oversimplified summary of the chapters in the book. Our primary purpose here is to briefly introduce each measurement topic and each analysis in the book, with special attention to the objective or purpose of the statistical procedure and the research context in which it is used.

Reliability and Generalizability Theory

Broadly speaking, *reliability* refers to the consistency of observations or measures. There are numerous ways in which observations can be consistent, and this gives rise to different types of reliability. For example, the term *interrater reliability* refers to the extent of agreement among observers, a concept often used in psychiatric diagnosis. A test designed to measure a construct that theoretically should be constant over time (e.g., extraversion) should yield the same score on repeated measurements, called *test–retest reliability*. When a test is constructed to measure one theoretical construct, we expect all of the items of the test to tap the construct. There should be consistency across the items because they are all suppose to be measuring the same thing or aspects of the same thing. Thus, scale developers are interested in the internal consistency of the test, and Chronbach's alpha is a common statistic that reflects the degree to which the items "hang together." Reliability is a cornerstone of scientific research. Without reliable measures, there is no way of answering some of the most important questions in science, which generally have the form of what attributes or events are related and what is the nature of the relationships. In emphasizing the importance of the reliability of measurement, psychometricians have coined the phrase "reliability precedes validity."

The purpose of a method of assessment is to tap a concept, but it is impossible to do so without error. Classical test theory holds that an observed score is influenced by what it is suppose to measure, yielding a true score, and by influences extraneous to the underlying concept, called an *error score*. Hence, observed score = true score + error score ($x = t + e$). One of the major weaknesses of classical test theory is that it informs us as to the potential sources of error variance but does not offer a way in which one can measure these sources. As Strube states in chapter 2, "the heart of generalizability theory, then, is the estimation of the components of variance that make up an obtained score with particular interest in estimating the sources of error variance." The multiway analysis of variance (ANOVA) is suited to the task because it can provide separate estimates for sources of variance, specifically, sources of error variance. One obvious benefit of knowing what contributes to measurement error is that the investigator can take steps to minimize error and, thus, increase the precision of observations. Strube provides a comprehensive introduction to the topic of generalizability theory that allows the reader to understand articles that use this approach to measurement. His chapter serves as an initial orientation for the reader who wants to conduct research using this method.

Item Response Theory

The development of *item response theory* (IRT) models was motivated by the inherent limitations of methods based on classical test theory. In chapter 3, Henard provides an introduction to the most commonly used IRT-based model, known either as the *Rasch model* or as the *one-parameter IRT model*. The fundamental hypotheses of this model include (a) the performance of an examinee on a test item can be explained by a set of predictors known as *traits* (or *abilities*), and (b) the relationship between the examinee's item performance and the traits underlying item performance may be described by a monotonically increasing function known as the *item characteristic curve* (ICC). Given a sample of individuals and a sample of items purporting to assess a single underlying dimension (e.g., arithmetic ability), how may an investigator decide which items are useful for measuring arithmetic ability, how much ability is required to solve each item correctly, and the relative arithmetic ability of individuals who complete the battery of items?

This chapter describes how to prune items that fail to provide useful information (e.g., everyone in the sample receives a correct answer and the item is too easy, or everyone receives an incorrect answer and the item is too difficult) and individuals (e.g., those who answered all of the items correctly and those who answered all of the items incorrectly). The reader then learns how to calibrate both item difficulties (think of this as obtaining item-relative difficulty weights by transforming the items onto a scale reflecting their relative difficulty) and person abilities (think of this as obtaining person-relative ability weights by transforming the people onto a scale reflecting their relative ability). The chapter then describes final calibration and evaluation of the resulting data in light of the assumptions of the IRT model.

A benefit of the IRT approach is that discrepancies between the data and the model may be evaluated in terms of whether the problem lies at the item level (in which case the item may be eliminated), or at the person level (in which case a second trait other than arithmetic ability, such as guessing, might be required to explain the discrepancy). Of course, the chapter explains the role of two- and three-parameter IRT models when an unacceptable number of discrepancies between the model and the data occur. The end result of using IRT is an instrument that may involve substantially fewer items than the original instrument, and that also provides scores for individuals (and items) that more accurately reflect their ability (and difficulty) on the trait being assessed. The improved accuracy of measurement may return scores that provide enhanced information (i.e., more "signal" and less "noise") for subsequent analyses in which they serve as either dependent or independent variables, versus the original raw scores.

Assessing the Validity of Measurement

The topic of *validity* of measurement is complex and potentially confusing because there are so many types of validities, including content, criterion, predictive or prospective, concurrent, construct, retrospective or postdictive, convergent and discriminant, internal, external, ecological, differential, single group, face, and statistical conclusion validity. Clearly, then, the frequently asked question "Is this test valid?" is meaningless in the face of so many different kinds of validity. In chapter 4, Bryant brings order and clarity to this bewildering array of validities. His discussion of validities includes numerous research examples from the literature and the types of statistical analyses used to assess each form of validity. Moreover, the reader who would like to pursue the topic in greater depth will appreciate the 75 citations found in the chapter.

A major focus of Bryant's chapter is an extended discussion and detailed example of the multitrait–multimethod approach to establishing construct validity. Where applicable, Bryant emphasizes the multivariate analyses involved in establishing various types of validity. The reader who masters the material of this chapter, along with chapters 2 (generalizability theory) and 3 (IRT), will obtain a thorough, conceptual grounding in the traditional and cutting-edge methods for establishing reliability and validity of assessment instruments.

Cluster Analysis

The purpose of *cluster analysis* (CA) is to identify natural groupings among a collection of observations; it is a procedure for identifying cases with distinctive characteristics in heterogeneous samples and grouping them into homogeneous groups. Clusters are formed so that observations that are similar form the basis of a single grouping and, as a group, are differentiated from other clusters that are, themselves, comprised of homogeneous observations. For example, Mowbray, Bybee, and Cohen (1993) had several measures of functioning for 108 people who were homeless and mentally ill. The results of a cluster analysis yielded four groups: hostile–psychotic (35%), depressed (19%), best functioning (28%), and substance abusing (19%). These findings can be used by policy makers to target specific programs for subgroups of the homeless population.

CA is *not* a hypothesis-testing procedure. Investigators most often use CA as an exploratory technique to detect natural groupings among those observations included in the study. There is a great deal of subjectivity in the decision as to which clusters are interpreted and retained for further analyses. The decision is based on what the investigator regards as conceptually meaningful, and no statistical index can answer that question.

Researchers who use CA often combine it with other data analytic strategies. For example, there are times in which the observations used as the basis for clustering have redundant variables, that is, variables that are intercorrelated. Before CA is used, factor analysis can be used to reduce the number of observations that may later be used in CA. Once clusters are selected, the investigator may choose to use analysis of variance (ANOVA) to see if the clusters differ with respect to other variables that were not used in the CA. For example, suppose a clinical psychologist identifies three clusters of drug abusers: one cluster is interpreted as comprising individuals who seem to be using drugs to control emotional problems (i.e., self-medicating), a second cluster appears to identify those individuals who use drugs as part of an antisocial pattern of behavior, and the third cluster looks like it is comprised of people who use drugs for recreation. An ANOVA can be used to see if these three groups differ in, for example, success in an intervention program. Discriminant analysis could be used to select which other variables can be combined so as to achieve separation among the groups, with the clusters serving as the dependent variable. Alternatively, suppose a researcher wants to know the odds that someone will be arrested within 2 years of an evaluation. Logistic regression can be used with arrest–no arrest as the dependent variable and cluster membership as the predictor. In other words, once the clusters are formed, they can be used as independent or dependent variables in other analyses.

In chapter 5, Hair and Black provide a thorough introductory treatment of this increasingly popular statistical technique. Writing from the vantage point of market researchers, the authors offer a nonmathematical discussion of when to use CA, the analytic decisions required of the researcher, and how to interpret the results.

Q-Technique Factor Analysis

One of the most widely used statistical methodologies in psychology has involved factor analysis of sets of items, or variables, on which a sample of people have been measured. The analysis returns to the investigator a set of factors, or dimensions, that together explain a preponderance of the variance in responses to the items. In chapter 6, Thompson describes *Q-technique factor analysis,* which is conceptually similar to factor analysis of variables but which instead involves factor analysis of people. This methodology allows an investigator to determine, for a given sample of people assessed on a given set of items, how many types (factors) there are, whether the people expected by theory to be represented by a given factor are indeed represented by the factor, and which variables were, versus were not, useful in differentiating the various person types from one another.

The chapter begins by describing the different measures of association that may be subject to factor analysis and the various types of factorable data matrices that may be created by crossing three measurement dimensions: variables, people, and occasions of measurement. The conditions under which this analysis performs well are then described and illustrated, for example, the selection of a small number of individuals carefully chosen for their "known" or presumed characteristics (e.g., extreme Type A vs. extreme Type B individuals); the use of structured versus unstructured responses reflecting items possessing known versus unknown underlying dimensionality, respectively; and the type of response scale (e.g., Q sort, mediated Q sort, normative measurement). An advantage of Q-sort factor analysis is that, rather than first factoring a set of variables to determine their underlying structure, and then attempting to differentiate types of people on the basis of the variable types, with Q-sort factor analysis investigators may directly differentiate person types, which often are more interesting than types of variables.

Structural Equation Modeling

Structural equation modeling (SEM) has emerged as one of the most important statistical methodologies in many areas of current psychology as well as general social science, and it is difficult to find quantitative journals that do not provide ample examples of this technique. The basic idea underlying SEM is the identification of a set of hypothetical underlying factors (latent factors) that relate to measured (observed indicator) variables, and the identification of the predictive (or causal) ordering of these factors in relation to each other. Using SEM, an investigator may be able to determine, for example, that three hypothetical factors (X, Y, and Z) underlie responses to the variables on which data were collected and that the predictive ordering of the hypothetical factors is $X \rightarrow Y \rightarrow Z$. Thus, SEM may be considered as a combination of a rigorous measurement model (factor analysis) and a rigorous structural model (path analysis). Because of its importance and widespread use in contemporary science, two chapters focus on SEM. In chapter 7, Klem describes the elements and language of SEM, discusses technical issues and assumptions and model identification and evaluation that must be considered, reviews the most useful types of SEM models for most research applications, and presents a guide to software for conducting SEM.

In chapter 8, Thompson first discusses key decisions that must be considered when SEM analyses are explained, such as the choice of measures of association to be analyzed, parameter estimation theory, model misspecification and specification searches, and sample size. This chapter next discusses two common misconceptions concerning the understanding of the results of SEM in the context of the role of measurement error and the notion of "causal modeling." The chapter concludes with a checklist of the most important precepts and principles to be adhered to in the conduct of SEM and the interpretation of its findings.

Canonical Correlation Analysis

Imagine that an investigator collected data for a sample on individuals assessed on two sets of variables (although any number of variable sets may be used, most investigators using this methodology typically use two). For example, one set of variables might constitute measures of patient functional status, that is, the ability to perform basic and intermediate tasks required by people for normal, everyday functioning, an index of the extent of their social interaction with friends, a depression measure, and so forth. The second set of variables might assess patient satisfaction with medical processes and outcomes: quality and quantity of communication with their health care providers, waiting times, perceived understanding of their particular condition, and so forth. A common methodology for assessing the relationship between sets of variables such as these involves conducting numerous multiple regression analyses, treating one of the variable sets as potential independent variables, and treating the individual variables in the other set as separate dependent measures. Not only does this methodology introduce rampant inflation in the Type I error rate, but it also runs the risk of "missing the big picture" in the data. As an alternative, imagine conducting an analysis in which (a) the most important hypothetical factor explaining variation in one set of variables is identified, (b) the most important hypothetical factor explaining variation in the other set of variables is identified, and (c) the simple correlation between scores computed for individuals on the two hypothetical factors is maximized. Then, return to the variation that remains after removing all variation identified in the first step, and repeat the analysis to determine whether secondary factors exist. This, of course, is (some of) what *canonical correlation analysis* (CCA) provides. Rather than conducting separate multiple regression analyses, CCA offers investigators more powerful and parsimonious multivariate regression analysis.

In chapter 9, Thompson provides a primer on CCA. He first explains the basic logic of CCA using a heuristic data set and then explains how CCA relates to a host of commonly used univariate and multivariate parametric analyses, such as t test, correlation, and ANOVA. He next discusses and illustrates the steps that are necessary to interpret the results and concludes by detailing common errors to avoid in conducting and interpreting CCA.

Repeated Measures Analyses

Repeated measures designs are defined by measures taken on each participant under each of several conditions. This research strategy has much to offer the researcher, as it is more powerful than between-subjects designs, meaning that there is a greater likelihood of rejecting a false null hypothesis. The reason for this is that participants serve as their own controls. More specifically, there is always variation among scores that is due, in part, to individual differences, or subject variables (those attributes of participants that are fixed and may influence scores on the dependent variable, such as personality traits and intelligence). Variation due to individual differences is one component of error variance. Error variance represents "noise" in a study, and it can drown out the "signal" or treatment effect. Researchers are always trying to find ways to decrease error variance, no matter what type of design is being used. The repeated measures methodology is perfectly suited to the task as it removes one potentially large source of error.

Not all research questions can be answered using a repeated measures design, which is why it is not as common as a between-subjects design. For example, there is no way to compare different therapy modalities by having participants receive each form of intervention. Because repeated measures designs are not as common as between-subjects designs, consumers of research may find themselves in unfamiliar territory when reading an article using this strategy. In chapter 10, Weinfurt provides a lucid discussion of the types of statistical analyses that are applied to various aspects of repeated measures. He walks the reader through increasingly complex designs, beginning with those having one independent variable, then factorial designs, multivariate repeated measures analysis, and finally longitudinal designs analyzed using hierarchical linear models. Throughout, Weinfurt pays special attention to the assumptions for each test, how effect sizes are computed, post hoc analyses, and planned comparisons. There is a section on the popular pretest–posttest, between-subjects design. Here Weinfurt compares and contrasts different approaches to analyzing the data of this design and provides a corrective to the common misconception that an analysis of covariance (ANCOVA) is the technique of choice when between-group means differ at pretest. Weinfurt's systematic presentation of simple to complex research situations and careful attention to assumptions and auxiliary analyses provide a thorough grounding in the basics of analyzing data from several types of repeated measures designs.

Survival Analysis

Consider a study in which two antidepressant medications are compared and of interest is whether one medication is more effective in preventing relapse after drug withdrawal. Patients are followed for 3 years after they stop taking the drug, and the duration to relapse is measured. It is found that many patients do not experience a relapse during the observation interval. What "duration score" should be recorded for these patients? Simply recording 36 months for all these patients is inaccurate because it suggests that there was a relapse at 36 months. *Survival analysis* (SA) was invented to handle just this kind of research situation. In SA, the dependent measure is the length of time to some critical event. In our example, the critical event is a reoccurrence of depression. Quality control engineers may be interested in the time it takes before a mechanical component fails. Personnel psychologists may want to know how long it takes for a promotion as a function of gender. A banker may be interested in how long customers maintain a savings account above a minimum account level. In each case, there is censored data, which are those instances in which the critical event has not occurred within the observation period.

SA is a family of techniques, with accommodations for the number of samples, and whether predictors are continuous, categorical, or both. In chapter 11, Wright discusses some of the most common applications, including Cox regression. In contrast to the Kaplan–Meier SA, used when there is one categorical predictor (e.g., treatment groups), Cox regression is used when the predictors are continuous or when there are categorical and continuous predictors. Unlike multiple regression, in which values of a continuous dependent variable are predicted, Cox regression predicts the rate of occurrence of a critical event. We expect articles that report SA will increase in number, especially in clinical and health psychology, as investigators become more interested in the long-term effects of interventions and the probability of emergence of interesting behavior patterns.

Conclusion

Considered together, the chapters in this volume describe powerful statistical analyses that can be, that should be, and indeed that are increasingly being used to model and study an enormous domain of research in the behavioral sciences, as well as in many other academic disciplines. In addition, as is typically true across the sciences, more powerful analyses address issues in ways that other types of analyses ignore, and they frequently provide more interesting and suggestive findings than less appropriate analyses. Our primary objective is to leave the reader with an understanding of several such new analyses, thereby opening a broader understanding of the state of the art appearing in published research. Our secondary objective is to encourage in readers an interest in pursuing such analytic methodologies in their own research and to steer them in the direction of some useful references to do so.

We close by noting again that the advent of statistical software packages has had a profound effect on behavioral researchers. New, more powerful and increasingly complex analytic techniques, which were relatively unknown or infrequently used because of the cumbersome computational mechanics, are now easily accessible. This computational ease creates a special dilemma, however, because it is possible to perform complex analyses without understanding them. It is our hope that this book (and its predecessor) will provide the kind of conceptual grounding that leads to further study and the deeper understanding required for responsible use of these powerful analytic tools.

References

Coombs, C. H. (1953). Theory and methods of social measurement. In L. Festinger & D. Katz (Eds.), *Research methods in the behavioral sciences* (pp. 471–535). New York: Dryden.

Gardner, P. L. (1975). Scales and statistics. *Review of Educational Research, 45,* 43–57.

Labovitz, S. (1970). The assignment of numbers to rank order categories. *American Sociological Review, 35,* 515–524.

Labovitz, S. (1972). Statistical usage in sociology: Sacred cows in ritual. *Sociological Methods & Research, 1,* 13–37.

Lord, F. M. (1953). On the statistical treatment of football numbers. *American Psychologist, 8,* 750–751.

Mowbray, C. T., Bybee, D., & Cohen, E. (1993). Describing the homeless mentally ill: Cluster analysis results. *American Journal of Community Psychology, 21,* 67–93.

Nunnally, J. (1978). *Psychometric theory* (2nd ed.). New York: McGraw-Hill.

Pedhazur, E. J., & Schmelkin, L. P. (1991). *Measurement, design, and analysis: An integrated approach.* Hillsdale, NJ: Erlbaum.

Siegel, S. (1956). *Nonparametric statistics for the behavioral sciences.* New York: McGraw-Hill.

Stevens, S. S. (1946). On the theory of scales of measurement. *Science, 103,* 677–680.

Stevens, S. S. (1951). Mathematics, measurement, and psychophysics. In S. S. Stevens (Ed.), *Handbook of experimental psychology* (pp. 1–49). New York: Wiley.

Tabachnick, B. G., & Fidell, L. S. (1989). *Using multivariate statistics* (2nd ed.). New York: HarperCollins.

Velleman, P. F., & Wilkinson, L. (1993). Nominal, ordinal, interval, and ratio typologies are misleading. *American Statistician, 47*(1), 65–72.

Wilkinson, L., & the APA Taskforce on Statistical Inferences. (1999). Statistical methods in psychology journals: Guidelines and explanations. *American Psychologist, 54,* 594–604.

Wilson, K. L. (1971). Critique of ordinal variables. *Social Forces, 49,* 432–444.

Zimmerman, D. W., & Zumbo, B.D. (1993). The relative power of parametric and nonparametric statistical methods. In G. Keren & C. Lewis (Eds.), *A handbook for data analysis in the behavioral sciences: Methodological issues* (pp. 481–517). Hillsdale, NJ: Erlbaum.

Reliability and Generalizability Theory

Michael J Strube

Just about everyone has an understanding of what *reliability* means, at least in its common, everyday usage. Most people know that a reliable friend will not let them down, a reliable car will not break down, and a reliable watch will not run down. Embedded in these everyday uses of the term is the expectation that when something is reliable, it can be depended on to act the same or be the same in the future, that it will be consistent in its characteristics. Reliable friends can be counted on to always be there when needed, reliable cars get people to work each morning, and reliable watches indicate the correct time whenever they are needed. When something is unreliable, however, expectations about its future performance are fraught with uncertainty. Indeed, reliability is often taken for granted until a formerly reliable friend lets a person down, a previously trustworthy car refuses to start, or a once precise watch now shows the correct time only twice a day. Reliability, then, is fundamental to an orderly and "error-free" everyday life.

Reliability is no less important to scientific life. In fact, reliability is fundamental to scientific progress. When systematic observations—the basic building blocks of science—are unreliable, one cannot trust those observations to provide insights into human behavior, the action of molecules, or the movement of planets. Scientific progress grinds to a halt. Scientists also seek to have orderly and error-free lives. Rather than taking reliability for granted, however, scientists make it a priority in their work.

In this chapter, I describe more formally the importance of reliability to science and show how scientists determine the reliability of their observations. I begin by describing more specifically the basic task that scientists try to accomplish in their work and show how reliability is crucial to accomplishing that task. Then I give a formal definition of reliability as it is typically used in scientific measurement—the so-called *classical theory*. An important extension of the classical view, called *generalizability theory*, is the focus of most of this chapter. Generalizability theory allows precise estimation of reliability, and it is a general-purpose approach to investigating broad questions about the consistency or dependability of measurement. Generalizability theory, as typically used, is a multivariable procedure. It makes use of repeated measures analysis of variance (ANOVA) in which multiple measures or observations are taken on the same units of measure (typically people, but the units could be any observable objects). Generalizability theory does, however, have a multivariate extension that is discussed later.

Reliability and Science

Although science uses a bewildering array of techniques and procedures and can take on many forms, the basic problem that most scientists must solve is surprisingly easy to understand. Most scientists seek to understand causality, that is, they want to know what causes what in the world. For example, medical scientists want to know what causes heart disease and cancer, behavioral scientists want to know what causes human aggression and lapses in memory, and astronomical scientists want to know what causes black holes. In many ways, scientists are like insistent two-year-olds, always asking "why?"

To answer those questions, scientists systematically observe events and, from those observations, draw inferences—educated guesses—about what causes what. If those observations are not reliable, however, then scientists cannot trust their observations and are unable to make confident conclusions about causality. Their world resembles that of the layperson with an unreliable friend, car, or watch: It is chaotic and prone to error.

Suppose, for example, that I am interested in testing the simple hypothesis that stress is related to psychological well-being: When people encounter stress, their psychological well-being declines. Focusing just on stress, I can see two problems that hinder testing this hypothesis. First, just what counts as "stress?" Second, once stress is defined, how is it measured? That is, can I be sure that I will see it when it occurs and not mistake other things for it instead?

To make their observations as precise and error free as possible, scientists carefully define the rules for observation, called *operational definitions,* or just *operations* for short. These rules dictate what "counts" and how to know it when it occurs. Stress, then, might be operationally defined as the number of daily annoyances from a list of 20 that a person claims to have endured during the past week. Psychological well-being might be operationally defined as self-reported anxiety on a simple 7-point rating scale. The effort to define precisely how observations are made, however, creates an important problem: The variables of primary interest are not observed directly. Rather, instances of them are observed and are used as proxies to inform the scientist about what might be true for the unobserved variables. The unobserved variables are known as *constructs,* and they are the variables about which scientists are most interested. These constructs can be glimpsed, however, only through the lens of the operational definitions. This places a heavy burden on the quality of the observations. On the basis of what a scientist records, inferences back to more general constructs are made. The quality of those observations then is crucial. A mistake or error made at the level of observations transfers back and creates errors of inference about constructs, leading to faulty scientific knowledge. This basic inferential problem is depicted in Figure 2.1.

Figure 2.2 gives the same construct–operation links, but in terms of the stress and well-being example. Stress is presumed to be causally related to psychological well-being, but making that inference rests on establishing a causal relation between daily annoyances and anxiety. Establishing causal relations is itself a complex task, but I do not discuss it further here (see, e.g., Cook & Campbell, 1979; and Whitley, 1996). Assuming that a causal link between daily annoyances and anxiety can be established, the key issue becomes the identification that means the same as "stress causes a decline in psychological well-being?" It does if two things are true. First, daily annoyances must be a good example or represent well the larger concept of stress. Likewise, self-reported anxiety must be a good example of the larger concept of psychological well-being. This is no small assumption, and it speaks to the validity of the observations. Second, daily annoyances and anxiety must be observed with little error; that is, the measures must be reliable. If they are not reliable, then the errors will prevent confident claims about the relation between daily annoyances and anxiety and, more importantly, about the

Figure 2.1

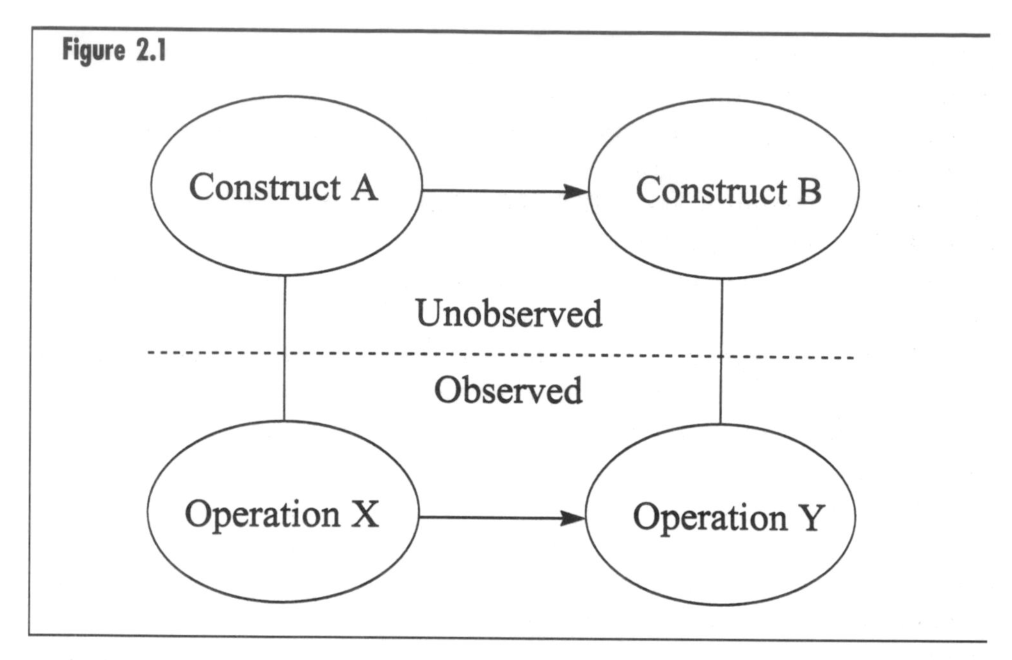

The basic inferential task in science.

Figure 2.2

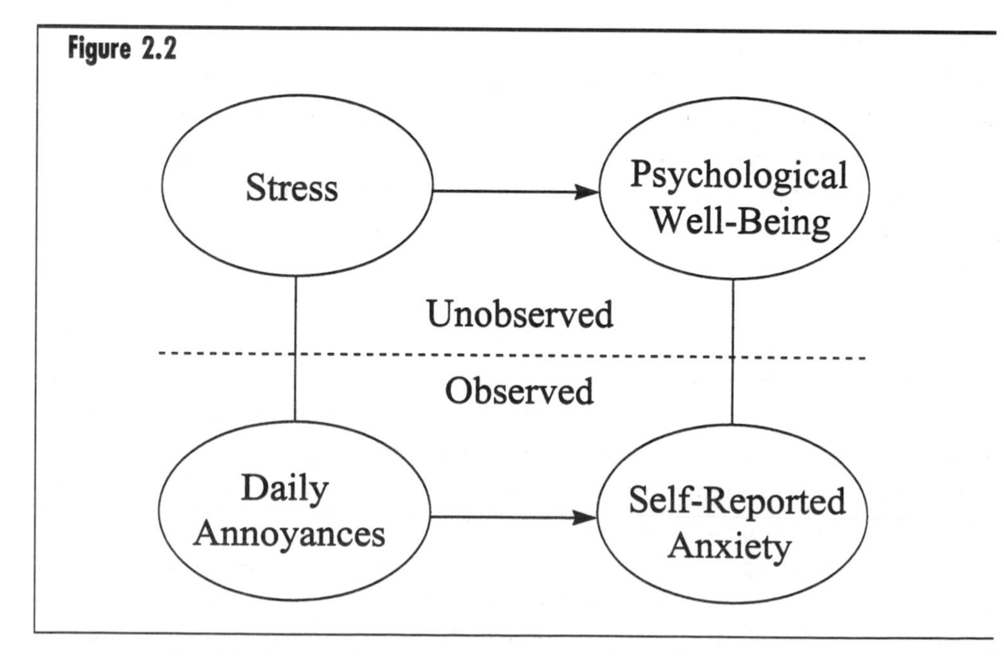

An example of the basic inferential task in science: Stress and psychological well-being.

relation between stress and psychological well-being from being made. Faulty or unreliable measures, then, obscure a scientist's ability to find the true relations between events and concepts (for a more formal treatment of reliability and statistical inference, see Bollen, 1989; Fleiss & Shrout, 1977; Humphreys & Dasgow, 1989; Kopriva & Shaw, 1991; Williams & Zimmerman, 1989; and Zimmerman & Williams, 1986). It is for this reason that researchers use the phrase, "reliability precedes validity."

Given the important role that reliability plays in science, two important questions come immediately to mind: (a) How does one know if observations are reliable? and (b) How can reliability be improved? Answering both questions requires the guidance of a theory about measurement. In the next section I describe the classical theory of reliability —a very simple and appealing way to think about measurement. That discussion sets the stage for my consideration of generalizability theory, a more complex but also more realistic way to think about measurement.

Classical View of Reliability

At the heart of reliability is the concept of error. Scientists know that observations are fallible and that error prevents one from finding the truth. But just what specifically is meant by error and truth? The most popular theory of measurement, known as the "classical theory," assumes that an observation of any given individual (symbolized as x and called an "observed score") comprises two simple pieces of information: a *true score* (t) and an *error score* (e). These are related in a very simple way:

$$x = t + e.$$

The true score is assumed to represent the standing of the individual on the construct of interest, that is, a person's real score. Of course, the true score is not observed directly; one can only get an estimate of it from the observed score. For example, I do not know a person's true intelligence from an IQ score; I only get an estimate of it. The observed score deviates from the true score due to the influence of error. The error component of an observed score reflects the random sources that can move scores up or down relative to the true score. These random sources can include a variety of idiosyncratic elements, but their influence is common: They produce noise against which it can be difficult to detect the signal of the true score. Good measures have little noise, that is, the observed score reflects mostly the person's actual standing on the construct (i.e., the true score).

So far I have considered the obtained score of a single person. Ordinarily, scores for many people are available. The variability of these scores can be described using a simple statistic called the *variance*, symbolized as σ^2, which indicates how much scores vary or deviate from the average or mean score for the sample. A large variance indicates that scores are more spread out. The square root of the variance, called the *standard deviation* (*SD*), is sometimes easier to understand. In a normal (i.e., bell-shaped) distribution of scores, approximately 68% of the scores will fall within a range between 1 *SD* below the mean and 1 *SD* above the mean. According to the classical view, if an obtained score comprises a true score and an error score, then the variability of a sample of obtained scores (symbolized as σ_x^2) likewise comprises true score variance (σ_t^2) and error variance (σ_e^2):

$$\sigma_x^2 = \sigma_t^2 + \sigma_e^2.$$

I am just a step away from answering my first question: "How does one know if a measure is reliable?" If the true score variability is good and error variability is bad, then clearly a good measure produces obtained score variability that contains much more true score variance than error variance. In other words, the obtained scores vary because the individuals are truly different on the construct being measured, not because of random differences. An intuitively appealing way to represent this idea is the proportion of obtained score variance that is due to true score variance:

$$\frac{\sigma_t^2}{\sigma_t^2 + \sigma_e^2}.$$

This, in fact, is the classical theory definition of reliability. The resulting *reliability coefficient* (r_{xx}) has a value between 0 and 1.0. It has a value of 0 when all of the observed score variance is random and does not reflect anything about a person's standing on the underlying construct. As the reliability coefficient approaches 1.0, more and more of the variability in observed scores is due to true score variability. In practice, reliability coefficients rarely exceed .95, and most often they are considered to be acceptable if they are above .80 (cf. Nunnally & Bernstein, 1994). Standards of "acceptable" reliability must be used with caution, however. Acceptable error in a measure depends on how that measure is used, a point made explicitly in generalizability theory.

I defined what reliability is, but is it estimated in practice? If the underlying characteristic being measured has not changed from one measurement occasion to another, then a reliable measure should give a similar score for each individual on each measurement occasion. Three common ways to assess reliability are based on this "consistency" idea. First, consider the same test given on two different occasions, known as *test–retest reliability*. Clearly, if a sample of scores differ on the two occasions that the tests are given and provided that the underlying true score has not changed, then I would be justifiably concerned about the reliability of the measure. It must be measuring other things. Similarly, two different measures of the same construct, called *parallel forms* or *alternative forms*, should provide similar answers. In addition, multiple items on a single test should converge on the same conclusions about true scores, called *internal consistency*. When they do not, the lack of consistency is revealed by a low reliability coefficient, and the ability to make much sense of the scores is reduced.[1]

Clearly by now measures are unreliable because they contain error, and estimates of reliability reflect the relative amounts of error that measures contain. Thus, the answer to the second question—"How can reliability be improved?"—should be obvious. Measures are made more reliable by reducing error. This is accomplished in two ways: standardization and aggregation. *Standardization* is simply the attempt to control the measurement conditions carefully so that no extraneous sources of error are allowed to influence scores. For example, the same instructions delivered in the same way should precede a measure. Otherwise individuals' obtained scores might vary slightly in response to differences in instructions; that would be an additional source of error that obscures the ability to detect the true score. Similarly, the same amount of time to complete the measure should be given to all respondents; otherwise, respondents' scores could vary depending on whether they felt rushed to complete the measure. The important point about standardization is that when measurement conditions are standardized, potential sources of variance become constants and so cannot influence the variability of obtained scores.

[1]The mathematics and assumptions underlying the classical theory derivations are not covered here; they are described well by Nunnally and Bernstein (1994). Also technically speaking, the measures of reliability described here are sensitive to changes in the relative standing of individuals. This point is made explicitly in the discussion of generalizability theory.

The second way to improve reliability is through *aggregation*, a fundamental principle in statistics and measurement. The importance of aggregation is clear to anyone who has ever take a test that matters. Ask students, for example, if they would like to have their grades in a course determined by a single exam question, and most would decline the offer. Instructors would similarly be hesitant to make the offer, despite the obvious savings in grading time. The concern for both parties is that too many irrelevant sources of error could prevent that one question from revealing the students' true ability. Intuitively, a test with many questions has a better chance of revealing ability than a test with few items.

Through aggregation the random sources of error that contaminate each observed score have a chance to cancel out and leave standing a better estimate of the true score. Think back to the definition of obtained scores in the classical view: $x = t + e$. If only one test question is available, then the obtained score is biased up or down to some extent because of *random error.* But imagine if a second test question were available that was also a measure of the same true score. It, too, would have an error component, but the direction in which this error influenced the obtained score would likely be different than for the first item (remember, error is random in the classical view). If the two items were added together, the total might provide a better measure because the two error parts might partly cancel each other out. They do not completely cancel each other out unless their error influences are directly opposite, but if more and more items are added, each with its own random error sources but measuring the same true score, then the odds of the error canceling out keep improving. This is a fundamental idea in measurement. Aggregating items that measure the same true score will allow the random error components to cancel and provide a more reliable measure.[2]

Although I emphasized r_{xx} as a measure of reliability, there is another way to view reliability that is also appealing and that anticipates one of the strengths of generalizability theory. One could just as easily think about reliability of measures in terms of the amount of error that those measures contain σ_e^2. The square root of this variance is an *SD*, but it is given a special name: the *standard error of measure* (σ_{meas}). It is appealing because it can be used to gauge the uncertainty inherent in an obtained score by providing an estimate of how much obtained

[2]The well-known Spearman–Brown prophesy formula is based on this idea (see Nunnally & Bernstein, 1994).

scores would be expected to vary on subsequent measurement occasions, even though an individual's true score does not change. Approximately 68% of an individual's obtained scores would be expected to fall within a range that is 2 σ_{meas} wide; approximately 95% of an individual's obtained scores would fall within a range that is 4 σ_{meas} wide. The σ_{meas} communicates reliability in terms of the original scale of measurement, which can be very revealing.[3]

I intentionally oversimplified matters a bit to introduce the key ideas surrounding reliability from the classical measurement view. Some of the complications, however, are important because they set the stage for introducing a more complete approach to measurement. For example, each of the ways of estimating reliability that I described previously defines error in a slightly different way. In other words, each is sensitive to different sources of error. For example, the internal consistency approach to reliability is used when a multiple-item measure is given on just one occasion. The inconsistency in item scores that can make a measure unreliable can occur because of random fluctuations that vary across responses to the items, but sources of error that might change from one day to another do not have any opportunity to show up as error. Indeed, on any one measurement occasion, such sources could masquerade as true score, acting with a constant influence on observed scores for a particular individual. My bad mood on the day of a test would influence my responses to all items on the test, but on a different day my mood could be different. However, test–retest reliability would treat the mood changes and other random shifts across individuals over time as error. This identifies an important problem with the traditional classical view. It considers error to be a single entity, but depending on the type of reliability assessment, the definition of that error can change considerably. A better approach would allow more precise estimation of the different pieces that make up *measurement error.*

Generalizability Theory

The major advance of generalizability theory (Cronbach, Gleser, Nanda, & Rajaratnam, 1972) over the classical approach is to allow for a more precise specification of error. This is accomplished by specifying more

[3]The standard error of measure is a simple algebraic translation of the reliability coefficient and the obtained score SD: $\sigma_{meas} = \sigma_x\sqrt{1 - r_{xx}}$. The actual calculation of confidence intervals can be complex, however (see Feldt & Brennan, 1989; and Nunnally & Bernstein, 1994).

carefully the conditions of measurement, and then testing the importance of those conditions of measurement as sources of error variance in the obtained scores. The conditions of measurement in generalizability theory are known as *facets.* The items on a test, for example, are a facet of measurement. The items are a facet of measurement because they might differ in their overall difficulty or in the relative performances of the individual respondents. In either case, item scores could vary by item, and one score would not necessarily provide a dependable indication of a person's real standing on the construct being measured. Time of day, form of test, and type of response format are all examples of other testing conditions that could affect obtained scores. Indeed the list of possible facets of measurement potentially is very long and depends on the specific measurement problem. The term *generalizability theory* essentially comes from the idea that the ability to generalize obtained scores across conditions of measurement requires explicitly specifying and testing the conditions of measurement that might affect obtained score variability. The specification of the facets in generalizability theory defines the *universe of admissible observations*; that is, the facets define the boundaries within which the investigator considers observations potentially to be interchangeable. Generalizability theory thus recognizes what classical theory somewhat vaguely assumes: Reliability depends on the conditions of measurement, and there are as many reliabilities for a measure as there are unique uses or conditions of measurement.

Generalizability theory resembles the classical view in assuming that obtained score variability is the sum of true score variability and error variability. It departs from classical theory in viewing error variance itself as having many possible sources. The heart of generalizability theory, then, is the estimation of the components of variance that make up an obtained score, with particular interest in estimating the sources of error variance. These efforts are known as *generalizability studies* or more simply as *G studies.* They attempt to specify more completely what the classical view oversimplifies.

The ability to estimate *variance components* hinges on the nature of the design that was used to collect the data. Some examples will illustrate the issues. First, imagine that I wanted to rate the aggressiveness of children on the playground using raters who observe the children and make an overall judgment on a numerical "aggressiveness" scale. The first key question that I ask is "What are the conditions of measurement that could make a difference in the scores that children get?"

Figure 2.3

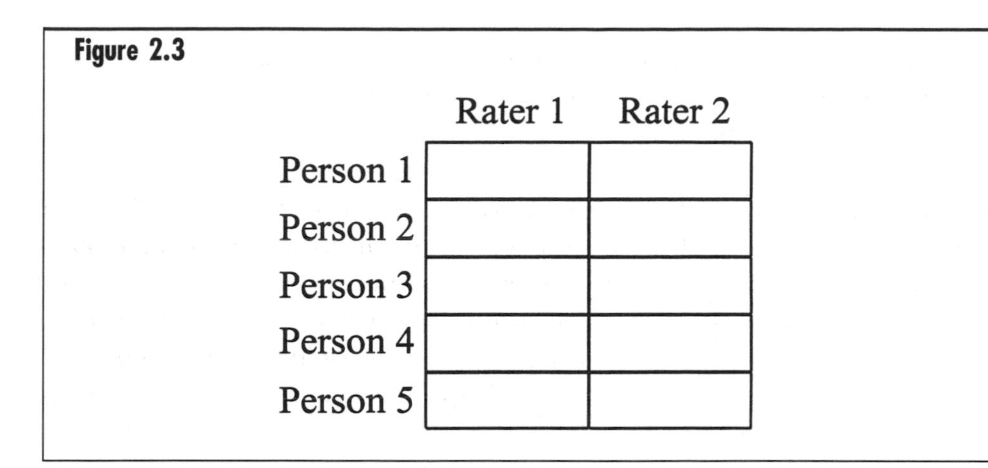

A completely crossed person × rater design.

Those need to be included explicitly in my data collection design so that their influence can be determined. One obvious source of variability is characteristics of the children themselves. This particular facet of measurement is known as the *object of measurement* in generalizability theory. Variability for this facet is desirable because it allows me to make the distinctions that I believe are meaningful; that is, children can be meaningfully distinguished in terms of their aggressiveness. I call this the p facet for people or participants. Another possible measurement condition that could matter would be the raters. I would attempt to train my raters, so that they are equal in their ability to judge aggression, but in generalizability theory, their status as interchangeable or exchangeable raters is an empirical question, not a given, so I specify raters (r) as a facet as well. Variability in the obtained scores due to rater differences would be undesirable, so I use generalizability theory to determine how big a problem this is. In the simplest G-study design, then, I would have each of r raters independently observe and rate the verbal aggressiveness of p children on the playground. The logistics might be a bit tricky because the raters have to observe the same behaviors simultaneously. This is known as a *crossed design* because each rater assesses each child, that is, each level of the rater facet is paired with each level of the participant facet. If I collected data from 5 children using 2 raters, the data that would be collected would provide information for each cell in Figure 2.3.[4]

[4]In actual practice, one would use more participants and perhaps more raters, so that the parameters estimated would have more precision.

The variance estimates in a G study are provided by *analysis of variance* (ANOVA), a statistical technique that ordinarily is used to test hypotheses about mean differences but that also provides the information needed for a G study. In this case, the ANOVA yields three distinct sources of variance: person (p), rater (r), and person × rater (pr). Actually, it is a bit more complicated than this. The ANOVA provides mean squares that must be transformed into the variance components (for details, see Cronbach et al., 1972; and Shavelson & Webb, 1991). In addition, the interaction term, pr, actually contains several sources of variance that cannot be separated. It not only contains information about the Person × Rater interaction, but also includes random unaccounted for sources of variance. This term is sometimes referred to as the *residual error* term. These sources of variance are shown in Figure 2.4 in a Venn diagram that depicts the three unique kinds of variance that arise from crossing people and raters.

The sources of variance refer to different systematic ways that the scores can vary. The person component (p) reflects real differences in the verbal aggressiveness of the children. The rater component (r) indicates systematic and overall differences in the way the raters view aggressiveness. If one rater simply provided consistently higher aggressiveness ratings than the other rater, that would show up in the rater component of variance. The person × rater component (pr) indicates that the relative aggressiveness of the children is different for the raters.

As mentioned previously, person variability is desirable and is analogous to the true score variance that was encountered in the classical view. In generalizability theory, this variance component is known as *universe score* variance to emphasize that it represents the variance of

Figure 2.4

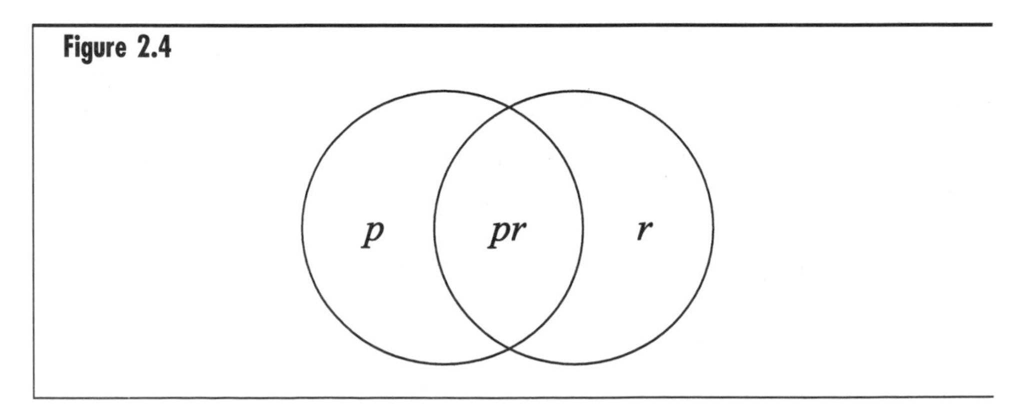

Sources of variance for a completely crossed person (p) × rater (r) design.

scores averaged over all conditions, defined by the universe of admissible observations. That leaves two other sources of variance (*r* and *pr*), whereas previously I only had one (*e*). The two sources must be error, but what kind of error are they? To understand their influence requires considering another important distinction made in generalizability theory: relative versus absolute decisions. Remember that the measures collected are used to make distinctions among the objects of measurement, people. There are two kinds of distinctions that can be made. First, I might wish to know the relative standing of the children in terms of their aggressiveness. This is known as a *relative decision.* In this case the absolute scores do not matter; the only interest is in a given child's aggressiveness relative to other children or relative to the sample mean. An alternative is that I may want to know if children exceed some criterion or cut-off score. In this case, the absolute value of the scores, not just their relative standing, is important. This is known as an *absolute decision.* The sources of error affect these decisions in different ways.

Consider first a relative decision in which my only concern is the relative differences among the children. Perhaps I want to identify the three most aggressive children. The top panel in Figure 2.5 shows that both raters provide the same rank ordering and that the answer to the question "Who are the three most aggressive children" would be the same for each rater. Note also that one rater provided consistently higher ratings than the other. In other words, there is an effect for raters, and the *r* component of variance would be sizeable. But this component does not affect the relative decision that I want to make. Now look at the bottom panel of Figure 2.5. This panel clearly shows that the relative differences among the children shifts across raters. This is a Person × Rater interaction, and it prevents me from making a clear statement about who are the three most aggressive children. I get a different answer to that question for each rater. This indicates that the *pr* component is a source of error for a relative decision. In fact, any interaction that involves the object of measurement (usually *p*) is a source of error for relative decisions because interactions involving the object of measurement reflect changes in relative standing across facet levels.

The sources of error for absolute decisions are a bit different. Figure 2.6 shows the same two data patterns but now with reference to a criterion score. This criterion represents a standard for defining aggressiveness, perhaps a normative standard, and I wish to identify as aggressive only those children who exceed this standard. As the top

Figure 2.5

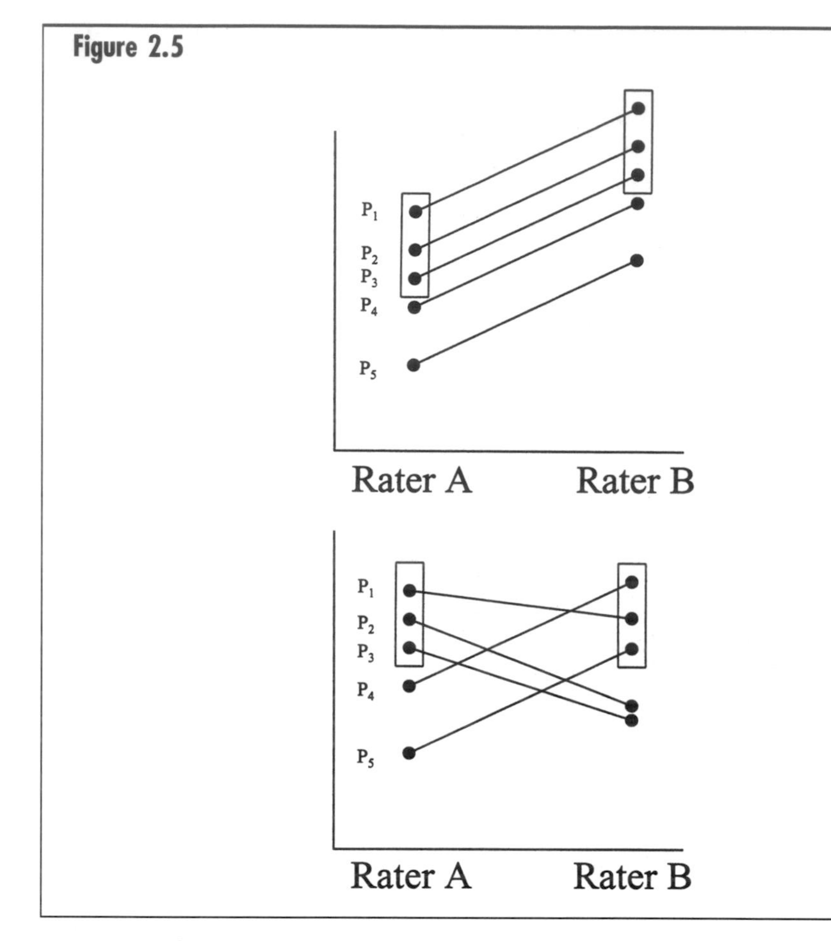

Sources of error for a relative decision.

panel indicates, the overall rater differences, estimated by the *r* component, influences whether a given child exceeds this criterion. Rater B has a tendency to see more aggressive behavior, perhaps because that rater is more observant or perhaps because the rating rule is being applied more liberally. In either case, the consequence is that more children would be labeled as aggressive if I relied on this rater's assessments. Ratings of aggressiveness would not be interchangeable. This indicates that components that are not interactions with the object of measurement affect absolute decisions, even though they do not affect relative decisions. The bottom panel indicates that interactions involving the object of measure affect absolute decisions as well.

In summary, relative decisions involve only relative standing of the objects of measurement, so only those error sources that are interac-

Figure 2.6

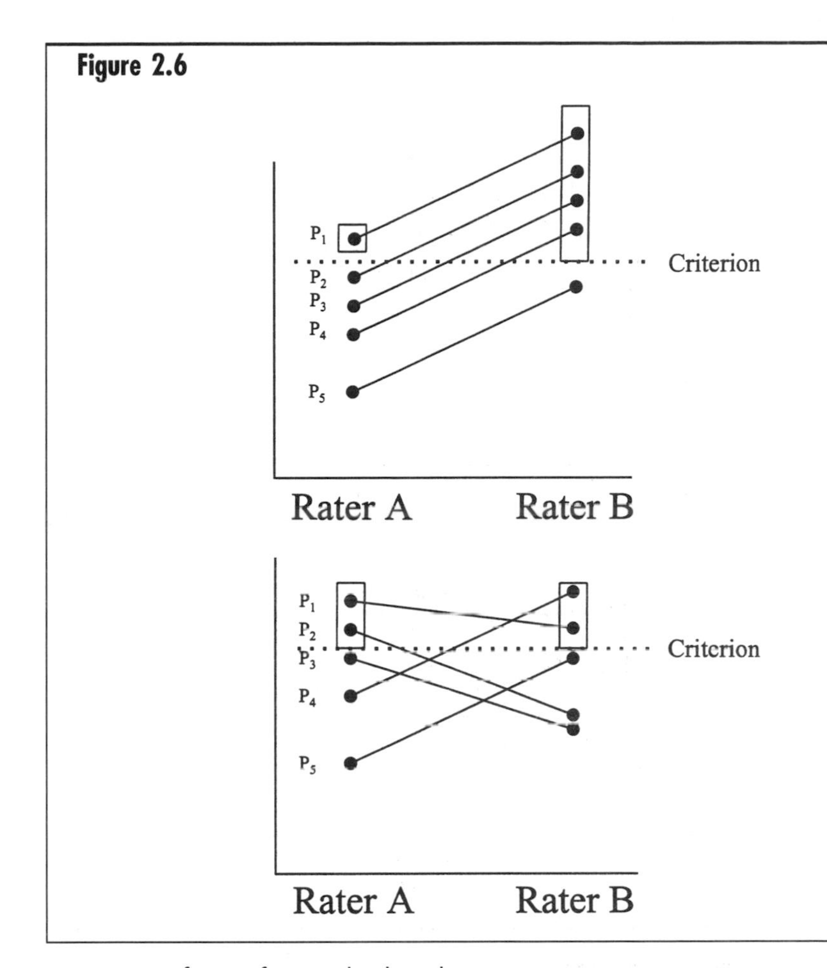

Sources of error for an absolute decision.

tions with the object of measurement affect relative decisions. Absolute decisions, however, are based on the absolute value of the obtained score, so any source that influences the value of the obtained score is a source of error. That means that all sources other than the object of measurement are a source of error for absolute decisions.

The calculation of reliability coefficients in generalizability theory follows the same logic as that for traditional classical theory, but two kinds of coefficients are calculated to reflect the two kinds of decisions that can be made. Reliability estimates for relative decisions are called *generalizability coefficients* (often $E\rho^2$ to indicate that it is the expected squared correlation between obtained scores and universe scores), and they represent the proportion of obtained score variance that is due to universe score variance. Remember that universe score variance is sim-

Figure 2.7

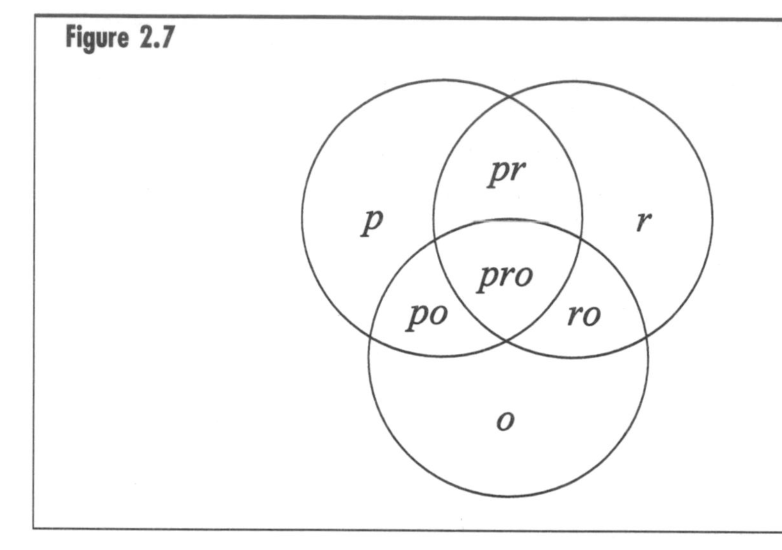

Sources of variance for a completely crossed person (p) × rater (r) × occasion (o) design.

ilar to true score variance and is estimated by the p component. Obtained score variance is the sum of universe score variance and error variance, as just defined. For absolute decisions, the reliability coefficient is given a different name—the *index of dependability* (ϕ; Brennan & Kane, 1977)—but is defined in an analogous way as the proportion of universe score variance to obtained score variance but where the obtained score variance includes additional sources of error.[5]

The person × rater design is the simplest, having only one facet other than the object of measurement. The power and complexity of generalizability theory increases when additional facets of measurement are included. I could extend the example to include different occasions of measurement. Perhaps I believe that levels of aggressiveness differ depending on the time of day that I assess the children. Adding an occasion (o) component to my design and crossing it with the other facets would produce a completely crossed person × rater × occasion design. In this design, all children are assessed by all raters on every occasion. Figure 2.7 shows the sources of variance that can be estimated

[5]The formula for estimating $E\rho^2$ is very similar to the classical theory definition of reliability: $E\rho^2 = \sigma_p^2/\sigma_p^2 + \sigma_{Rel}^2$, in which σ_{Rel}^2 is the error variance for a relative decision. For a simple person × rater design, relative error is σ_{pr}^2/n_r, in which n_r is the number of raters in the design. The formula for estimating ϕ takes a similar form: $\phi = \sigma_p^2/\sigma_p^2 + \sigma_{Abs}^2$, in which σ_{Abs}^2 refers to the error variance for an absolute decision. For a simple person × rater design, absolute error is $\sigma_r^2/n_r + \sigma_{pr}^2/n_r$. Cronbach et al. (1972) contains an extensive listing of error sources for more complex designs (see also Shavelson & Webb, 1991).

from this design. Again, the person (*p*) component is desirable variance and the rest are error, but their influence depends on the type of decision that will be made. For relative decisions, *pr, po,* and *pro* sources all obscure my ability to rank order the children reliably in terms of their aggressiveness. The *pr* component reflects, as it did in the simpler person × rater design, different relative ratings for the different raters. If the *po* component is sizeable, it indicates that the relative aggressiveness of the children is different across occasions, meaning that I cannot generalize about the aggressiveness of the children from one occasion to another. For example, the most aggressive children in the morning may not be the most aggressive children in the afternoon. The *por* component indicates that the relative aggressiveness of the children varies simultaneously with rater and occasion. This component also contains other random sources of error.

Completely crossed designs are the most typical and familiar way to collect data, but there are other options that might be required due to practical considerations. Going back to the simpler design, it might not be possible for the same pair of raters to assess each child. Perhaps different pairs of observers are asked to each assess a different child. This data collection design is depicted in Figure 2.8 and is known as a *nested design* because raters are nested within people. The most notable feature of a nested design is that each level of each facet does not appear in combination with each level of all other facets. Although there are still two raters for each child, it is a different pair of raters for each child (Raters 1 and 2 assess Person 1 but not any other child). The major consequence of this type of design is that some variance components cannot be separately estimated. This is clear from the Venn diagram in Figure 2.9, which shows that the person × rater (*pr*) effect cannot be separated from the rater (*r*) effect. These effects are said to be *confounded.* The overlap of the person component and rater component (*pr*) is completely contained in the rater (*r*) component. Consequently, there are only two components of variance that can be estimated from the ANOVA of these data: a *p* component and a combined component that contains both the *r* source of variance and the *pr* source of variance (this combined source is often symbolized as *r:p* to indicate "*r* nested within *p*").

Matters can get a bit more complex. It is possible for designs to have both crossed and nested features. For example, the person × rater × occasion design might not be realistic. It might be possible to assess all children on all occasions (people and occasions are crossed), but it

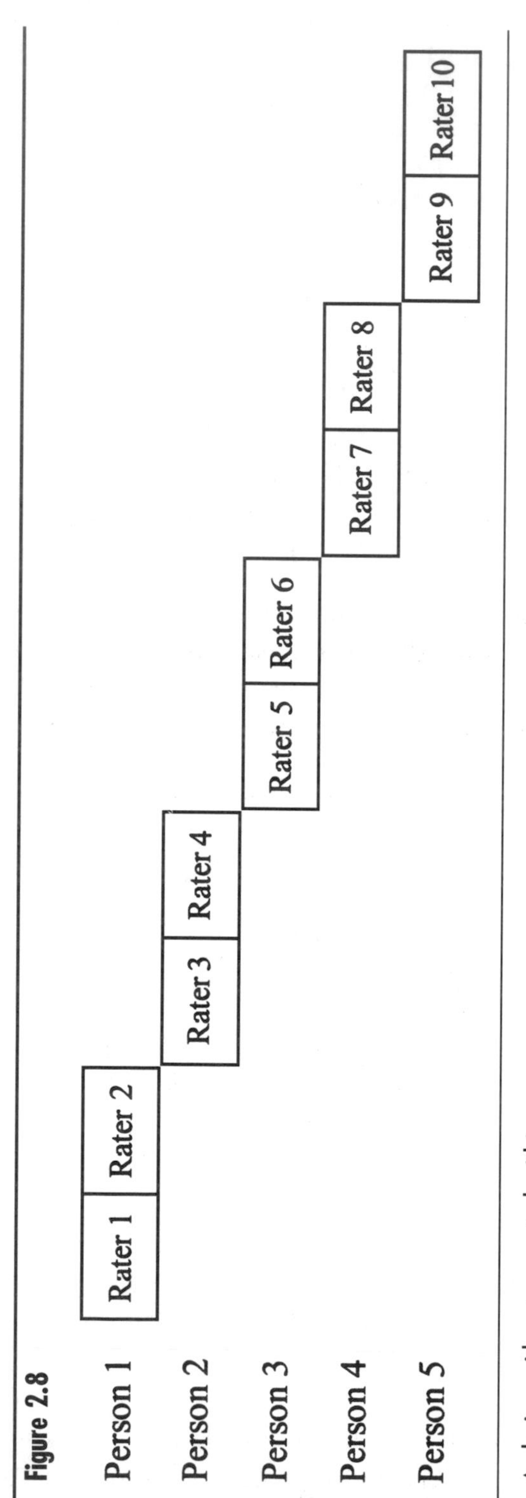

A design with raters nested within persons.

Figure 2.9

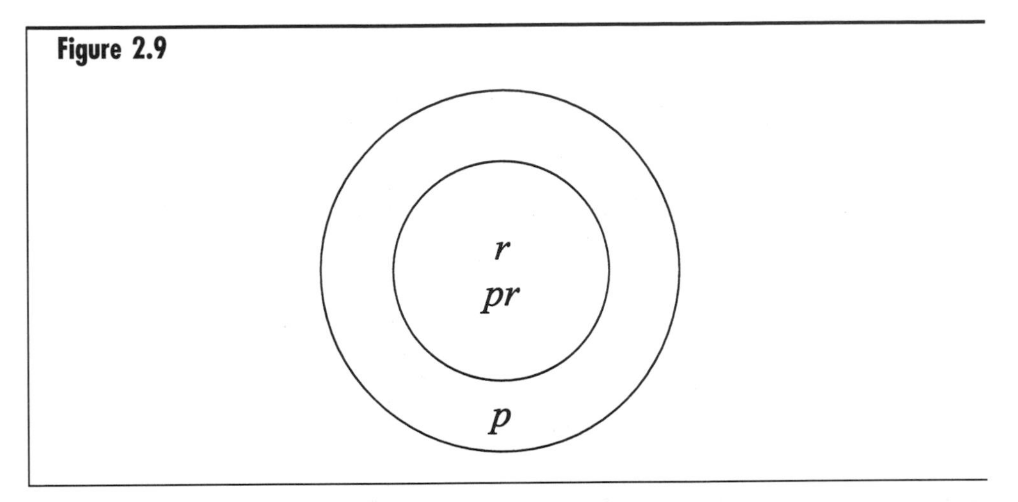

Sources of variance for a design with raters (*r*) nested within persons (*p*).

may not be possible to use the same raters on all occasions. If raters are nested within occasions, then the design is a partially nested design, as indicated in Figure 2.10. With raters nested within occasions, the rater (*r*) source of variance is confounded with the rater × occasion (*ro*)

Figure 2.10

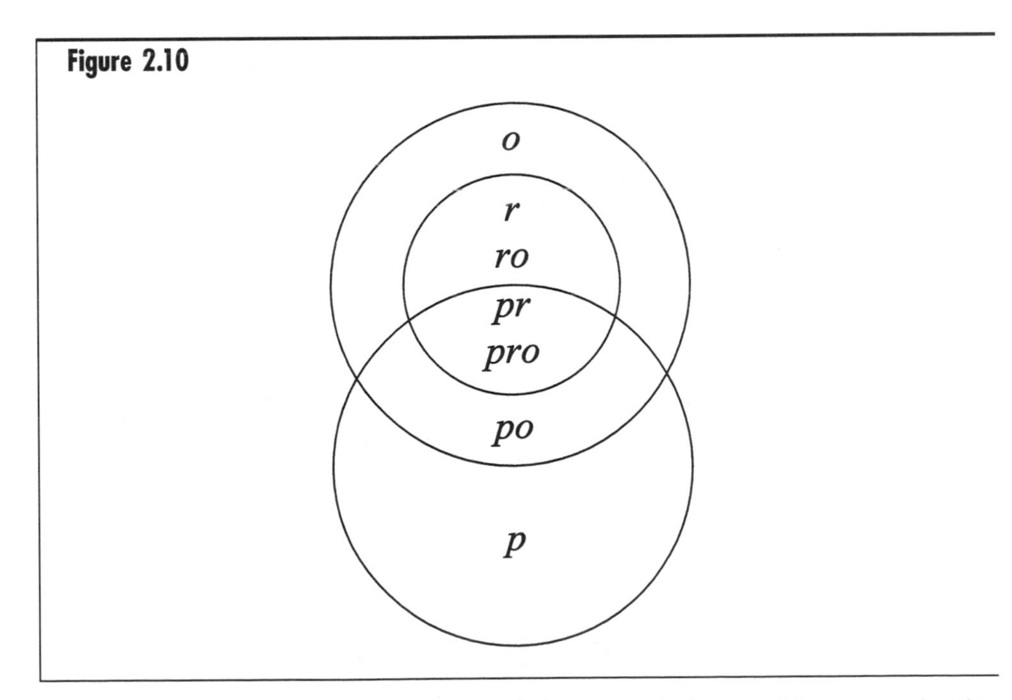

Sources of variance for a partially nested design in which raters (*r*) are nested within occasions (*o*), and both are crossed with persons (*p*).

source of variance. Also confounded are sources of variance that interact with the nested effect (i.e., *pr* and *pro*). The ANOVA for this design would yield only 5 sources of variance rather than 7.

It should now be clear that a completely crossed design allows the most information about sources of error, and nesting reduces that information through confounding. A completely nested design is less informative than a partially nested design, which is less informative than completely crossed design. Figure 2.11 shows the consequences of a completely nested design. In this case, raters are nested within occasions, which are, in turn, nested within people. This might be necessary if the children cannot be observed on all occasions and raters are only available to assess one child on one occasion. In this design, there are only three sources of variance that can be estimated, with two composed of *confounded effects*. Given the less informative nature of nested designs, one might justifiably wonder why they would be used. As I suggested in the examples, one common reason is that it is impossible to collect the data as a completely crossed design or, if possible, it might be very costly to do so. Another reason will be clear shortly when I discuss the aggregation principle as it applies to generalizability theory.

Figure 2.11

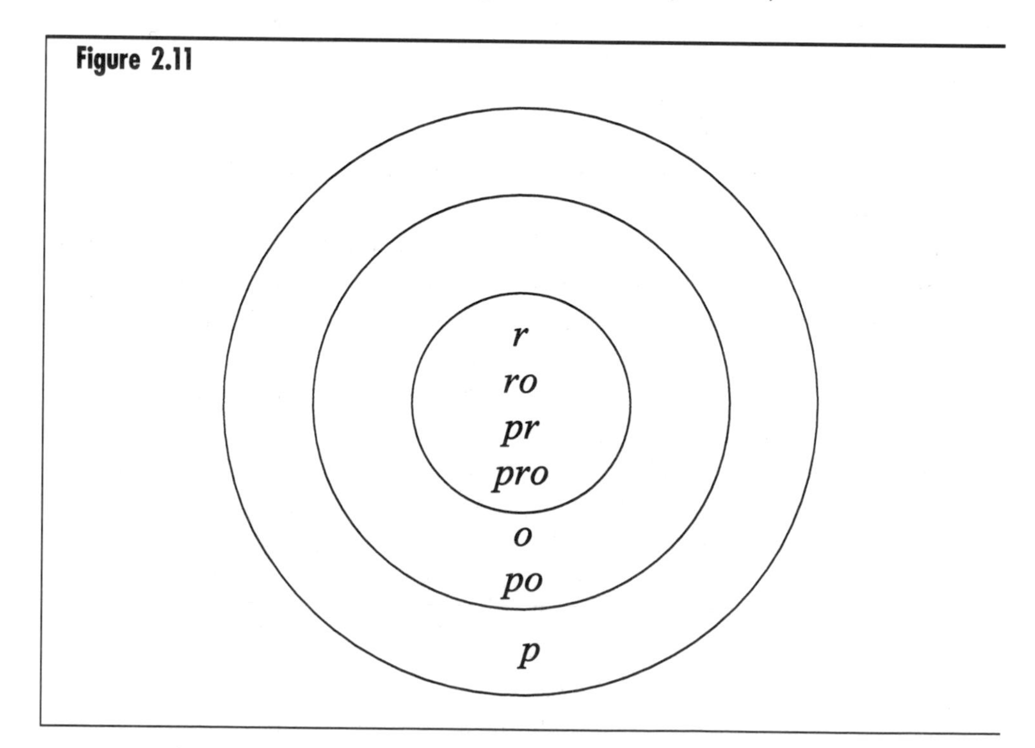

Sources of variance for a completely nested design. o = occasion; p = person; r = rater.

There is one further distinction often made in generalizability theory: random effects versus fixed effects. To this point, I described the facet levels as though they were exchangeable with any other facet levels; that is, they are assumed to be *random effects.* Ideally, they would be randomly sampled from the population of possible levels. In theory, this population would be infinite, but in practice, it is usually merely very large. Furthermore, the levels actually used in a generalizability study may not be truly random samples; it is sufficient to assume that they are if there is no particular interest in the levels used and any other levels would work equally well. By contrast, when a facet is a *fixed effect,* the levels used are the only ones of interest. There is no intent to generalize to some larger population, either because there is no larger population (all facet levels are included) or because the ones included in the study are the only ones that the researcher cares about. This distinction between fixed and random effects has important consequences for the estimation of variance components in generalizability theory. Two approaches have been recommended (see Shavelson & Webb, 1991). First, if each level of a fixed effect was selected because there was particular interest in that level, then it might make conceptual sense to conduct a separate generalizability analysis for each level. Alternatively, if a researcher wants an overall average of the generalizability, then averaging across the fixed effects is another option. Note, however, that if the variance components that involve the fixed effect are quite large, then averaging across fixed facet levels obscures an important difference.

As seen, then, generalizability theory provides a more complete way to assess the sources of error and to estimate the reliability from a design. Another powerful use of generalizability theory is directed toward that second question I posed some time back: "How can reliability be improved?" In fact, generalizability theory provides a very powerful and sophisticated way to answer that question using what is known as a *D study* (for decision study). D studies pose "what-if" questions about future possible ways to collect data; they forecast the expected reliability under those different hypothetical scenarios. D studies use the variance components from the G study to make these projections. D studies essentially use the aggregation principle to ask what the reliability would be for different combinations of facet levels and for different kinds of designs. The reliability coefficients that can be estimated in a G study refer to the number of facet levels and their particular arrangement (e.g., crossed or nested) as they actually existed in the data collection (i.e., in the universe of admissible observations). The parallel idea in a D study is known as the *universe of admissible generalization*.[6] In a D study, the number of facet levels and their arrangements are modified to determine how reliability changes with different kinds of designs. Each design defines a different universe of admissible generalization. This allows careful planning that can optimize reliability while taking design cost into account. If two designs produce essentially the same reliability, the researcher can choose the one that is logistically easier or is less costly to conduct. Alternatively, for a fixed cost a researcher can identify the design that maximizes reliability.

Two Examples

I have covered a lot of ground; I now show how all of this actually works. An excellent example of how generalizability theory can be used is provided by Brennan, Gao, and Colton (1995). They examined the reliability of a measure designed to assess listening skills.[7] In their study, 50 examinees listened to 12 tape-recorded messages, taking notes while each message was played. After each message was completed, the examinees were told to consult their notes and write a summary of the message. The written summaries were then scored by three raters, each of whom independently assigned a score from *low* (0) to *high* (5) representing the holistic judgment of the accuracy and completeness of each summary. The researchers thus used a person (p) × task (t) × rater (r) crossed design, with 50 levels of p, 12 levels of t, and 3 levels of r. These facets define the universe of admissible observations and yield seven components of variance. Table 2.1 gives the variance components for these sources.

Figure 2.12 shows the percentages of variance in the listening skills scores that are attributable to the different sources of variance. Several features of these data are notable. One of the larger components of

---

[6]A D study defines $E\rho^2$ and ϕ in a way that resembles a G study, except the components of error are defined by the projected design and the values for n represent possible levels in those designs, not the levels actually present in the G study. Thus, $E\rho^2$ for a D study, assuming a nested design (raters nested within people), would be estimated as $E\rho^2 = \sigma^2_p/\sigma^2_p + \sigma^2_{r,pr}/n'_r$. The accent mark over n_r indicates that it need not be the same as in the original G study. The effect of aggregation is apparent in this formula.

[7]Brennan et al. (1995) also investigated a writing-skills test. For simplicity, only their analyses of the listening-skills test are described here. The results for three forms of the listening test were averaged. Brennan et al. also conducted more complex analyses of their data than is described in this chapter. Their article provides a good example of the multivariate extension of generalizability theory.

Table 2.1

Variance Components from Brennan et al.'s Generalizability Study of Listening Skill

Source of variance	Variance component
Person (*p*)	.324
Task (*t*)	.127
Rater (*r*)	.012
pt	.393
pr	.014
tr	.022
ptr	.317

Note. Results represent averages from three test forms. From *Educational and Psychological Measurement* (p. 161), by R. L. Brennan, X. Gao, and D. A. Colton, 1995, Thousand Oaks, CA: Sage. Copyright 1995 by Sage. Adapted with permission.

variance is *p*, the estimate of universe score variation. Remember that in this context this is a desirable source of variance, and it indicates that much of the variability in the examinees' listening skills scores is due to true or real differences in their underlying talent (at least to the extent that the tasks validly measure what most people think of as lis-

Figure 2.12

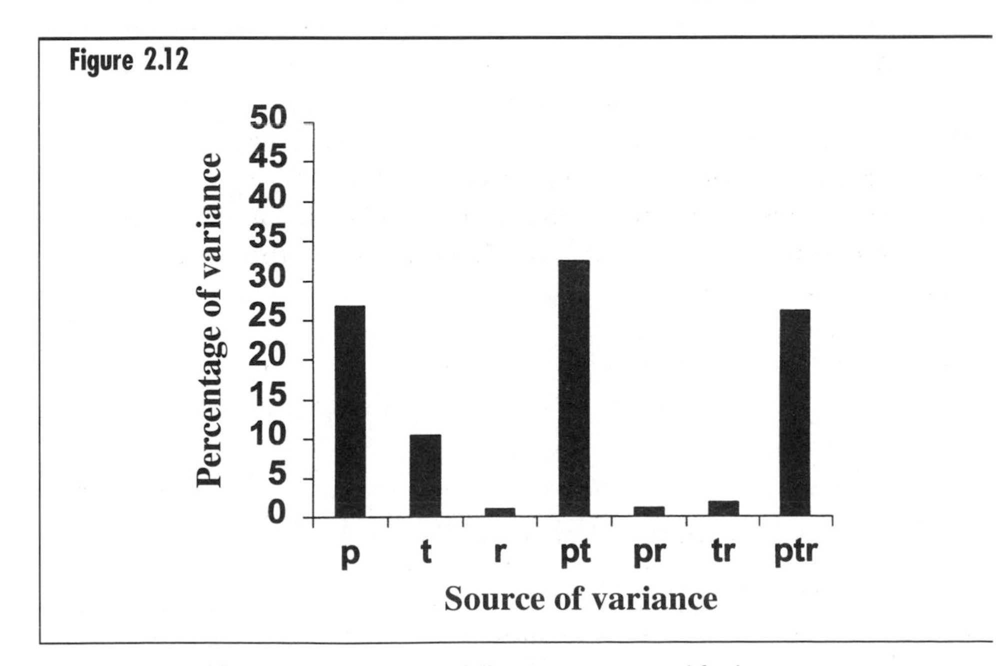

Percentages of variance in listening skills scores accounted for by sources in Brennan et al.'s (1995) generalizability study. p = person; r = rater; t = task.

tening skills). But another large component of variance, *pt*, is not desirable, and it indicates that the relative standing of examinees changes from task to task. In other words, the answer to the question "Who are the better listeners?" would, on the basis of this analysis, yield the answer "It depends to some extent on the particular task or message." Another sizeable source of undesirable variation is the *ptr* component. This source is not so easily interpreted because it contains both systematic and random sources. The systematic part reflects shifts in the relative standing of examinees across combinations of tasks and raters. Finally, what is not a considerable source of variation is equally notable. For example, the *pr* source is fairly small, indicating that raters provided similar relative ratings of the examinees.

The real power of generalizability theory is that it allows a way to ask "what-if?" questions about future use of a measure. Such D studies allow a researcher to tailor a measure to produce optimal reliability given cost and pragmatic considerations. How many tape-recorded messages should be used in future applications of the listening test? How many raters should be used? Remember that aggregation suppresses error, so increasing the number of tasks and increasing the number of raters reduces error and increases the reliability of the measure. But does increasing the number of tasks have the same effect as increasing the number of raters? The answer is "not necessarily." Examining Figure 2.12 provides some guidance about likely effect. If I were concerned only about making relative decisions, then only interactions involving *p* would be sources of error. The *pr* interaction is already very small, so I might guess that increasing the number of raters will not help me much. By contrast, the *pt* interaction is quite large, so increasing the number of tasks should suppress error more substantially. These intuitions are correct, as indicated in Figure 2.13. The generalizability coefficient increases with greater numbers of raters and tasks, but the rate of increase is much more substantial for tasks. Faced with limited resources, I would be better off using a large number of tasks with relatively fewer raters.

Remember that there are two ways to view reliability. The gains in reliability arise from reductions in error. Figure 2.14 shows how the standard error of measure decreases with increasing numbers of tasks and raters. The same conclusions would be drawn from either Figure 2.13 or Figure 2.14 because they represent two different ways of viewing the same phenomenon. One advantage to Figure 2.14 is that it represents reliability in the metric of the measure that is used. In other words, I can get a good sense of the uncertainty that I have about a person's

Figure 2.13

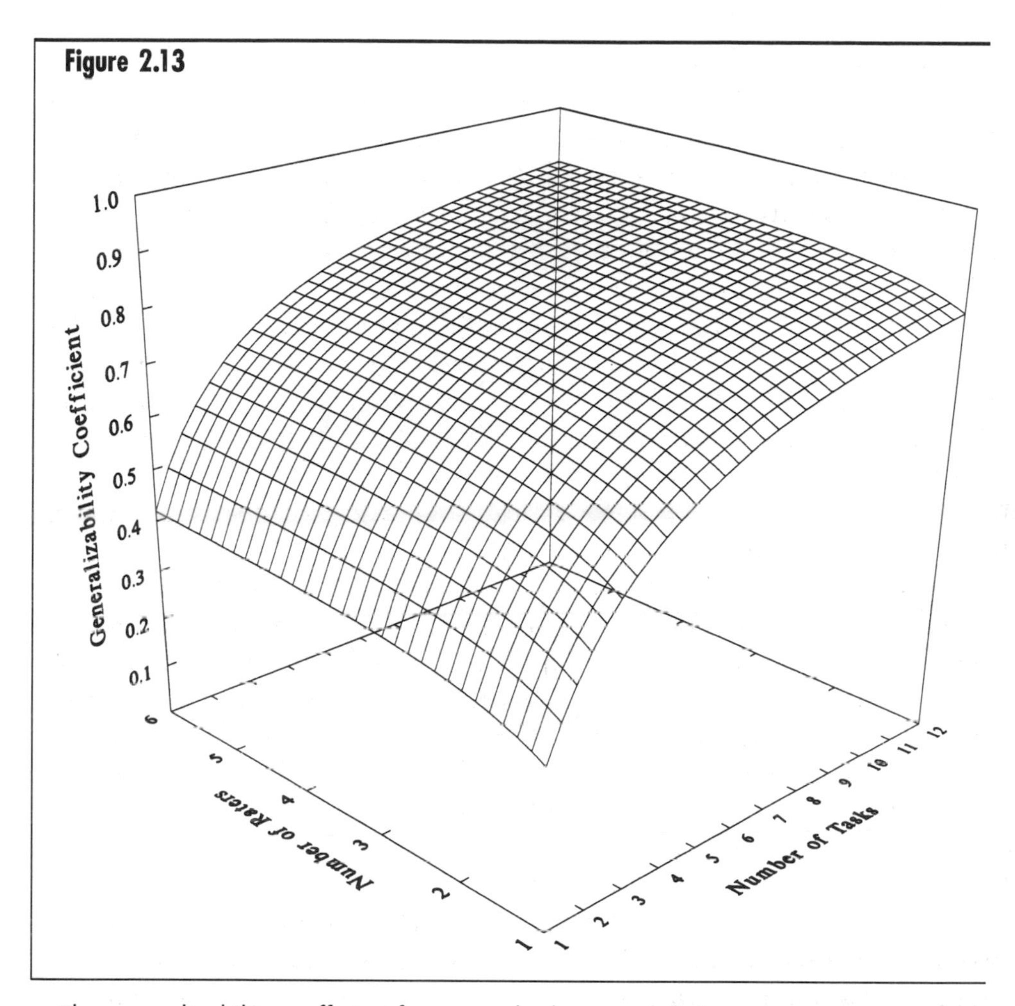

The generalizability coefficient for a completely crossed design using variance components from Brennan et al.'s generalizability study of listening skill. From *Educational and Psychological Measurement* (p. 167), by R. L. Brennan, X. Gao, and D. A. Colton, 1995, Thousand Oaks, CA: Sage. Copyright 1995 by Sage. Adapted with permission.

listening skills score by examining the *standard error*. For example, with eight tasks and two raters, the standard error for a relative decision is .28, and the same universe score would be consistent with obtained scores that differ by as much as 1 point (with 95% confidence). This might be more informative to a researcher or practitioner than knowing that the generalizability coefficient under these conditions is .81.

The power of generalizability theory is evident in another feature of this D study. By specifying errors of measurement more specifically, generalizability theory allows me to ask "trade-off" questions. I can ask, for example, what numbers of raters and tasks will produce a reliability

Figure 2.14

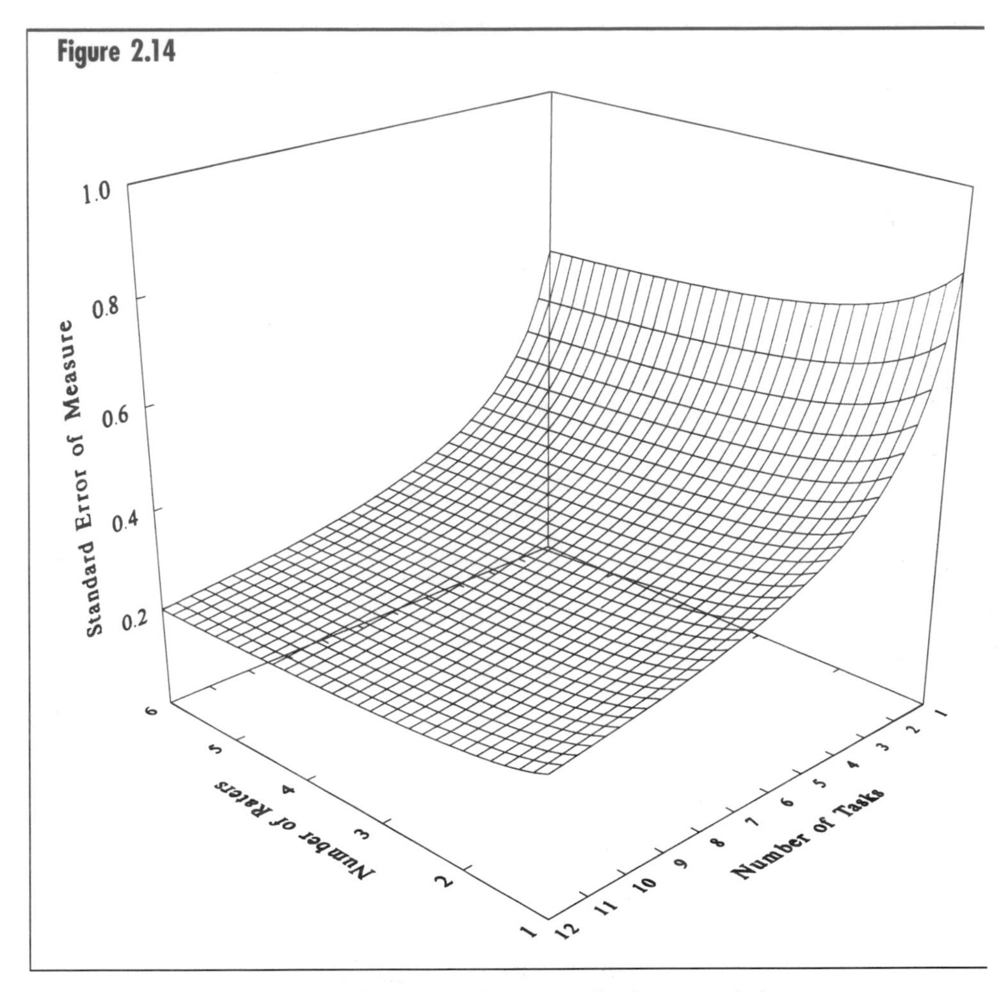

Projected standard errors of measure for a completely crossed design using variance components from Brennan et al.'s generalizability study of listening skill. From *Educational and Psychological Measurement* (p. 165), by R. L. Brennan, X. Gao, and D. A. Colton, 1995, Thousand Oaks, CA: Sage. Copyright 1995 by Sage. Adapted with permission.

of a given amount. Figure 2.15 displays the results of asking this question for a generalizability coefficient of .80. Several conclusions are obvious. First, a minimum of 6 tasks is required; reliability of at least .80 cannot be achieved with 5 or fewer tasks no matter how many raters are used. Second, as the number of raters is reduced, the number of tasks that must be added to compensate goes up, a little at first but then dramatically. If 4 raters are available, only 6 tasks are needed. If 2 raters are used, then 7 tasks are necessary. Dropping the number of raters to 1 requires an even greater trade off; at least 11 tasks are necessary to

Figure 2.15

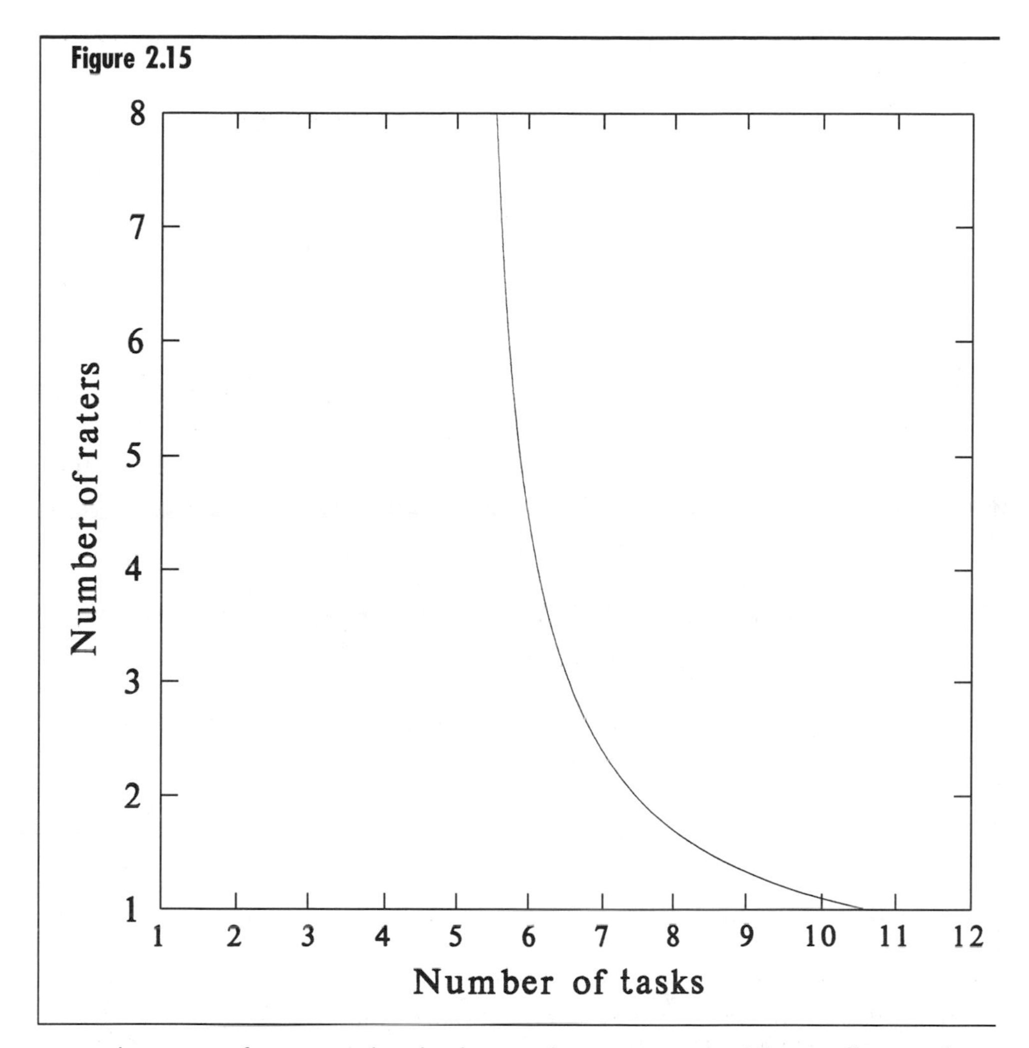

Combinations of raters and tasks that produce a generalizability coefficient of .80 (based on variance components from Brennan et al.'s (1995) generalizability study of listening skill).

achieve a reliability of at least .80. Depending on the costs of training raters or administering tasks, a sensible combination of raters and tasks can be determined in any D study.

Figures 2.13 and 2.14 are based on the assumption that all raters evaluate the summaries of all examinees but that is not the only design option. D studies can pose "what-if" questions about designs other than the one that defines the original G study. For example, pragmatic constraints might dictate that in future applications a different set of 3 raters are required for each examinee. This requires a D study assuming a partially nested design: $(r{:}p) \times t$. A D study, appropriately combining variance components from the G study that would be confounded (i.e., *r* and *pr*, *tr*, and *prt*), could determine if this alternative design would produce noticeably different estimates of reliability. In this case, it would not because the additional sources of variance that are added to the error variance due to confounding (i.e., *r* and *tr*) are quite small.

Another question that I might ask is how changing the type of decision affects the reliability. The results in Figures 2.13, 2,14, and 2.15 assume that a relative decision is made, that is, that all that matters is the relative standing of the examinees. What if the absolute value of the score mattered, perhaps because some criterion score was going to be used to select people with outstanding listening skills for a particular job that required such talent? A D study can address that question by using the G-study variance components but combining them to produce the appropriate definition of error for an absolute decision. Figure 2.16 shows the index of dependability that would result from different combinations of facet levels in a completely crossed design. The shape of Figure 2.16 resembles Figure 2.13 very closely, a result that could have been anticipated from Table 2.1 in which clearly the additional error components for the absolute decision (i.e., *t*, *r*, and *tr*) are small compared with the error components that defined the relative decision.

Although D studies can be used to explore numerous questions about reliability, they cannot address questions about measurement facets that were not included in the original G study. Likewise, D studies cannot estimate the consequences of different designs if the variance components are not available from the G study. Imagine, for example, that Brennan et al. (1995) designed their G study as an $(r{:}t) \times p$ design instead of a $p \times r \times t$ design. This would mean that the *r* source of variance and *rt* source of variance could not be estimated separately; they would be confounded. Consequently, it would not be possible to conduct a D study in which their separate effects are forecasted. For example, it would not be possible to explore the consequences of a completely crossed design because all of the separate variance components that exist in that design cannot be estimated from the results of an $(r{:}t) \times p$ G study. Accordingly, the widest range of D-study questions can be posed if the G study is completely crossed.

To this point, I discussed generalizability theory within its traditional boundaries—as a means of investigating the reliability of measures and forecasting reliability under different measurement conditions. Researchers have made creative use of generalizability theory to

Figure 2.16

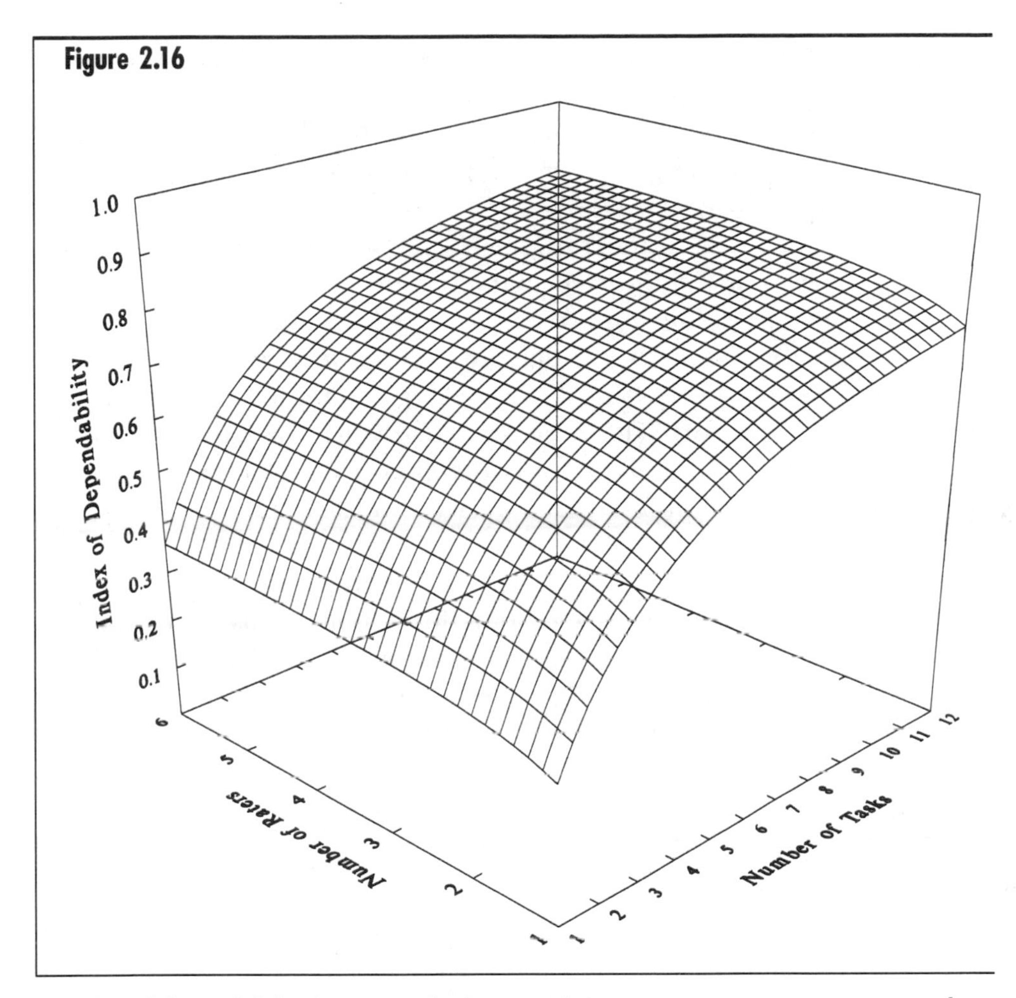

Index of dependability for a completely crossed design using variance components from Brennan et al.'s generalizability study of listening skill. From *Educational and Psychological Measurement* (p. 167), by R. L. Brennan, X. Gao, and D. A. Colton, 1995, Thousand Oaks, CA: Sage. Copyright 1995 by Sage. Adapted with permission.

address questions of more substantive interest as well (e.g., Shrout, 1993). Lakey, McCabe, Fisicaro, and Drew (1996) provided a good example of the additional powerful ways that generalizability theory can be used. Lakey et al. were interested in perceptions of social support, particularly the extent to which perceptions of support are determined by characteristics of the supporters and the characteristics of the perceivers. The question is an important one because different theories of social support place different emphasis on the locus of perceived support. Some claim that it reflects actual or enacted support and thus should be determined largely by the characteristics of the supporters;

that is, some people are better than others at providing social support. Other theories claim that support is in the eye of the beholder and thus should be determined to a large extent by the characteristics of the perceivers; that is, some people are more inclined to view the behavior of others as supportive. Of course a mixture of the two makes sense as well, that is, perceived support might be determined by the joint influence of perceiver and supporter characteristics.

Lakey et al. (1996) conducted three studies using a very similar methodology. In each study, participants (perceivers) rated the supportiveness of a set of targets (supporters) using several items meant to tap perceived social support. The design of each study was thus a perceiver (p) × supporter (s) × item (i) completely crossed design. Averaging the results from the three studies provides a clear picture about the sources of perceived support. As Figure 2.17 shows, the most important source of variance was the Perceiver × Supporter interaction, accounting for 41% of the variance in support judgments. This indicates that perceptions of supportiveness were determined by the joint combinations of perceiver and supporter characteristics. In other words, some behaviors are viewed as supportive by some but not all perceivers. Lakey

Figure 2.17

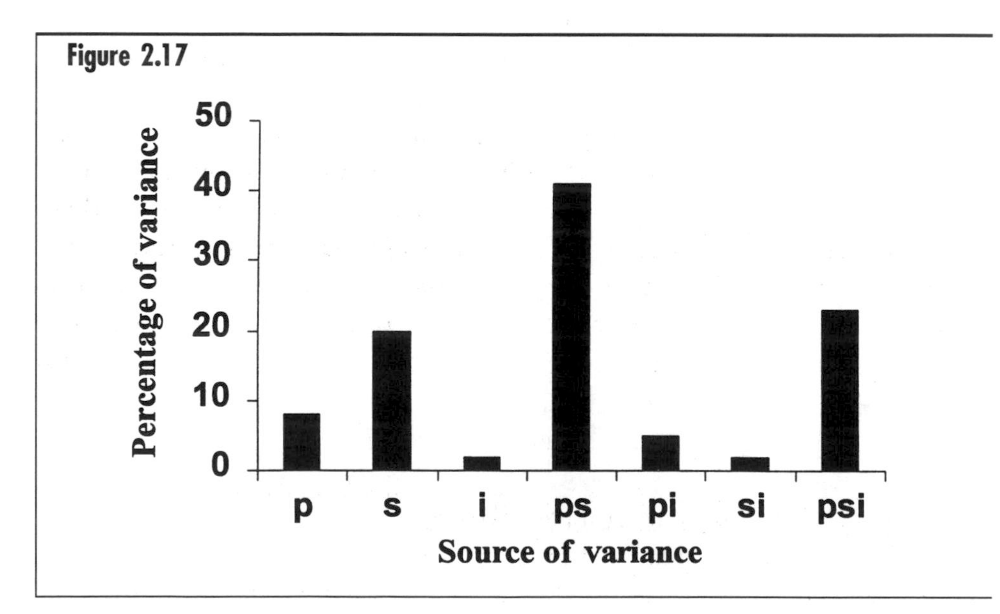

Percentages of variance accounted for in support judgments in Lakey et al.'s generalizability study. i = item; p = perceiver; s = supporter. From "Environmental and Personal Determinants of Support Perceptions: Three Generalizability Studies," by B. Lakey, K. M. McCabe, S. A. Fisicaro, and J. B. Drew, 1996, *Journal of Personality and Social Psychology,* p. 1276. Copyright 1996 by the American Psychological Association. Reprinted with permission.

et al. provided several interpretations. For example, people may exert varying amounts of influence in shaping the support provided by others. Alternatively, recipients and providers of support may have preferred styles of enacted support, such that support judgments reflect the match between these styles. Still another explanation rests on the likely individual variation in the interpretation of supportive behaviors. As Lakey et al. (1996) pointed out,

> people who make light of serious situations might be seen as supportive by some people but as unsupportive by others. In this case, judging supportiveness would be similar to judging art. In the same way that cubism is beautiful to some but grotesque to others, the same personal qualities that are supportive to some will be insensitive to others. (p. 1277)

However, Figure 2.17 also indicates an agreement among perceivers as to whom is most supportive; the supporter component accounted for 20% of the variance in support judgments. By comparison, the perceiver component contributed relatively little to support judgments, as did the item component. In other words, it was not the case that some people were substantially more inclined to report greater amounts of social support from the sources they viewed, nor was it the case that support judgments depended much on the particular way the judgment was rendered. Note, however, that the perceiver effect, although small, has been found in other research (see Lakey et al., 1996).

As these examples show, generalizability theory can be used to address both measurement and substantive issues. Its application is bounded only by the imagination of creative scientists and methodologists. The common thread that ties these applications together is the recognition that variability in observations is multifaceted, and careful study of the components of variability that define observed variability can led to important insights into the quality of measures and the truth of theoretical propositions.

Multivariate Generalizability Theory

Generalizability theory has a multivariate extension that considerably expands the potential application of the approach (Brennan, 1992; Brennan et al., 1995; Cronbach et al., 1972; Marcoulides, 1994; Nussbaum, 1984; Webb & Shavelson, 1981; Webb, Shavelson, & Maddahian, 1983). The basis for the extension is best understood by remembering that the basic building blocks in *univariate* generalizability theory are the variance components; the obtained score variance for a single measure is separated into components of variance that reflect universe score variability and a variety of error variances. When more than one measure is examined in generalizability theory, the procedure becomes multivariate and an additional important piece of information can be obtained and separated into components: the obtained covariance between two measures. A *covariance* is simply a measure of the direction and magnitude of relation between two measures. The more familiar correlation coefficient is simply a covariance divided by the standard deviations of each measure.

When multiple measures are included in the analysis, the basic information for *multivariate generalizability theory* is contained in the *variance–covariance matrix* of obtained scores. This is a square, symmetrical matrix in which the variances for the measures are the main diagonal elements, and the covariances between pairs of measures are the elements that are not on the main diagonal. This variance–covariance matrix is separated into a universe score variance–covariance matrix and additional variance–covariance matrices that represent the different error sources. The procedures for making that separation are precisely those used when variances for a single measure are considered. The only additional complications in multivariate generalizability theory are that variances for more than one measure and covariances between the measures need to be considered.

The approach is best understood by returning to one of my examples. Recall that Brennan et al. (1995) examined the reliability of a measure designed to assess listening skills. In that study, 50 examinees listened to 12 tape-recorded messages, taking notes while each message was played. After each message was completed, the examinees were told to consult their notes and write a summary of the message. The written summaries were then scored by three raters, each of whom independently assigned a score from *low* (0) to *high* (5) representing the holistic judgment of the accuracy and completeness of each summary. In that same study, Brennan et al. had a different set of three raters score the summaries for writing skill. Thus, two measures—listening skill and writing skill—were obtained and can form the basis for a multivariate generalizability analysis. Of course, the writing skills measure can be examined using univariate generalizability theory too; Brennan et al. (1995) reported those results. As with the univariate approach, there are seven basic components in the multivariate analysis, but these components are variance–covariance matrices instead of single variances. These component matrices are listed in Table 2.2.

The diagonal entries in each matrix are variances, and for the listening skills measure, those variances duplicate the entries in Table 2.1. The corresponding variances for the writing skills measure could be used to carry out a univariate generalizability theory analysis. The new elements—covariances—are the off-diagonal elements.[8]

One particularly useful variance–covariance matrix is that for universe scores (*p*). The covariance indicates the degree to which the universe scores for listening skills and writing skills are related. When that covariance is divided by the standard deviations for each measure (i.e., the square roots of the diagonal elements in the matrix), the result is the correlation between listening skills universe scores and writing skills universe scores:

$$\rho_{LW(p)} = \frac{\sigma_{LW(p)}}{\sqrt{\sigma_{L(p)}\sigma_{W(p)}}} = \frac{.356}{\sqrt{.324(.691)}} = .75.$$

Because this correlation involves universe scores, it represents a disattenuated correlation, that is, a correlation free from measurement error. It indicates that the two skills—listening and writing—are closely related to each other.

The other matrices are also quite informative. For example, the covariance for the *t* facet indicates the degree to which the rank ordering of task means replicates across the two skills (these task means are the averages across people and raters). The covariance (.039) is fairly small, but so are the variances. The corresponding correlation is .69, suggesting that tasks that produce a high listening-skill mean also tend to produce a high writing-skill mean. Finally, consider the variance–covariance matrix for the *pt* component. The variance components indicate the degree to which the rank ordering of people changes over tasks. The covariance component indicates the degree to which the rank ordering differences are consistent across the two tests. The covariance component is quite small compared to the variances, indicating the rank ordering differences are not consistent across the two tests.

Multivariate generalizability theory has a D-study feature as well; it can be used to pose "what-if" questions about the influence of different study designs and facet levels on observable correlations in the presence

[8]All covariances involving the rater facet are zero in this analysis because different raters were used for listening and writing assessments.

Table 2.2

Variance and Covariance Components From Brennan et al.'s Generalizability Study of Two Skills

Source of variance	Component	Listening	Writing
p	Listening	.324	.356
	Writing	.356	.691
t	Listening	.127	.039
	Writing	.039	.025
r	Listening	.012	.000
	Writing	.000	.010
pt	Listening	.393	.030
	Writing	.030	.159
pr	Listening	.014	.000
	Writing	.000	.047
tr	Listening	.022	.000
	Writing	.000	.008
ptr	Listening	.317	.000
	Writing	.000	.218

Note. Results represent averages from three test forms. p = person; r = rater; t = trainer. From *Educational and Psychological Measurement* (p. 169), by R. L. Brennan, X. Gao, and D. A. Colton, 1995, Thousand Oaks, CA: Sage. Copyright 1995 by Sage. Adapted with permission.

of measurement error (recall that the previously calculated universe score correlation is an "errorless" estimate). For example, I could estimate how the correlation between listening and writing scores would vary if I used different numbers of tasks and raters. The fewer the number of tasks and raters, the greater the error and the more the correlation would be attenuated. Figure 2.18 shows this relation, assuming that the same task is used to produce both listening and writing scores (see Brennan et al., 1995, for an elaboration). Note that the correlation never exceeds .75. That value is the upper limit when the correlation is based on the average of all scores in the universe of admissible observations; it is the same value that I established earlier when I calculated the universe score correlation. Note also that when few raters or tasks are used, the attendant measurement error reduces the correlation that one can find between obtained listening and writing skill scores.

There is one final advantage to multivariate generalizability theory that deserves mention. It can be used to determine the generalizability of composite variables formed by constructing linear combinations of

Figure 2.18

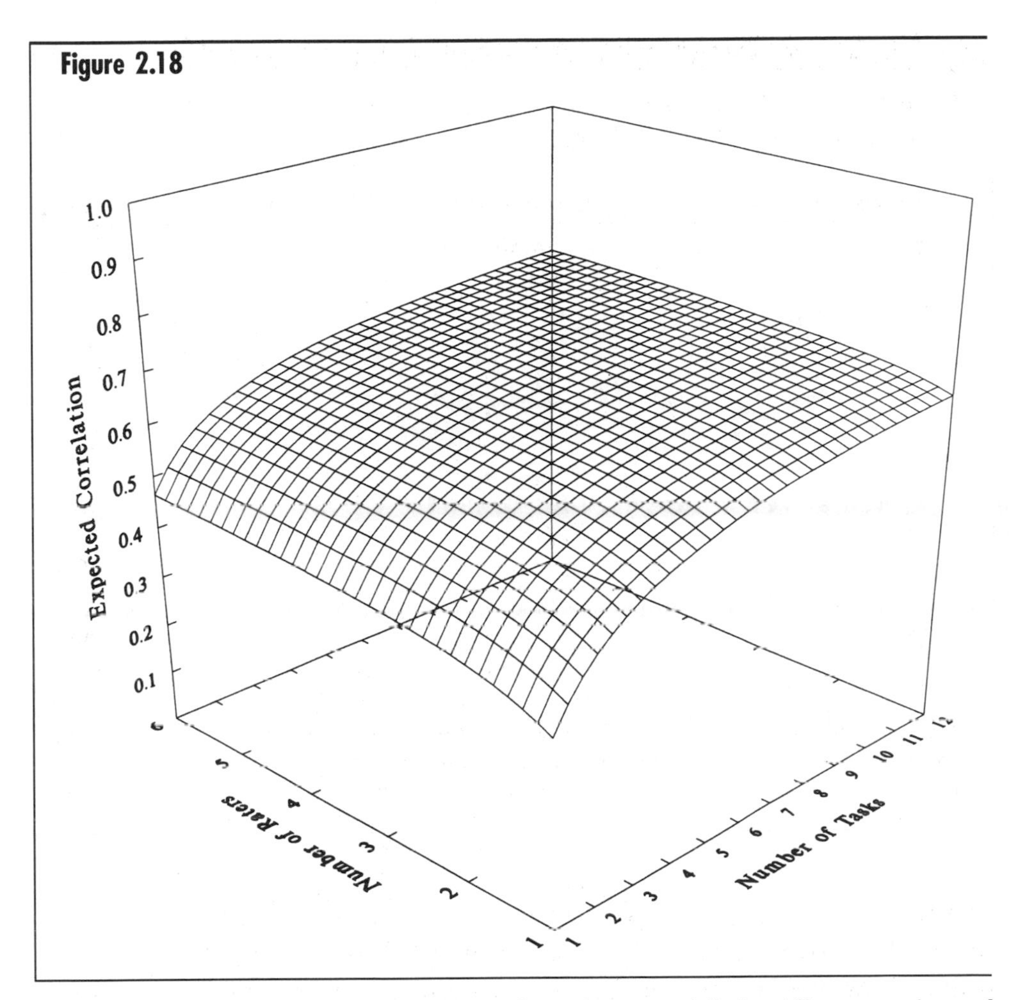

Expected correlation between listening skills and writing skills for different numbers of tasks and raters. From *Educational and Psychological Measurement* (p. 171), by R. L. Brennan, X. Gao, and D. A. Colton, 1995, Thousand Oaks, CA: Sage. Copyright 1995 by Sage. Adapted with permission.

the multiple measures. Two approaches can be taken. One approach is data driven and asks "What weighted combination of the measures provides the largest possible generalizability coefficient?" The weights are determined empirically, so the resulting generalizability coefficient is as large as possible given the specified design and number of facet levels (e.g., see Webb & Shavelson, 1981). The approach is similar in spirit to discriminant analysis in which weighted linear combinations of variables are derived to maximize group separation. A second approach is theory driven and asks "What is the generalizability of a linear combination with a specified set of weights?" This approach may not produce the highest possible generalizability coefficient, but it provides the generalizability coefficient for a linear combination that may be of more substantive interest. For example, in Brennan et al.'s study, there might be substantive interest in the difference between listening and writing skills.

This brief summary of multivariate generalizability theory certainly does not do justice to the full range and complexity of the approach, but it should underscore the additional and powerful ways that generalizability theory can be used to explore the reliability and measurement consequences of different designs when multiple measures are available.

Caveats, Complications, and Unfinished Business

Like any statistical procedure, the appropriate use and interpretation of generalizability theory rests on some assumptions. First, the data examined in a generalizability theory analysis should be interval in nature or at least strongly ordinal, so that interpretations of variance components are clear (but see Maxwell & Delaney, 1990). Most numerical rating scales (e.g., Likert scales, bipolar ratings, and semantic differential scales) are not, strictly speaking, interval scales. It is routine to treat them as interval scales in statistical analyses, however, and no harm seems to have resulted from that practice. No particular distributional form for the data is assumed to estimate variance components from ANOVA, although the use of maximum likelihood procedures does require a normality assumption (Brennan, 1992; Cronbach et al., 1972). Less an assumption than a caution is the reminder that variance components are estimates that are subject to sampling variability. This means that variance components will be more precise for generalizability studies that use large samples of both people and facet levels. Procedures for estimating confidence intervals for variance components have restrictions (e.g., normality), but distribution-free procedures such as bootstrapping may prove useful in this regard (see also Smith, 1981).[9]

[9]Bootstrapping (e.g., Mooney & Duval, 1993) is a resampling technique that builds a sampling distribution for a statistic from the empirical data rather than assuming some theoretical sampling distribution that requires assumptions that may not be true. In bootstrapping, a random sample size N is drawn, with replacement, from the obtained sample. The statistic (e.g., a mean, a correlation or a regression coefficient) is calculated. Then the procedure is repeated, typically at least 1,000 times. This produces a bootstrap sampling distribution for the statistic. The mean or median of that distribution is the best estimate of the population value. The upper and lower tails of the distribution can be used for significance testing by establishing whether the null hypothesis value falls below or above, for example, the 2.5% or 97.5% values.

The application of generalizability theory can also encounter some special problems. For example, the designs described in this chapter are all relatively "clean" and do not create unusual problems for estimating variance components. These balanced designs do not have missing data or unequal levels of facets nested within other facets. So-called unbalanced designs require careful attention to ensure that variance components are derived correctly (see Brennan, 1992; and Cronbach et al., 1972).

Another complication is the occasional appearance of a negative variance component. Although this is a conceptual impossibility, it can arise in practice, and it signals one of two problems, one benign, the other serious. The benign cause of negative variance components is sampling error. In these cases it is customary to set the variance component to zero (see Shavelson & Webb, 1991, for a discussion of when in the estimation process this should be done). Sometimes, however, a variance is negative and so large in absolute value that sampling error seems an unlikely explanation. The more likely problem is a misspecified measurement model (e.g., a missing facet).

A third problem that may arise frequently is the presence of correlated effects (Smith & Luecht, 1992; see also Cronbach et al., 1972; and Cronbach, Linn, Brennan, & Haertel, 1997). Generalizability theory assumes that all of the effects in the measurement model are uncorrelated, a condition that may be violated with repeated measurement trials over brief periods or situations in which repeated observations are made for a fixed task. Smith and Luecht (1992) showed that variance components can be biased under these conditions.

A fourth limitation of generalizability theory is the frequent assumption that standard errors are the same at all score levels. In other words, the same standard error of measure is often applied to all objects of measurement, regardless of the underlying universe score. It seems unlikely that this assumption is true (Kolen, Hanson, & Brennan, 1992), and attempts to estimate separate standard errors for different score levels is a topic of keen interest in the measurement field (e.g., Brennan, 1996).

Finally, incomplete understanding of error sources can lead to faulty interpretations (see Schmidt & Hunter, 1996, for an excellent discussion of this general problem). One common problem is to ignore "hidden" facets. For example, a researcher might administer multiple items to a group of respondents on one occasion and forget that, although occasion is a constant in the study, it likely will be a source of variability and considered as error in application. The problem is that occasion variance (what Schmidt & Hunter, 1996, called "transient error") acts like a constant for each individual on a given occasion and is therefore counted as true score or universe score for each individual. Its transient nature would only be revealed and estimated separately if multiple occasions were included in the measurement design. Likewise, error specific to a particular test form (what Schmidt & Hunter, 1996, referred to as "specific error") masquerades as true score variance unless test forms are explicitly included in the measurement design and their contribution to error is estimated. The general principle here is simple: Universe score variability and error variability depend on the conditions of measurement and the conditions of application. Careful attention must be given to what should count as error, and then the measurement design must be constructed so that all sources of error can emerge and be estimated correctly.

The limitations that I described might seem rather severe and call into question the utility of generalizability theory. A better conclusion would be that they represent minor challenges likely to be solved in time. Despite these limitations, generalizability theory offers such clear advantages over the classical model that it should be the method of choice for understanding, developing, and applying measurements in the behavioral sciences.

Suggestions for Further Reading

There is much more to reliability and generalizability theory than is described in this chapter. Developments continue at a rapid clip, and novel applications continue to emerge. For example, one useful extension is the recognition that generalizability theory is symmetrical, that is, any facet can be used as the object of measurement (Cardinet, Tourneur, & Allal, 1976, 1981). One common example of this feature is the use of groups rather than individuals as the focus of decision making (Brennan, 1995; see also Cronbach, Linn, Brennan, & Haertel, 1997). More generally, this aspect of generalizability theory emphasizes that what counts as error in one measurement context could well be the object of measurement in another. This greatly extends the potential uses of generalizability theory.

These and other advances are described in more complete descriptions of the generalizability theory approach (e.g., Brennan, 1992; Cronbach et al., 1972; Shavelson, & Webb, 1991; Shavelson, Webb, & Rowley, 1989). Additional general references for psychometric theory are Feldt and Brennan (1989), Ghiselli, Campbell, and Zedeck (1981) and Nunnally and Bernstein (1994). Because generalizability theory uses ANOVA to estimate the variance components, additional background in that area may be necessary to conduct generalizability studies or to interpret complex designs. Useful sources include Maxwell and Delaney (1990), Myers and Well (1991), and Winer (1971). Finally, generalizability theory represents an extension of the classical approach to measurement, and the classical approach represents just one way of thinking about measurement. Another modern measurement model—item response theory—holds considerable promise because it does not make some of the restrictive assumptions that are embedded in the classical view. For an introduction to item response theory, see Hambleton and Swaminathan (1991) and chapter 3 of this volume.

Glossary

ABSOLUTE DECISION A decision in which the absolute score of a respondent, and not just the respondent's relative standing, is used to make a decision. The decision typically involves a criterion or cut-off score.

AGGREGATION The statistical principle in which random error is suppressed through replication (e.g., test items, raters, or occasions).

ALTERNATIVE FORMS RELIABILITY Reliability estimated from the correlation between two versions of the same test administered to the same respondents.

ANALYSIS OF VARIANCE (ANOVA) A statistical procedure for partitioning the variability of measurements into systematic and random components.

CLASSICAL THEORY The approach to measurement that views an obtained score as being the linear combination of a true score and an error score. Also known as "classical test theory" and "classical measurement theory."

CONFOUNDED EFFECTS Sources of variance that cannot be estimated separately. They arise in nested designs.

CONSTRUCT An unobservable entity that is estimated by observations.

COVARIANCE A measure of the direction and magnitude of relation between two measures. Sometimes called an "unstandardized correlation" because it is the numerator of the more familiar correlation coefficient.

CROSSED DESIGN A design in which each level of each facet is combined with each level of every other facet. The number of discrete combinations of conditions in a crossed design is equal to the product of the numbers of levels of all facets.

D STUDY A study in which the components of variance from a G study are used to estimate the measurement error or reliability under different proposed conditions of measurement. The proposed conditions of measurement define the universe of admissible generalization.

ERROR SCORE (e) See measurement error.

FACET A condition of measurement in a generalizability study. Facets define the universe of admissible observations.

FIXED EFFECT A condition of measurement in which all levels of interest are included in the design.

GENERALIZABILITY COEFFICIENT ($E\rho^2$) An estimate of reliability from generalizability theory that is appropriate for relative decisions.

G STUDY A study in which the variance components corresponding to different facets of measurement are estimated.

GENERALIZABILITY THEORY The extension of classical theory in which sources of error variance are separately estimated and used to forecast the measurement implications for different measurement conditions.

INDEX OF DEPENDABILITY (ϕ) An estimate of reliability from generalizability theory that is appropriate for absolute decisions.

INTERNAL CONSISTENCY An estimate of reliability based on the intercorrelations among multiple items of a test administered on one occasion to the same respondents.

MEASUREMENT ERROR The component of an observed score that reflects uncertainty about a respondent's standing on the true score.

MULTIVARIATE GENERALIZABILITY THEORY The extension of generalizability theory in which more than one measure is examined, and the sample variance–covariance matrix is separated into universe score components and error variance components.

NESTED DESIGN A design in which different levels of one facet occur at each level of another facet. Nested designs produce confounded effects.

OBJECT OF MEASUREMENT The facet in a generalizability study about which decisions will be made. The variance component for this facet is called "universe score variance."

OBSERVED SCORE (x) The number resulting from the operation designed to assess an underling construct. The observed score will assess both true score (the construct) and error. The terms *test* and *measure* are often used interchangeably with the term *observed score.*

OPERATIONAL DEFINITION The specific rules for obtaining systematic measurement corresponding to some underlying construct.

PARALLEL FORMS RELIABILITY See *alternative forms reliability.*

RANDOM EFFECT A condition of measurement in which the levels are assumed to be randomly sampled from the population of levels and are considered exchangeable with any other random sample.

RANDOM ERROR An error that is unpredictable in its influence.

RELATIVE DECISION A decision based on the relative standing of respondents.

RELIABILITY The trustworthiness of a measure. Common terms to describe reliability include *consistency, generalizability, repeatability,* and *dependability.*

RELIABILITY COEFFICIENT (r_{xx}) An estimate of the amount of obtained score variance that is due to true score variance.

RESIDUAL ERROR The variance remaining after all other components of variance have been estimated.

STANDARD DEVIATION (*SD*) A measure of variability equal to the square root of the variance. In a normal distribution, approximately 68% of the sample will fall within a range of scores extending from 1 *SD* below the mean to 1 *SD* above the mean.

STANDARD ERROR The standard deviation of a sampling distribution for a statistic.

STANDARD ERROR OF MEASURE (σ_{meas}) The standard deviation of error scores.

STANDARDIZATION The procedure by which all aspects of the measurement situation are carefully controlled so that extraneous sources of error cannot influence test scores.

SYSTEMATIC ERROR A source of error that produces a constant influence on obtained scores.

TEST–RETEST RELIABILITY An estimate of reliability based on the correlation between scores from the same measure administered to the same respondents on two different occasions.

TRUE SCORE (t) The component of an obtained score that corresponds to the respondent's standing on the underlying construct.

UNIVARIATE Statistical analyses that are conducted on a single outcome variable.

UNIVERSE OF ADMISSIBLE GENERALIZATION The conditions of measurement in a D study defined by the facets of measurement.

UNIVERSE OF ADMISSIBLE OBSERVATIONS The conditions of measurement in a G study that are defined by the facets of measurement.

UNIVERSE SCORE The generalizability theory analog to the true score in classical theory. It represents the average of all observations in the universe of admissible observations.

VALIDITY The extent to which a measure is labeled appropriately, that is, the extent to which it measures what it purports to measure.

VARIANCE (σ^2) A measure of variability in a distribution of scores. The variance is the average of the squared deviations of individual scores from the mean of those scores.

VARIANCE COMPONENT The variability in obtained scores that corresponds to a particular source of variability in a generalizability study.

VARIANCE–COVARIANCE MATRIX The square symmetric matrix containing the *variances* (diagonal elements) and *covariances* (off-diagonal elements) for a collection of measures.

References

Bollen, K. A. (1989). *Structural equations with latent variables.* New York: Wiley.

Brennan, R. L. (1992). *Elements of generalizability theory* (rev. ed.). Iowa City, IA: American College Testing.

Brennan, R. L. (1995). The conventional wisdom about group mean scores. *Journal of Educational Measurement, 32,* 385–396.

Brennan, R. L. (1996). *Conditional standard errors of measurement in generalizability theory* (ITP Occasional Paper No. 40). Iowa City: University of Iowa, Iowa Testing Programs.

Brennan, R. L., Gao, X., & Colton, D. A. (1995). Generalizability analyses of work keys listening and writing tests. *Educational and Psychological Measurement, 55,* 157–176.

Brennan, R. L., & Kane, M. T. (1977). An index of dependability for mastery tests. *Journal of Educational Measurement, 14,* 277–289.

Cardinet, J., Tourneur, Y., & Allal, L. (1976). The symmetry of generalizability theory: Application to educational measurement. *Journal of Educational Measurement, 13,* 119–135.

Cardinet, J., Tourneur, Y., & Allal, L. (1981). Extension of generalizability theory and its application in educational measurement. *Journal of Educational Measurement, 18,* 183–204.

Cook, T. D., & Campbell, D. T. (1979). *Quasi-experimentation: Design and analysis issues for field settings.* Chicago: Rand McNally.

Cronbach, L. J., Gleser, G. C., Nanda, H., & Rajaratnam, N. (1972). *The dependability of behavioral measurements: Theory of generalizability of scores and profiles.* New York: Wiley.

Cronbach, L. J., Linn, R. L., Brennan, R. L., & Haertel, E. H. (1997). Generalizability analysis for performance assessments of student achievement or school effectiveness. *Educational and Psychological Measurement, 57,* 373–399.

Feldt, L. S., & Brennan, R. L. (1989). Reliability. In R. L. Linn (Ed.), *Educational measurement* (3rd ed., pp. 105–146). New York: Macmillan.

Fleiss, J. L., & Shrout, P. E. (1977). The effect of measurement error on some multivariate procedures. *American Journal of Public Health, 67,* 1184–1189.

Ghiselli, E. E., Campbell, J. P., & Zedeck, S. (1981). *Measurement theory for the behavioral sciences.* New York: Freeman.

Hambleton, R. K., & Swaminathan, H. (1991). *Fundamentals of item response theory.* Newbury Park, CA: Sage.

Humphreys, L. G., & Dasgow, F. (1989). Some comments on the relation between reliability and statistical power. *Applied Psychological Measurement, 13,* 419–425.

Kolen, M. J., Hanson, B. A., & Brennan, R. L. (1992). Conditional standard errors of measurement. *Journal of Educational Measurement, 29,* 285–307.

Kopriva, R. J., & Shaw, D. G. (1991). Power estimates: The effect of dependent variable reliability on the power of one-factor ANOVAs. *Educational and Psychological Measurement, 51,* 585–595.

Lakey, B., McCabe, K. M., Fisicaro, S. A., & Drew, J. B. (1996). Environmental and personal determinants of support perceptions: Three generalizability studies. *Journal of Personality and Social Psychology, 70,* 1270–1280.

Marcoulides, G. A. (1994). Selecting weighting schemes in multivariate generalizability studies. *Educational and Psychological Measurement, 54,* 3–7.

Maxwell, S. E., & Delaney, H. D. (1990). *Designing experiments and analyzing data: A model comparison perspective.* Belmont, CA: Wadsworth.

Mooney, C. Z., & Duval, R. D. (1993). *Bootstrapping: A nonparametric approach to statistical inference.* Newbury Park, CA: Sage.

Myers, J. E., & Well, A. D. (1991). *Research design & statistical analysis.* New York: HarperCollins.

Nunnally, J. C., & Bernstein, I. H. (1994). *Psychometric theory* (3rd ed.). New York: McGraw-Hill.

Nussbaum, A. (1984). Multivariate generalizability theory in educational measurement: An empirical study. *Applied Psychological Measurement, 8,* 219–230.
Schmidt, F. L., & Hunter, J. E. (1996). Measurement error in psychological research: Lessons from 26 research scenarios. *Psychological Methods, 1,* 199–223.
Shavelson, R. J., & Webb, N. M. (1991). *Generalizability theory: A primer.* Newbury Park, CA: Sage.
Shavelson, R. J., Webb, N. M., & Rowley, G. L. (1989). Generalizability theory. *American Psychologist, 44,* 922–932.
Shrout, P. E. (1993). Analyzing consensus in personality judgments: A variance components approach. *Journal of Personality, 61,* 769–788.
Smith, P. L. (1981). Gaining accuracy in generalizability theory: Using multiple designs. *Journal of Educational Measurement, 18,* 147–154.
Smith, P. L., & Luecht, R. M. (1992). Correlated effects in generalizability studies. *Applied Psychological Measurement, 16,* 229–235.
Webb, N. M., & Shavelson, R. J. (1981). Multivariate generalizability of general educational development ratings. *Journal of Educational Measurement, 18,* 13–22.
Webb, N. M., Shavelson, R. J., & Maddahian, E. (1983). Multivariate generalizability theory. In L. J. Fyans (Ed.), *New directions for testing and measurement: Generalizability theory: Inferences and practical application* (No. 18, pp. 67–82). San Francisco: Jossey-Bass.
Whitley, B. E., Jr. (1996). *Principles of research in behavioral science.* Mountain View, CA: Mayfield.
Williams, R. H., & Zimmerman, D. W. (1989). Statistical power analysis and reliability of measurement. *Journal of General Psychology, 116,* 359–369.
Winer, B. J. (1971). *Statistical principles in experimental design.* New York: McGraw-Hill.
Zimmerman, D. W., & Williams, R. H. (1986). Note on the reliability of experimental measures and the power of significance tests. *Psychological Bulletin, 100,* 123–124.

Item Response Theory

David H. Henard

How does one measure a psychological quantity, such as arithmetic ability? One method is to devise an instrument that one thinks will measure the ability by an ungoverned method, such as is often seen in the popular press and some published research. Classical test theory (CTT) represents an improvement over the lay approach, and discussions of CTT derivations of reliability and validity coefficiencies are ubiquitous. An important limitation of CTT methods, however, is that examinee characteristics cannot be separated from test characteristics. Item response theory (IRT) overcomes this limitation and allows the researcher to develop test questions that, theoretically, are free from both test item and examinee bias. IRT allows the researcher to obtain better methods for "scoring" psychological scales used to assess people on a theoretical dimension or factor. When they were first introduced, IRT or latent trait measurement models were heralded as "one of the most important methodological advances in psychological measurement in the past half century" (McKinley & Mills, 1989, p. 71). However, the pluses and minuses of these models have been debated despite their widespread use in various applications, such as test equating, item selection, and adaptive testing.

On reviewing this chapter, it is hoped that the reader will have a greater understanding of IRT and its inherent advantages and weaknesses. It is anticipated that the material contained in this chapter will

---

I wish to thank Bruce Thompson for his insightful comments on an earlier version of this chapter and also Paul Yarnold and Larry Grimm for their helpful suggestions. Portions of the material presented in this chapter were originally presented at the 1998 annual meeting of the Southwest Psychological Association, New Orleans, LA.

not only allow the reader to engage in an intelligent conversation regarding IRT but also will provide sources of additional or extended information on the subject for those seeking a deeper understanding on the matter. In keeping with the tenets of this book, the goal of this chapter is to induce broad and general subject knowledge. Those readers interested in a more complete IRT discussion should consult the publications listed in the Suggestions for Further Reading section at the end of this chapter.

As mentioned initially, IRT arose out of the acclaimed limitations of CTT. Because one way to define something is to explain what it is not, I begin with a synopsis of CTT and the limitations therein.

Classical Test Theory

An encompassing examination of CTT is beyond the scope of this work. The following discussion is designed to briefly acquaint the reader with classical theory and to lay the groundwork for why more theoretically grounded models, such as item-response models, were initially developed. Readers desiring greater discourse on the subject are directed to Crocker and Algina (1986) and Nunnelly and Bernstein (1994) as excellent starting points.

CTT and its related methodologies (e.g., see chap. 2, Reliability and Generalizability Theory) have dominated the field of psychological measurement to date. CTT's use is evidenced in the reporting of coefficients of internal consistency and of test–rest reliability, such as Cronbach's alpha, by many authors. Among its key benefits are that it is relatively simple to execute and that it can be applied to a broad range of measurement situations. CTT is primarily test oriented by nature, meaning that the measurement statistics derived from it are fundamentally concerned with how individuals perform on a given test as opposed to any single item on that test. At an item level, CTT simply calculates the success rate of test-taking individuals on any given test item as the ratio of individuals successfully answering that item to the total number of individuals participating in the test. The lower the ratio, known as the p value, the more difficult the test item is presumed to be.

In CTT, the observed score variance is partitioned into two parts —true score and error score—and is concerned with determining the reliability of test scores across different testing situations. The observed score is what the individual actually scored on the measure. Reliability estimates are simply the relationship of the true score to the observed score. Hence, a "perfectly reliable" test would be one in which the true score equaled the observed score.

CTT estimates three types of error variance by calculating reliability estimates for internal consistency, stability, and equivalence. *Internal consistency* is concerned with the consistency across items within a given test measure. *Stability* refers to the performance of the measure when given to the same subjects across multiple testing situations, and *equivalence* simply refers to the difference in an individual's scores on variations of the same test. Each of these reliability estimates allows one to evaluate a single source of error variance. One measurement difficulty, however, is that one is not able to evaluate the effect of error variance resulting from the distinct and independent interaction effects among these three sources.

CTT has other limitations that have been noted by many psychometricians. For example, comparison across examinees is limited to situations in which the participants of interest are administered the same (i.e., parallel) test items. Also, a false presumption of CTT is that the variance of errors of measurement is the same for all examinees. Reality dictates that some people perform tasks more consistently than do others and that consistency varies with ability (Hambleton & Swaminathan 1985). Two major CTT limitations of note are the following:

1. Examinee characteristics (i.e., observed scores) cannot be separated from test characteristics.
2. CTT is test oriented rather than item oriented.

The first limitation above can be summarized as a situation of circular dependency. The examinee statistic is item sample dependent, whereas the item statistics (i.e., item difficulty or item discrimination) are examinee sample dependent. Stated simply, when the test is "difficult," examinees appear to have lower ability, and when the test is "easy," they appear to have higher ability. Likewise, the difficulty of a test item is determined by the proportion of examinees who answer it correctly and is thus dependent on the abilities of the examinees being measured (Hambleton, Swaminathan, & Rogers, 1991). This circular dependency poses some theoretical difficulties in CTT's application in measurement situations, such as test equating and computerized adaptive testing.

The second major limitation listed is a question of orientation. The CTT model fails to allow one to predict how an examinee, given a stated ability level, is likely to respond to a particular item (Hambleton et al.,

1991). Predicting how an individual examinee or a group of examinees will perform on a specific item is relevant to a number of testing applications. Consider the difficulties facing a test designer who wishes to predict test scores across multiple groups, or to design an equitable test for a particular group, or possibly to compare examinees who take either different tests or the same test at different times. Such inherent limitations of CTT led psychometricians to develop models that not only overcame these limitations but also led to improved bias detection, enhanced reliability assessment and increased precision in ability measurement. IRT provides a framework within which to accomplish these desired features.

Item Response Theory

Concepts

As with the CTT section, a comprehensive discussion of IRT is beyond the bounds of this chapter (see Hambleton et al., 1991; and Wright and Stone, 1979, for more depth). The following discussion, however, should instill a solid grounding for understanding the theory.

IRT primarily rests on two basic postulates:

1. The performance of an examinee on a test item can be explained (or predicted) by a set of factors called "traits," "latent traits," or "abilities."
2. The relationship between examinees' item performance and the traits underlying item performance can be described by a monotonically increasing function called the "item characteristic curve" (ICC).

Several IRT models exist, including the three-parameter, two-parameter, and one-parameter models. The one-parameter model (often referred to as the "Rasch model" after George Rasch, the Danish mathematician who pioneered it) is the most commonly used of the three and are the dominant focus of this chapter. Two- and three-parameter models are discussed later in the chapter after the Rasch model has been more fully covered. The various models differ principally in the mathematical form of the ICC and the number of parameters specified in the model. The differences and similarities across the three models will become clear shortly.

When an IRT model fits the test data of interest, many of the aforementioned limitations of CTT are resolved. For example, examinee latent trait estimates are theoretically no longer test dependent, and item indices are no longer group dependent. Ability estimates derived from different groupings of items will be the same, barring measurement error, and item parameter estimates derived from different groups of examinees will also be the same, barring sampling error.

Fundamental Assumptions

Unidimensionality and local independence are two assumptions that are fundamental to IRT. The *unidimensionality* assumption requires that only one ability or latent trait is measured by the various items that make up the test. Intuitively, this assumption cannot be strictly satisfied because of the reality that multiple factors normally affect the test-taking performance of an examinee. Exogenous factors, such as generic cognitive ability, test anxiety, and motivation levels, are likely to affect test performance as well. For a set of test data to satisfy the assumption of unidimensionality, a "dominant" factor influencing performance must be present (Hambleton et al., 1991). This dominant factor is referred to as the ability or latent trait measured by the test.

The assumption of *local* independence requires an examinee's responses to the various items in a test to be statistically independent of each other. This implies that an examinee's response to any one item will not affect their response to any other item in the test. Simply put, the latent trait specified in the model is the only factor believed to be influencing the respondent's answer to the test items, and one item does not hold clues for subsequent or previous items in the test. It is important to note that the assumption of local independence does not imply that the test items are uncorrelated across the total group of examinees. Whenever there is variation among the examinees on the measured ability, positive correlations between pairs of items are the result.

There are three primary advantages to using item response models:

1. Assuming the existence of a large pool of items, each measuring the same latent trait, the estimate of an examinee's ability is independent of a particular sample of test items that are administered to the examinee.
2. Assuming the existence of a large population of examinees, the descriptors of a test item (e.g., item difficulty, item discrimina-

tion) are independent of the sample of examinees drawn for the purpose of item calibration.

3. A statistic indicating the precision with which each examinee's ability is estimated is provided.

Thus, the primary argument for using IRT methods is that the resulting analyses are both person-free and item-free measurements. Note that not all researchers agree that IRT offers such rich benefits. Lawson (1991) subjected three test data sets to both classical and Rasch procedures and found "remarkable similarities" between the results. Findings for both examinee abilities and item difficulties yielded "almost identical information." More recently, Fan (1998) replicated Lawson's work and arrived at strikingly similar conclusions.

Given these two study results and the mathematical intricacies inherent to IRT that are not required of classical methods, one may arguably question the necessity of the Rasch procedure; once misfitting items and people are removed from the analysis, IRT and CTT models seem to yield highly correlated person-ability and item-difficulty estimates. These empirical studies do not question the effectiveness of IRT models, yet they do call into question whether there is any substantial benefit to using the IRT methods versus the classical methods. Despite these concerns, the Rasch model continues to be widely used by psychometricians, and its use is likely to broaden. The relatively recent rise in computerized adaptive testing bears testament to the continued use of IRT, and it may prove to be the biggest beneficiary yet of the item-level approach.

To this point in the chapter, the focus has been on outlining and contrasting the fundamental tenets of both CTT and IRT with an emphasis on how IRT models capitalize on the inadequacies of classical models. I now turn attention to the specific calculations detailed in the one-parameter IRT model. In practice, sophisticated computer software programs are used to perform the Rasch model calculations. For my purposes, the model calculations to follow are simplified in an effort to efficiently portray the multiple calculations entailed in the model.

Calculating the One-Parameter Model

The Rasch model calculations can appear daunting to many observers. Although extremely powerful in its applications, the fundamentals of IRT are actually quite straightforward and should not be viewed as a black box process. The following discussion facilitates a conceptual grasp of the subject. Later in the chapter, I provide an annotated example of how IRT has been applied in published research on self-esteem.

Initial Model Calibration

In the following data set example, assume that 13 people were tested on a 13-item exam. The reader should bear in mind that this heuristic example is a gross oversimplification of a real-life testing occasion, and one should apply the understanding of the example to a practical situation that is personally relevant. Table 3.1 details the example test results. In it, the 13 people taking the test are listed with their respective outcome on each of the 13 items. Because items were dichotomously scored as either right or wrong, a "1" in the table indicates that the item was answered correctly by the examinee, while a "0" denotes an incorrect response to the item. The person score denotes how many items (out of 13) that each individual answered correctly. The item score indicates how many individuals (out of 13) answered each item correctly.

Because the object of the item response model is to predict performance based on item calibrations that are independent of the people generating the data (i.e., person free) and examinee ability estimates that are independent of the items used in the measurement (i.e., item free), all items that are answered either correctly or incorrectly by everyone (i.e., the extremes) must be removed from further analysis. Likewise, any person who answered either 0% or 100% of the items correctly must also be removed because neither can be calibrated against the group, and thus they provided no usable statistical information.

Such items and people provide no information to facilitate the model-estimation process in much the same manner that the addition of a single data point to a data set that is equal to the mean of the existing data set has no statistical effect on the data set. In this *model calibration* example, the person with all of the items correct may be exactly smart enough to accomplish that or may have any of the infinite ability levels that lie above the ability that is just sufficient to yield this perfect score. One has no way to reasonably know which assumption is correct.

These items and individuals are graphically partitioned in Table

Table 3.1

Responses of Examinees to Test Items

Person	Test item no. 1	2	3	4	5	6	7	8	9	10	11	12	13	Score
1	1	1	1	1	1	1	0	0	0	0	0	0	0	6
2	1	1	1	0	1	1	1	0	0	0	0	0	0	6
3	1	1	0	1	1	1	1	0	1	1	0	0	0	8
4	1	1	1	1	1	1	1	1	0	1	1	0	0	10
5	1	1	0	1	0	1	1	1	1	0	0	0	0	7
6	1	1	1	1	1	1	1	0	0	0	0	1	0	8
7	1	1	1	1	0	1	1	0	0	0	0	0	0	6
8	1	1	0	1	1	0	1	0	0	0	0	0	0	5
9	1	1	1	0	1	1	1	0	0	0	1	0	0	7
10	1	1	1	1	1	1	1	0	0	0	0	0	0	7
11	1	1	1	1	1	1	1	0	0	0	2	0	0	8
12	1	1	0	0	0	0	0	0	0	0	0	0	0	2
13	1	1	0	0	0	0	0	0	0	0	0	0	0	2
Item score	13	13	8	9	9	10	10	2	2	2	3	1	0	

Note. 1 = a correct response; 0 = an incorrect response.

3.1 by the dashed lines. Note that test Items 1 and 2 were answered correctly by all 13 examinees and that Item 13 was answered incorrectly by all. Once these three test items have been removed from further consideration, note that Persons 12 and 13 incorrectly answered each of the remaining items. As a result, these individuals are eliminated from the model for the reasons mentioned previously. Before continuing, re-examine Table 3.1 until the logic behind the removal of these items and individuals becomes clear.

The revised data set, after this initial cut of the information, can be seen in Table 3.2, which is laid out so that examinees are sorted in increasing order of the number of items answered correctly, whereas items are sorted in decreasing order of examinees that correctly answered the item. Thus, the items that individuals who answered the most correctly appear at the bottom of the table, while items that were answered correctly by most of the examinees are located to the left in the table. The proportion correct for people (out of 10 items) and items (out of 11 people) has also been added to the table and are used in calculations.

In looking at Table 3.2, one finds that Person 8 answered the fewest number of questions correctly (three), while Person 4 answered the greatest number of items correctly (eight), with the remaining individuals falling somewhere in between these two. Remember that any examinee who scored either 100% or 0% has been removed. Any extreme items have also been removed. In this data set, Items 1, 2, and 13 were removed, and Persons 12 and 13 were also removed. This editing of the data continues in this manner until no extreme items or people remain in the model.

Further Model Refinement

Given this initial editing of the data, the next step in the process is to calibrate both the item difficulties and the person abilities. To make valid assessments and predictions from the Rasch model, both of these statistics (difficulties and abilities) must be continuous and in the same metric. In IRT, this is accomplished by converting the values into logits. *Logits* for item difficulties are calculated as the natural logarithm of the proportion of items incorrect $(1 - P_i)$ divided by the proportion correct (P_i). Conversely, the logit calculation for person ability is the natural log of the proportion of items that an examinee correctly answered (P_i) divided by the proportion answered incorrectly $(1 - P_i)$. Thus,

Table 3.2

Revised Data Set After Initial Adjustments

	Test item											
Person	**6**	**7**	**5**	**4**	**3**	**11**	**8**	**10**	**9**	**12**	**Revised score**	**Proportion correct (of 10)**
8	0	1	1	1	0	0	0	0	0	0	3	.30
1	1	0	1	1	1	0	0	0	0	0	4	.40
2	1	1	1	0	1	0	0	0	0	0	4	.40
7	1	1	0	1	1	0	0	0	0	0	4	.40
5	1	1	0	1	0	0	1	0	1	0	5	.50
9	1	1	1	0	1	1	0	0	0	0	5	.50
10	1	1	1	1	1	0	0	0	0	0	5	.50
3	1	1	1	1	0	0	0	1	1	0	6	.60
6	1	1	1	1	1	0	0	0	0	1	6	.60
11	1	1	1	1	1	1	0	0	0	0	6	.60
4	1	1	1	1	1	1	1	1	0	0	8	.80
Revised item score	10	10	9	9	8	3	2	2	2	1		
Proportion correct (of 11)	.91	.91	.82	.82	.73	.27	.18	.18	.18	.09		

Note. 1 = a correct response; 0 = an incorrect response.

$$\text{item difficulty logits } (\delta) = \ln[(1 - P_i)/P_i], \quad (1)$$

$$\text{person ability logits } (\theta) = \ln[P_i/(1 - P_i)]. \quad (2)$$

Table 3.3 shows the distribution of test item scores from the example and details how item scores are transformed into proportions, and subsequently into item difficulty logits (δ). For example, Item 3 was answered correctly by 8 of the 11 people. This yields a proportion correct of 0.73 for that item and, obviously, a proportion incorrect of 0.27. Using logit Equation 1 above, I arrive at a δ for Item 3 of −0.98. Take note that the easier items have the lower item difficulty value and vice versa.

Table 3.4 denotes the distribution of person scores, as well as the pertinent calculations entailed in obtaining person ability or latent trait ability (θ) in logits. The number of items correctly answered by examinees can vary between 1 and 9, inclusive. Remember that scores of 0 and 10 have been removed from the model. From Table 3.4, one sees that three individuals answered four items correctly, two answered six items correctly, and so on. The proportion correct (P_i) is simply calculated as the number of possible correct answers divided by the total number of questions in the revised model. Equation 2 above is then used to calculate θ. Here, one sees that negative logit values are associated with "lower" latent ability by the examinee. Although this runs counter to the scaling seen in Table 3.3, an examination of the inverse relationship of equations (1) and (2) leads one to expect this result.

Table 3.3

Distribution of Item Scores and Calculation of Item Difficulty in Logits

Test item	Item score	Proportion correct (P_i)	Proportion incorrect ($1 - P_i$)	Item logit ($\ln[(1 - P_i/(P_i)]$)
6	10	.91	.09	−2.30
7	10	.91	.09	−2.30
5	9	.82	.18	−1.50
4	8	.73	.27	−0.98
3	8	.73	.27	−0.98
11	3	.27	.73	0.98
8	2	.18	.82	1.50
10	2	.18	.82	1.50
9	2	.18	.82	1.50
12	1	.09	.91	2.30

Table 3.4

Distribution of Scores and Calculation of Ability in Logits

Possible correct	Person frequency	Proportion correct (P_i)	Proportion incorrect ($1 - P_i$)	Person logit ($\ln[(1 - P_i/(P_i)]$)
1	0	.10	.90	−2.20
2	0	.20	.80	−1.39
3	1	.30	.70	−0.85
4	3	.40	.60	−0.41
5	3	.50	.50	0.00
6	2	.60	.40	0.41
7	0	.70	.30	0.85
8	1	.80	.20	1.39
9	0	.90	.10	2.20

Once the logit values for both the people and items are calculated, one has overcome another key weakness associated with CTT. Namely, although item difficulty and person ability levels realistically range from negative infinity to positive infinity, the proportion correct/incorrect found in CTT is bound by the values of 0 and 1. Conversion to logits transforms the values into a ± infinity (∞) scale. One further step involves calculating the mean and standard deviation of the data and converting the logits to a standardized (z) scale, arbitrarily assigning a center-point value of 0. Although the scale theoretically runs from negative infinity to positive infinity, values realistically tend to vary between −3.00 and +3.00 logits, as evidenced in my example.

Final Calibrations

Two final calibration steps remain. The initial measurement of item difficulty must be corrected for the difficulty dispersion of the items. Additionally, the initial measurement of person ability needs to be corrected for the natural ability dispersion of people. The calculations are modeled in Tables 3.5 and 3.6 and result in item calculations that are corrected for sample spread and in person calculations that are corrected for test width. These final calibrations are necessary because the calculated difficulty level of the items depends on the variance in the ability levels of the people sampled, and the calculation of the apparent person ability levels depends on the variance in test item difficulty.

In essence, if there is a large variance in the ability level of people

Table 3.5

Estimates of Item Difficulty

Possible correct	Test item no.	Initial item logit (*d*)	Spread expansion factor (*Y*)	Final item logit (*d* × *Y*)
1	12	2.30	1.80	4.14
2	8, 9, 10	1.50	1.80	2.70
3	11	0.98	1.80	1.76
4	–	–	–	–
5	–	–	–	–
6	–	–	–	–
7	–	–	–	–
8	3, 4	−0.98	1.80	−1.76
9	5	−1.50	1.80	−2.70
10	6, 7	−2.30	1.80	−4.14

Note. In our example data set, there were no test items that were answered correctly by either 4, 5, 6, or 7 out of 11 examinees. Therefore, these fields (–) are blank above. In a practical situation, all of the fields will likely be filled and, as such, will likely have logit calculations for each.

Table 3.6

Estimates of Ability

Possible correct	Person number	Initial person logit (*T*)	Spread expansion factor (*Y*)	Final person logit (*T* × *Y*)
1	–	–	–	–
2	–	–	–	–
3	8	−0.85	1.05	−0.89
4	1, 2, 7	−0.41	1.05	−0.43
5	5, 9, 10	0.00	1.05	0.00
6	3, 6, 11	0.41	1.05	0.43
7	–	–	–	–
8	4	1.39	1.05	1.46
9	–	–	–	–

Note. In our example data set, there were no examinees that answered either 1, 2, 7, or 9 items out of 10 correctly. Therefore, these fields (–) are blank above. In a practical situation, all of the fields will likely be filled and, as such, will have logit calculations for each.

taking the test, then the item difficulty levels appear similar in difficulty across items. Likewise, the greater the variance in item difficulty levels on the test, the more similar the ability levels of people appear. The effects of sample spread and test width must be eliminated from the estimates for person ability and item difficulty for the final estimates to realistically be considered both person free and item free. The process is highly mathematical and, thus, is not elaborated on here. Wright and Stone (1979, pp. 21–24) offered a comprehensive discussion of the calculations. This calibration step is known as the *calculation of expansion factors,* and it is crucial given the premise of the one-parameter IRT model that the achievement of any person on a given item is solely dependent on that person's latent ability and the difficulty of the specific item.

The final step in the modeling process is to fit the model, obtained via the preceding steps, to the data and evaluate the model's goodness-of-fit. One cannot merely assume that the preceding steps are sufficient in developing an effective model. If one re-examines Table 3.2, a pattern in responses should emerge. Because items are ordered by increasing level of difficulty and examinees are progressively ranked according to correct responses, one would intuitively expect to see more incorrect responses to the top and right of the table and vice versa. Given that the heuristic example used here is overly small, the pattern is somewhat more difficult to show visually. Yet the logic concerning one's expectations should be clear. In other words, one would expect a person who is higher on a latent trait (θ) to have a greater chance of answering a difficult question than a person who is lower on that same latent trait. Table 3.7 accentuates the different expectations.

Although those people or items that do not fit well with the model are statistically identified by the software program used to calculate the Rasch model, some of the potential misfits are circled here to visually highlight points where the model may not perfectly fit the data. It should be noted that both items and people can be identified as aberrant. For example, Person 5 answered Items 8 and 9 correctly when the expectation (given that person's other responses) would be that the items would be answered incorrectly. This same individual missed Items 3 and 5, which also seems to indicate aberrant results.

Item 3 was answered incorrectly by Person 3 when the expectation might be a correct response by a person of that latent ability level on an item of that difficulty level. Also note that Person 8 incorrectly answered Item 6—the easiest item remaining in the data set. Although

Table 3.7

Comprehensive Data Set

	Person	Easier items					Harder items					Revised score	Proportion correct	Person ability logits
		6	7	5	4	3	11	8	10	9	12			
Lower ability	8	⓪	1	1	1	0	0	0	0	0	0	3	.30	−0.89
	1	1	0	1	1	1	0	0	0	0	0	4	.40	−0.43
	2	1	1	1	0	1	0	0	0	0	0	4	.40	−0.43
	7	1	1	0	1	1	0	0	0	0	0	4	.40	−0.43
	5	1	1	⓪	1	⓪	0	①	0	①	0	5	.50	0.00
	9	1	1	1	⓪	1	1	0	0	0	0	5	.50	0.00
	10	1	1	1	1	1	0	0	0	0	0	5	.50	0.00
	3	1	1	1	1	⓪	⓪	0	1	1	0	6	.60	0.43
	6	1	1	1	1	1	⓪	0	0	0	1	6	.60	0.43
Higher ability	11	1	1	1	1	1	1	0	0	0	0	6	.60	0.43
	4	1	1	1	1	1	1	1	1	0	0	8	.80	1.46
	Revised item score	10	10	9	9	8	3	2	2	2	1			
	Proportion correct	.91	.91	.82	.82	.73	.27	.18	.18	.18	.09			
	Item difficulty logit	−4.14	−4.14	−2.70	−1.76	−1.76	1.76	2.70	2.70	2.70	4.14			

Note. Circled numbers indicate potential items that do not fit well with the model. 1 = a correct response; 0 = an incorrect response.

Person 8 seemingly has the lowest latent ability ($\theta = -.89$) in the group of individuals, one might still expect him to answer Item 6 correctly given that every other person answered it correctly and that it is the easiest of the remaining 10 items. If other people of similar ability levels were present in the data, the question of aberrance in this specific case would be clearer. In other words, a much larger data set would likely contain more people with latent ability levels similar to Person 8, and the incorrect response to Item 6 would then not appear as irregular to the remainder of the data set as it does in this small set example.

Once an IRT computer software program identifies discrepancies between the specified model and the existing data, the source of the variance (item or person based) is explored. It may be helpful to think of the process as a sophisticated χ^2 test in which the goodness-of-fit between the model and the data is tested and sources of misfit are identified and subsequently eliminated from the data. In examining Table 3.7, Persons 5 and 3, among others, appear to post responses that are aberrant to expectations. Likewise, Items 3, 5, 9, and 11 do not appear to totally fit with expectations. Bear in mind that in practice, the aberrant sources of variance are identified statistically by the software program and that, in this example, I am merely examining the data set visually in an effort to more fully understand the process.

As mentioned previously, the source of the inconsistencies can originate at either the item or the person level. Table 3.8 simulates how the software program would investigate the irregularities caused by Persons 5 and 3. Each test item has a calculated difficulty level (δ) in logits, and each person has a calculated ability level (θ) in logits. The first step in the analysis of fit is to determine the difference between the ability level and the difficulty level for each person and each item. Table 3.8 highlights the values specific to Persons 5 and 3.

When the difference in the two values (i.e., $\theta - \delta$) is a positive number, it is an indication that this particular item should be "easy" for that particular examinee and that it, therefore, should theoretically be answered correctly. The higher the number, the greater the likelihood of a correct response. Conversely, the more negative the difference, the greater the likelihood that the item difficulty theoretically exceeds that person's latent ability. Remember, I am discussing the one-parameter model, so I am stating that only one trait (e.g., knowledge level on a subject) is being tested and captured by the researcher. Other sources of variance in person–item responses are discussed shortly.

Table 3.8

Fit Analysis for Persons 5 and 3

Person 5					Person 3				
Test item	Item difficulty (d)	Ability logits (T)	$T - d$	Item response	Test item	Item difficulty (d)	Ability logits (T)	$T - d$	Item response
6	−4.14	.00	4.14	1	6	−4.14	.43	4.57	1
7	−4.14	.00	4.14	1	7	−4.14	.43	4.57	1
5	−2.70	.00	2.70	⓪	5	−2.70	.43	3.13	1
4	−1.76	.00	1.76	1	4	−1.76	.43	2.19	1
3	−1.76	.00	1.76	⓪	3	−1.76	.43	2.19	⓪
11	1.76	.99	−1.76	0	11	1.76	.43	−1.33	0
8	2.70	.00	−2.70	①	8	2.70	.43	−2.27	0
10	2.70	.00	−2.70	0	10	2.70	.43	−2.27	①
9	2.70	.00	−2.70	①	9	2.70	.43	−2.27	①
12	4.14	.00	−4.14	0	12	4.14	.43	−3.71	0

Note. The "Ability logits" columns here represent the "Final person logit" columns from Table 3.6 after adjustment for expansion factors. Circled numbers indicate potential items that do not fit well with the model. 1 = a correct response; 0 = an incorrect response.

Looking at Table 3.8, it appears that Person 5 missed Items 3 and 5, when both items should theoretically have been answered correctly, while correctly answering Items 8 and 9, when the probability was that those items would be missed by a person with a theta (person ability logit) equal to 0. Person 3 missed Item 3 while getting Items 9 and 10 correct—all opposite of theoretical expectations. Viewing this information in isolation may lead one to conclude that Persons 5 and 3 are hampering the effort to develop a solid theoretical model. Again, however, the source of variance can result from item irregularities and person irregularities.

Table 3.9 further facilitates understanding of how item-response patterns are examined for misfitting results. The process here is similar to the aforementioned one, yet it concerns the fit analysis for items. This table illustrates the examination of Items 5 and 9 for all people. Please note, this model-fitting process occurs for all items and all people, not simply those selected in my heuristic example.

From Table 3.9, one sees two aberrant responses for each selected item. Recall that when one subtracts the item difficulty logits from the person ability logits one arrives at a theoretical, linear definition of model performance. A positive difference score indicates that the item should theoretically be answered correctly by that person. A negative difference score implies a theoretical expectation of an incorrect response. For Item 5, one would expect that both Persons 5 and 7 would have answered it correctly. This is not what one observes. Furthermore, one would expect, given $\theta - \delta$, that no one from my sample would answer Item 9 correctly because the item difficulty in logits exceeds all of my individual abilities in logits. Therefore, the possibility exists that Items 5 and 9 are contributing to model misfitting. Note that when logits are negative, lesser values indicate greater ability (for person logits) or difficulty (for item logits).

From analyzing only the test-item fit analyses in Table 3.9, it appears that Items 5 and 9 may not be causing irregularity between the model and the data, but rather that Persons 3 and 5 are. This is most likely what is occurring. The removal of these people from the data set would eliminate most of the irregularity associated with the two items. Look again at Table 3.8 and compare these results to that observed in Table 3.9. Even though I eliminated the extreme items (i.e., 0% or 100% of which were answered correctly) and the extreme people (those who answered 0% or 100% of all items correctly) from my model in the initial revision of the data set, I am still left with some model vari-

Table 3.9

Fit Analysis for Test Items 5 and 9

	Item 5				Item 9				
Person	Ability logits (T)	Item difficulty (d)	$T - d$	Item response	Person	Ability logits (T)	Item difficulty (d)	$T - d$	Item response
8	−0.89	−2.70	1.81	1	8	−0.89	2.70	−3.59	0
1	−0.43	−2.70	2.27	1	1	−0.43	2.70	−3.13	0
2	−0.43	−2.70	2.27	1	2	−0.43	2.70	−3.13	0
7	−0.43	−2.70	2.27	⓪	7	−0.43	2.70	−3.13	0
5	0.00	−2.70	2.70	⓪	5	0.00	2.70	−2.70	①
9	0.00	−2.70	2.70	1	9	0.00	2.70	−2.70	0
10	0.00	−2.70	2.70	1	10	0.00	2.70	−2.70	0
3	0.43	−2.70	3.13	1	3	0.43	2.70	−2.27	①
6	0.43	−2.70	3.13	1	6	0.43	2.70	−2.27	0
11	0.43	−2.70	3.13	1	11	0.43	2.70	−2.27	0
4	1.46	−2.70	4.16	1	4	1.46	2.70	−1.24	0

Note. Circled numbers indicate potential items that do not fit well with the model. 1 = a correct response; 0 = an incorrect response.

ance or "noise," that precluded one from developing a solid item response model. That noise resides mainly with Persons 3 and 5.

On the removal of these individuals from the data set, the calibration process reiterates, and a new evaluation of fit is calculated for the remaining distributions. Again, all combinations of people and items are examined. By removing Persons 3 and 5, Item 9 would fit perfectly with my theoretical expectations (see Table 3.9). Likewise, a large portion of the noise associated with Item 5 would also be removed. The result would be an improved model-to-data fit over what it would have been with these individuals included.

At the point that no further removal of either items or people enhances the goodness-of-fit, the model is said to fit the data, and the result is items that are theoretically both unidimensional and independent—the two core assumptions of IRT. By eliminating both items and individuals that deviate from expectations, I can begin to develop a test bank of items that should optimally fit the individual person ability levels for subsequent test takers. Stated differently, one can develop items that should indicate a person's ability level on the latent trait being tested simply by whether the person answered the items correctly or not.

One benefit to this is that far fewer questions are needed to determine individual ability level or proficiency because the ability or inability to answer an individual test item indicates the level of examinee ability on the trait being tested. Furthermore, the accuracy of test-based conclusions is improved over CTT approaches. This concept is the cornerstone of computerized adaptive testing procedures. The improved time efficiencies for both the examiner and examinee is obvious. Still, imagine the difficulties facing an instructor who must explain how a person's ability level in an area of interest has been captured from only 5 to 10 questions and that the person has not passed the test?

Item Characteristic Curves

One of the key terms in and elements of IRT is the ICC. It is basically a plot (a logistic ogive) of the probabilities of a correct response to a specific question across various latent ability levels. Hence, after all of the preceding modeling calibrations have been completed and given the assumptions of IRT, the ICC is a graphic portrayal of an item's difficulty level (δ) and the expectations of how people of varying ability levels will perform on that item.

Figure 3.1 illustrates one-parameter item characteristic curves for four hypothetical items, noted as A, B, C, and D. Late ability (θ) is represented, in logits, along the x axis. The probability of a correct response is located on the y axis. In the one-parameter IRT model no traits other than ability (e.g., guessing) are assumed to affect responses, so the curves are asymptotic to the 0 and 1 endpoints of the probability distribution. The difficulty level (δ) of any specific item is defined as the logit point at which the probability of answering the item correctly is 50% (p = 0.50). For example, the item difficulty for Item A is approximately −1.80, while the item difficulty for Item C is approximately 0.00.

The dotted lines in Figure 3.1 are meant to facilitate the visual grasp of how one can determine the item difficulty level by simply viewing an ICC. Those items with curves that are toward the right side of the x axis, and thus associated with higher logit ability values, are relatively more difficult than those items to the left on the axis. Therefore, a person with an ability (θ) equal to 0.0 would probably answer Items A and B correctly and would likely miss Item D. There is a 50% chance of the person answering item C correctly. An examination of Figure 3.1 will help make these statements clear.

Figure 3.1

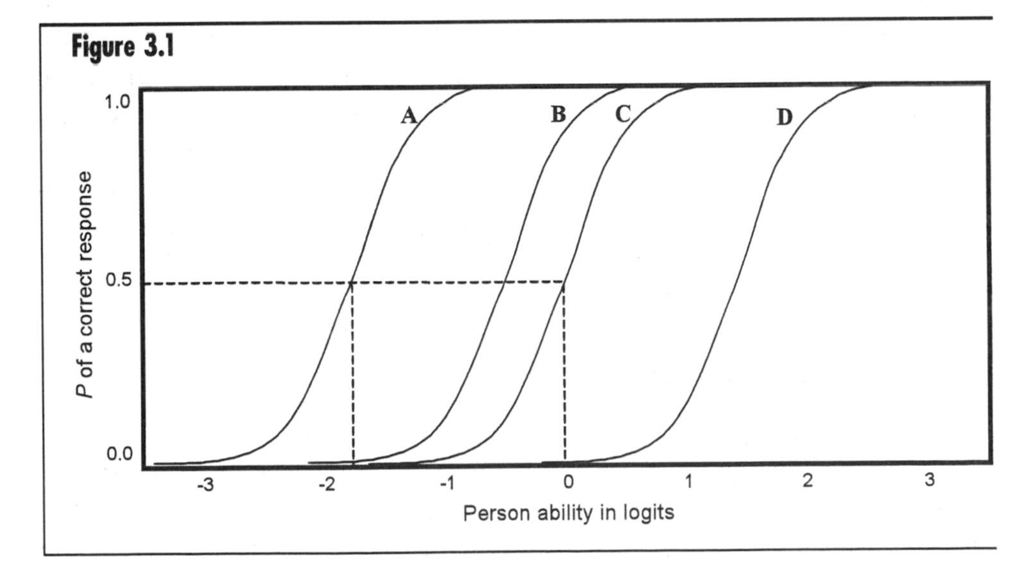

Item characteristic curves for a one-parameter item response theory model.

Multiple Parameter Models

The process for calibrating two- and three-parameter IRT models is similar to that entailed in the one-parameter model, albeit more mathematically complex given that additional parameters are introduced into the model. A comprehensive discourse on the multiple parameter extensions of the Rasch model would be of limited usefulness in this chapter, given the overwhelming use of the one-parameter model. Some elaboration on them, however, may be beneficial. Basically, the information garnered from the two- and three-parameter models enhances understanding of how well each test item can discriminate among numerous individuals.

A simple and straightforward way to illustrate the two- and three-parameter IRT models is through the use of representative item characteristic curves for each model. Figures 3.2 and 3.3 depict these points. Figure 3.2 represents four ICCs for a two-parameter IRT model. The two-parameter model assumes that there are two parameters affecting examinee responses: latent ability and item discrimination. The curve endpoints remain asymptotic as with the one-parameter model, as answers are still only classified as correct or incorrect. With the two-parameter model, the slope of the curve indicates how well the item is able to differentiate between people with varying latent abilities. For instance, Item B in Figure 3.2 has a much flatter slope than Item C. It then follows that Item C is the better discriminating item of the two

Figure 3.2

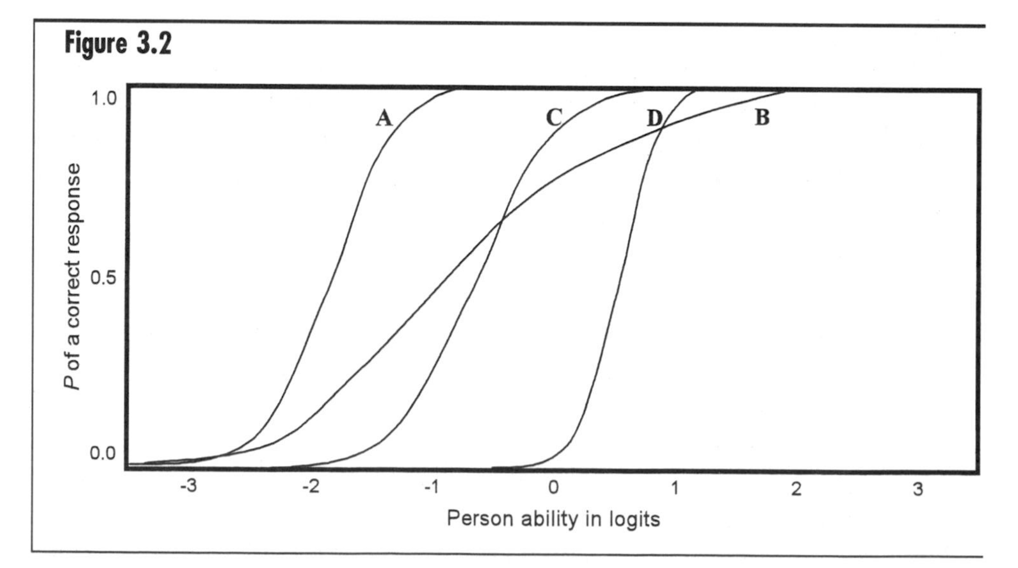

Item characteristic curves for a two-parameter item response theory model.

Figure 3.3

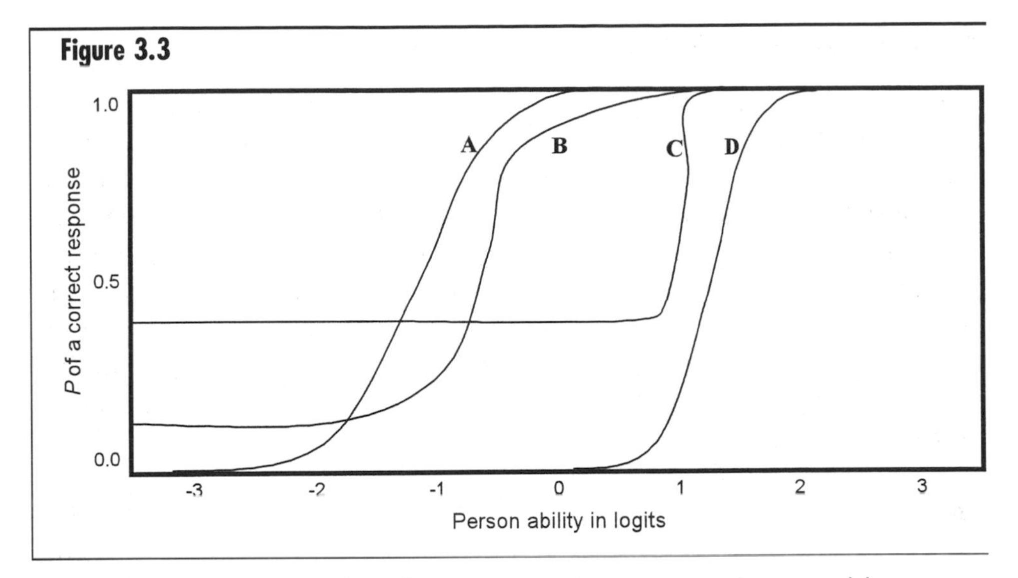

Item characteristic curves for a three-parameter item response theory model.

because C's curve is spread across a more narrow range of person ability levels. Look at the range of ability levels covered by the ICCs for Items B and C. The item difficulty level for each curve is determined in the same manner as illustrated in Figure 3.1. Hence, Item A would be the easiest of the four items, with Item D being the hardest. Likewise, Item B would be the worst item with regard to discriminating ability because the ICC covers such a broad range of ability levels (e.g., −3 to +2 logits).

Figure 3.3, the three-parameter model curves, adds a third variable to the modeling equation—the effect of guessing. In this model, the curve endpoint may begin at a value other than 0 because the effect of correctly guessing an item is taken into account—regardless of latent ability level. The evaluation guidelines that apply to the other two ICCs apply here as well; however, the location of the initial endpoint gives the researcher an indication as to how effective the item distracters (e.g., incorrect alternative choices in a multiple-choice test) may be. For example, Items B and C appear to have the potential to be guessed correctly whereas Items A and D do not. Specifically, there is a 40% probability that a person with a ability level of −3 logits would answer Item C correctly—which is unlikely, given his low ability level and the difficulty level of the item. One should now be able to examine Figure 3.3 and develop some conclusions regarding the four items as depicted by their respective ICC. For instance, Item C, in addition to having 40%

of test takers able to guess the correct answer, is an extremely poor discriminator of people of low ability levels. One should also notice that Item D would likely be missed by many individuals and that it is a better discriminating item than either Items A or B.

A Practical Example

At this point in the chapter, an annotated example of how IRT has been applied at a practical level proves useful. Although there are many capable articles that could serve as a practical example, the investigation by Gray-Little, Williams, and Hancock (1997) of the Rosenberg Self-Esteem (RSE) Scale illustrates the insight provided by the application of IRT analytical methods.

The RSE scale, a self-reporting instrument that is used to quantify the self-esteem level of an individual, is arguably the most widely used self-report scale of its kind. It has been the focus of numerous substantive investigations and has likewise undergone a number of psychometric assessments to assess its reliability and validity. One of the selling points of the scale is that it claims to be a unidimensional construct. In other words, each of the 10 items that make up the RSE scale are purported to measure self-esteem and no other underlying personality dimension. Recall that IRT rests on two assumptions, one of which is unidimensionality. Citing the popularity of the scale, Gray-Little et al. (1997) noted that the RSE scale carries a "heavy burden" and, as such, "should be the focus of further psychometric evaluation" (p. 443).

The authors had two dominant goals in conducting their analysis of the RSE scale. The first was to assess the dimensionality of the scale. Although Rosenberg had posited that the scale was unidimensional, researchers have debated this assumption with some studies supporting the unidimensional view and others obtaining a two-factor solution. Verification of the unidimensional view was obviously necessary, given the assumptions of IRT, before IRT analysis could rationally be performed. The second goal was to investigate the discriminating ability of each of the 10 individual items of the scale and of the RSE scale as a whole. IRT analysis is suited for this purpose and is capable of providing researchers with the ability to assess the measurement precision of the RSE or any other scale.

Exploratory factor analysis was used to assess the dimensionality of the scale. The results indicate that the RSE scale is unidimensional as represented by the one-factor solution. Although a second factor was statistically supported in the analysis (i.e., an eigenvalue greater than 1.0; see chap. 6 on factor analysis for greater detail), it accounted for little additional variance beyond that accounted for by the first factor, and it failed to enhance the interpretability of the results. Hence, the scale is presumed to globally measure self-esteem as it is purported to. Table 3.10 reports the factor loadings for the Gray-Little et al. (1997) analysis.

Accepting the conclusion that the scale is unidimensional still leaves certain questions about the RSE scale unanswered. These are, namely, how well do each of the items discriminate across individual levels of self-esteem and how does the scale as a whole discriminate across these varying levels. To answer these questions, Gray-Little et al. (1997) used IRT methods. Given that the RSE scale items are scored on a Likert-type scale (e.g., a 5-point scale, such as 1 = *strongly agree* to 5 = *strongly disagree*), the authors used a polychotomous IRT model. I illustrated a simple dichotomous model (i.e., answers are either 0 or 1) in this chapter to facilitate understanding, but the underlying dynamics of the two approaches are the same. An understanding of the model presented in this chapter provides the ability to comprehend more complex IRT models.

Although support for the global nature of the scale was provided

Table 3.10

Factor Loadings for the Items of the Rosenberg Self-Esteem Scale

Scale item	Factor 1
1	.7227
2	.7121
3	.7781
4	.6193
5	.7396
6	.7727
7	.7147
8	.6128
9	.6067
10	.6679

Note. From "An Item Response Theory Analysis of the Rosenberg Self-Esteem Scale," by B. Gray-Little, V. S. L. Williams, and T. D. Hancock, 1997, *Personality and Social Psychology Bulletin, 23,* p. 449. Copyright 1997 by the Society for Personality and Social Psychology, Inc. Adapted with permission.

by the factor analysis, IRT analysis revealed that the 10 items of the scale did not equally discriminate between people of varying levels of self-esteem. Specifically, Items 8, 9, and 10 were shown to be less effective than the other items in distinguishing among individuals. Because the RSE scale score is calculated by summing the 10 item scores, this heterogeneity in item discrimination ability indicates that the existing method of calculating the total scale score is somewhat suboptimal. Gray-Little et al. (1997) proposed an IRT-scoring strategy in which each item's score was weighted by its item discriminating ability as a solution to the suboptimal situation. Subsequent analysis, however, revealed that there was insufficient improvement in the IRT-weighted total scale score over the existing total scale score to justify the extra steps involved in scale score calculations. The authors subsequently derived a test characteristic curve—which is analogous to the ICC discussed in this chapter—from the data in an effort to illustrate the discriminating ability of the RSE scale as a whole. This curve is illustrated in Figure 3.4 and can be interpreted in the same manner as discussed for the ICCs.

The test characteristic curve depicted in Figure 3.4 reflects the summed scores of the 10 items of the RSE scale. Note that the slope of the curve is somewhat flat and, thus, indicative of poor scale discrimination across varying levels of the trait of interest. In this example, the

Figure 3.4

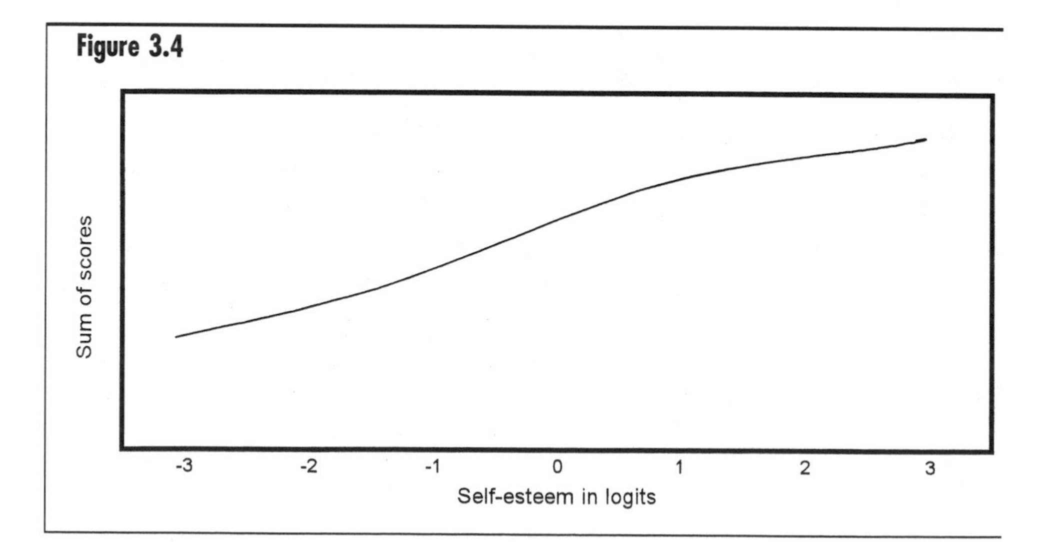

Test characteristic curve for the Rosenberg Self-Esteem Scale. From "An Item Response Theory Analysis of the Rosenberg Self-Esteem Scale," by B. Gray-Little, V. S. L. Williams, and T. D. Hancock, 1997, *Personality and Social Psychology Bulletin, 23,* p. 449. Copyright 1997 by the Society for Personality and Social Psychology, Inc. Adapted with permission.

trait of interest (θ) is self-esteem. The use of IRT analyses by the authors provided readers with valuable insight into the RSE scale. Gray-Little et al. (1997) were able to move beyond test-oriented questions concerning scale dimensionality, reliability, and validity and actually investigate the individual scale items. One insight from their work is that the scale can perhaps be shortened (e.g., by dropping Items 8, 9, and 10) without jeopardizing the validity of the instrument. Although it was ultimately deemed inefficient, the authors were also able to develop a more precise method for calculating the total scale score through an IRT-weighting approach. Similar IRT analyses to these can be applied to a variety of scales or tests and thus can provide researchers with similarly unique insights. IRT analyses can contribute to the understanding of personality processes, improve instrument assessment abilities, and ultimately enhance the validity of research instruments (Gray-Little et al., 1997).

Conclusion

IRT models arose from the inherent limitations of CTT methods of test analysis. Chief among the CTT limitations is that examinee characteristics cannot be separated from test characteristics. IRT overcomes these limitations and rests on two major assumptions: (1) the performance of an examinee can be explained by a set of factors (traits), and (2) the relationship between an individual's item performance and the traits underlying that performance can be described by a monotonically increasing function (ICC).

IRT allows the researcher to develop test questions that are theoretically both person free and item free. IRT stresses maximizing the test information function over the range of abilities that are of interest, rather than maximizing reliability, as does classical psychometrics. Although the usefulness of IRT continues to be debated, IRT continues to hold many benefits. Among these is a more accurate ability to detect item or test bias; the ability to administer customized, individualized, computer-adaptive tests; and the ability to construct more effective testing instruments in general.

Suggestions for Further Reading

Although this chapter was intended as a broad yet comprehensive discussion of CTT and IRT measurement models, space does not permit a full discourse on either topic. Excellent references exist for both topic areas. For those readers interested in CTT, I suggest initially consulting both Crocker and Algina (1986) and Nunnally and Bernstein (1994). Those interested in a more encompassing IRT discussion should consult Wright and Stone (1979), Lord (1980), and Hambleton, Swaminathan, and Rogers (1991). The Lawson (1991) and Fan (1998) articles discuss the pros and cons of the two measurement approaches and may be of interest to some readers.

Glossary

ABILITY One of the focal areas of emphasis in IRT that refers to the level of competence of an individual on a certain task or aptitude measurement. Also referred to as *trait* or *latent trait.*

BLACK BOX PROCESS Jargon that refers to any system where the inputs and outputs are known but where the intermediate process is a mystery to the observer.

EQUIVALENCE The difference in an individual examinee's scores on taking different versions of the same text.

ERROR SCORE The difference between a person's observed, or actual, score on a test and his or her true, or expected, test score.

EXOGENOUS FACTORS Any variable outside of those being tested for or theorized about that may have an effect on the subject of interest.

EXPANSION FACTORS Final calibration steps in the IRT model where item difficulty levels are adjusted for the variance in person ability (i.e., adjusting for "sample spread"), and person ability levels are adjusted for the variance in item difficulty (i.e., adjusting for "test width"). The aim is to produce a test or scale that is both person and item free.

GENERIC COGNITIVE ABILITY The base intelligence level of an individual.

INTERNAL CONSISTENCY The degree that all individual scale items within a measure are collectively capturing the construct of interest.

ITEM DIFFICULTY LOGITS (δ, d) The natural logarithm of the proportion of items that a person answers incorrectly divided by the proportion of items answered correctly.

ITEM DISTRACTERS Answer options on a multiple choice test, for example, that are incorrect choices.

ITEM CHARACTERISTIC CURVE (ICC) A graphical plot of the probabilities of a correct response by a person to a specific test or scale item across varying levels of ability. It is a simple, graphic portrayal of an item's difficulty level and the corresponding latent ability level that is theoretically required to correctly answer the item.

ITEM ORIENTED Having the focus of an analysis on the individual items that make up the test (e.g., IRT) as opposed to on the test as a whole (e.g., CTT).

ln The natural logarithm mathematical calculation.

LATENT TRAIT One of the focal areas of emphasis in IRT that refers to the level of competence of an individual on a certain task or aptitude measurement. Also referred to as *ability* or *trait*.

LOCAL INDEPENDENCE One of the two key tenets of IRT. It assumes that a person's response to any one item on a test or scale is independent to their response to any other item on the same test or scale. This implies that only the factor influencing a person's response to an item is the trait of interest to the researcher.

LOGISTIC OGIVE The distribution curve of a logarithmic frequency distribution.

LOGIT The mathematical unit of measure used to denote individual ability and item difficulty. It is the natural logarithm of the odds of an individual succeeding on a chosen item or of an item inducing failure from an individual with "zero" ability, respectively.

MODEL CALIBRATION The refinement of the measurement instrument of interest such that its predictive capability becomes increasingly more accurate with each refinement step.

MODEL-TO-DATA FIT The statistical goodness-of-fit between the theoretical model and the test or scale data. Elimination of nonessential items and people during the model calibration steps leads to a good model-to-data fit and the indication that the remaining items are both unidimensional and independent.

MONOTONIC A particular set of values of a mathematical function where each value is related to the preceding values. The trend in values can be either increasing or decreasing.

ONE-PARAMETER MODEL An IRT model that specifies that only one parameter is affecting a person's response to an item—namely, a person's ability level on the trait of interest to the researcher. Also known as the *Rasch model.*

PERSON ABILITY LOGITS (θ, T) The natural logarithm of the proportion of items that a person answers correctly divided by the proportion of items answered incorrectly.

POLYCHOTOMOUS The division of something into more that two parts. Here, it refers to the Likert-type scaling method where the scale comprised five response choices (1 through 5) for each individual scale item.

TEST ORIENTED Having the focus of the analysis on the test as a whole (e.g., CTT) as opposed to on the individual items (e.g., IRT) that make up the test.

THREE-PARAMETER MODEL An IRT model that specifies that three parameters are affecting a person's response to an item, namely, a person's ability level on the trait of interest, the discriminating ability of the particular item, and the effect of guessing by the person.

TRAIT One of the focal areas of emphasis in IRT that refers to the level of competence of an individual on a certain task or aptitude measurement. Also referred to as *ability* or *latent trait.*

TRUE SCORE A person's expected test score over repeated administrations of the same test (or parallel forms of the test). The relationship of the true score to the observed (actual) score is commonly referred to as the *reliability estimate.*

TWO-PARAMETER MODEL An IRT model that specifies that two parameters are affecting a person's response to an item—namely, a person's ability level on the trait of interest and the discriminating ability of the particular item.

UNIDIMENSIONALITY One of the two key tenets of IRT. It assumes that only one trait or ability level is being measured by the various items that compose a test or scale. For example, the 10 items of the Rosenberg Self-Esteem Scale are each assumed to only measure self-esteem and no other trait.

References

Crocker, L., & Algina J. (1986). *Introduction to classical and modern test theory.* New York: Holt, Rinehart & Winston.

Fan, X. (1998). Item response theory and classical test theory: An empirical comparison of their item/person statistics. *Educational and Psychological Measurement, 58*(3), 357–381.

Gray-Little, B., Williams, V. S. L., & Hancock, T. D. (1997). An item response theory analysis of the Rosenberg Self-Esteem Scale. *Personality and Social Psychology Bulletin, 23*(5), 443–451.

Hambleton, R. K., & Swaminathan, H. (1985). *Item response theory: Principles and applications.* Boston: Kluwer.

Hambleton, R. K., Swaminathan, H., & Rogers, H. J. (1991). *Fundamentals of item response theory.* Newbury Park, CA: Sage.

Lawson, S. (1991). One parameter latent trait measurement: Do the results justify the effort? In B. Thompson (Ed.), *Advances in education research: Substantive findings, methodological developments* (pp. 159–168). Greenwich, CT: JAI Press.

Lord, F. M. (1980). *Applications of item response theory to practical testing problems.* Hillsdale, NJ: Erlbaum.

McKinley, R., & Mills, C. (1989). Item response theory: Advances in achievement and attitude measurement. In B. Thompson (Ed.), *Advances in social science methodology* (pp. 71–135). Greenwich, CT: JAI Press.

Nunnally, J. C., & Bernstein, I. H. (1994). *Psychometric theory.* New York: McGraw-Hill.

Wright, B. D., & Stone, M. H. (1979). *Best test design.* Chicago: MESA Press.

4 Assessing the Validity of Measurement

Fred B. Bryant

A clinical psychologist is interested in developing a cognitive–behavioral therapy for the treatment of arachnophobia (fear of spiders). Because she wants to assess the efficacy of this new treatment, she designs a physiological instrument to gauge blood pressure, heart rate, and muscle tension as a measure of fear. When exposed to a crawling spider, the clients who are most afraid of spiders should register the highest readings on the new instrument.

* * *

A social psychologist travels to a small Caribbean island to conduct cross-cultural research on superstition. Rather than creating a questionnaire in the native language, he develops an observational checklist for measuring superstitious behaviors in public settings. He predicts that such behaviors will occur more frequently over time as tropical storms approach the island.

* * *

Based on a prominent theory of personality, an industrial–organizational psychologist develops a self-report inventory to measure accident proneness. According to the theory, accident proneness involves three psychological "risk factors": thrill seeking, mindlessness, and impatience. Individuals with high scores on all three of these risk factors should be most strongly predisposed to accidents.

How accurate are the new instruments described above? Are they appropriate measures of what they are intended to assess? How can a

---

This chapter is respectfully and affectionately dedicated to the memory of Donald T. Campbell, whose wisdom, ingenuity, and kindness have inspired countless researchers in the social sciences.

researcher be sure that an instrument actually measures what it is supposed to measure? Social scientists have devised a variety of strategies to answer such questions about validity.

Consider the industrial–organizational psychologist who developed the measure of accident proneness. To evaluate the test's *content validity*, the researcher could use exploratory or confirmatory factor analysis to assess how thoroughly the test taps each of the three content areas that it is supposed to include (i.e., thrill seeking, mindlessness, and impatience).

To evaluate the test's *criterion validity*, the researcher could assess how accurately it predicts the experience of accidents, using several approaches. To evaluate *predictive validity* or *prospective validity*, the researcher could follow the same individuals over time and use multiple regression analysis to see if people who score higher on the test initially experience more future accidents, compared with people who score lower on the test initially. To evaluate *concurrent validity*, the researcher could measure a sample of respondents at a single point in time and use logistic regression analysis to see if people's test scores relate to whether or not they are currently recovering from an accident. To evaluate *retrospective validity*, the researcher could administer the test to a sample of respondents and then ask them to recall the number of accidents they have had in the past, to see if higher test scores correlate with the number of reported past accidents. As a stronger test of *postdictive validity*, the researcher could administer the test to a sample of respondents and then check these individuals' archived medical records to see if higher test scores correlate with the number of actual treated accidents.

To evaluate the test's *construct validity*, the researcher could examine whether the test actually measures what it is supposed to measure, using several approaches. To assess *convergent validity*, the researcher could examine the correlation between scores on the new test of accident proneness and scores on other similar measures known to assess predisposition to accidents, to see if there is convergence or agreement among these multiple forms of measurement. To assess *discriminant validity*, the researcher could see if scores on the new test of accident proneness diverge from scores on measures of other, distinct phenomena, such as the Type A coronary-prone behavior pattern or proneness to illnesses. Alternatively, the researcher could construct a *multitrait–multimethod matrix* (MTMM) and analyze the correlations among multiple measures using confirmatory factor analysis to assess convergent and discriminant validity simultaneously. In addition, the researcher could administer the new test to two groups of respondents—one known to have experienced many accidents and one known to have experienced few accidents—and use discriminant analysis to see if test scores can be used to correctly classify people into the different criterion groups.

This chapter explores the many forms of validity prominent in the social sciences and considers the prevailing multivariate statistical approaches to validity assessment. I begin with a broad definition of validity and its general subtypes, and then focus more narrowly on the validity of measurement and how researchers evaluate it. I introduce the major types of test validity—content, criterion, and construct—and describe research examples that have used multivariate statistics to investigate each of these. Finally, I discuss the MTMM matrix as an integrative framework for assessing convergent and discriminant validity, describe commonly used multivariate approaches to the analysis of MTMM data, and consider a concrete MTMM example.

A General Definition of Validity

What is *validity*, and how are multivariate statistics used to evaluate it? Simply stated, validity is truth. In the context of research, validity concerns whether a particular inference or conclusion that one wishes to make is accurate, reasonable, or correct. Stated more generally, validity is an overall evaluative judgment of the degree to which empirical evidence and theoretical rationales support the adequacy and appropriateness of conclusions drawn from some form of assessment (Messick, 1989, 1995).

Although people sometimes consider conclusions to be partly true (or partly false), in judging validity it is more accurate to consider empirical evidence as either supporting validity or failing to support validity. Thus, a conclusion may be judged to be either valid or invalid, although the evidence in support of this judgment may be either weak or strong. Comparing the evaluation of validity with the American legal process helps to clarify these points.

In a civil trial, the jury weighs evidence to reach a verdict of guilty or not guilty, based on the preponderance of evidence, and the burden of proof is on the plaintiff, who is accusing the defendant of wrongdoing. Likewise, in the field of science, the observer weighs evidence to decide whether a particular research conclusion is valid or invalid, based on the bulk of the evidence, and the burden of proof is on the one who is claiming validity. If the majority of evidence supports the conclusion, then the conclusion is deemed valid; if the majority of evidence fails to support the conclusion, then the conclusion is deemed invalid.

There is at least one major difference, however, between legal decisions regarding guilt and scientific decisions regarding validity. In a court of law, defendants are presumed innocent until proven guilty. In the field of science, in contrast, research conclusions are presumed invalid until proven otherwise. Unless the bulk of the evidence supports a particular conclusion, one must assume that the conclusion is not valid. This principle reflects the essential role of falsification in drawing conclusions using the scientific method (see Popper, 1959).

Furthermore, it is inappropriate to judge an instrument as being either valid or invalid in an absolute sense without knowing the particular application in which it is to be used (Cronbach, 1990). Validity cannot be judged independent of the circumstances in which generalizations are to be made. Without an understanding of the specific job involved, for example, it would be improper to conclude that a test of ability is necessarily a valid measure of job-related skills.

The process of assessing or establishing validity is known as *validation.* It is important to recognize that validation involves a gradual accumulation of research evidence from a variety of sources (Anastasi, 1988). In principle, just as no one study alone can prove an experimental hypothesis true, no one study by itself can definitively establish validity.

In planning validation studies, investigators strive first to develop a *nomological net* that identifies the key concepts or *constructs* associated with the particular phenomena being studied and the interrelationships among these key constructs. They then use this integrative theoretical framework to make critical decisions about methodological design and measurement in validation research. The general question of how closely the obtained data patterns fit the theoretical predictions about such data patterns is known as *nomological validity* (Cronbach & Meehl, 1955).

General Varieties of Validity

Validity has multiple forms, depending on the research question and on the particular type of inference being made. In their pioneering treatise on social science methodology, Cook and Campbell (1979) distinguished among four general types of validity in research. In the context of experimentation, one may first wish to know whether or not some aspect of the experimental manipulation (or "treatment") influenced responses to the dependent measures (or "outcomes"). At issue here are both the accuracy of the cause-and-effect conclusions drawn from statistical tests (i.e., *statistical conclusion validity*), and the appropriateness of concluding that the treatment, and not some extraneous factor, caused the observed results (i.e., *internal validity*). If an experimental effect is observed, then the question arises whether one can confidently label the specific cause and the specific effect in theory-relevant, conceptual terms (i.e., construct validity). Can the researcher precisely identify the conceptual variables responsible for the treatment's causal effects, as well as the conceptual variables being measured as outcomes? A final issue in experimentation concerns whether the conclusions drawn about cause and effect can be generalized to or across other populations, settings, and time periods (i.e., *external validity*). Attempts to assess the degree to which research results generalize across different samples, settings, or times are known as *cross-validation.* A related issue concerns *ecological validity:* whether the environment, or situational context, of a research study is representative of the natural, "real-world" social environment (see Cook & Campbell, 1979, for a more extensive discussion of these forms of validity).

This chapter focuses primarily on the role of multivariate statistics in assessing the validity of measurement. Although the essence of statistical conclusion validity is the proper use of statistics to draw inferences, each chapter in this volume, and in its predecessor, already addresses the valid use of specific multivariate statistical techniques. Moreover, statistics have little to do with evaluating internal validity, which largely involves logically ruling out alternative methodological explanations for obtained results. Although statistics are useful in assessing selection bias (i.e., whether preexisting differences between experimental and control groups are responsible for apparent treatment effects) and maturation effects (i.e., whether changes in participants that occur naturally over time explain observed results), most threats to internal validity are typically controlled through methodological design. Whereas external validity is best established by demonstrating the generalizability of results statistically, multivariate strategies for accomplishing this task have already been covered in the previous volume (Durlak, 1995; Licht, 1995). Because multivariate statistics play a major role in evaluating the validity of measurement instruments, this chapter addresses different forms of test validity and how researchers use multivariate statistics to assess them.

The specialized research area concerned with assessing and establishing the validity and reliability of measurement instruments is called *psychometrics* (Nunnally, 1978). Psychometric issues of test validity are distinct from issues of test reliability, which concern the consistency of scores within subjects across either (a) time (i.e., test–retest, or temporal reliability), (b) items tapping the same construct (i.e., internal consistency), (c) equivalent forms of the test (i.e., parallel-forms reliability), or (d) multiple raters (i.e., interrater reliability; for a discussion of the role of statistics in assessing reliability, see chapter 2 in this volume). Test validity, in contrast, concerns what the test measures and how well—but not how consistently—it does so. A measure can be reliable, but not valid; an instrument can provide consistent responses, although it may have nothing to do with what one really wishes to measure. The validity of an instrument concerns how thoroughly (content validity) and accurately (construct validity) it measures a specific theoretical concept of interest, and how useful it is in predicting important outcomes (criterion validity).

Specific Types of Test Validity

Content Validity

Content validity concerns the degree to which an instrument assesses all relevant aspects of the conceptual or behavioral domain that the instrument is intended to measure, or how thoroughly it samples the relevant target domain. For example, if an instrument is supposed to measure the coronary-prone lifestyle of Type A behavior, then the instrument would have content validity if it measures all essential components of Type A behavior, including competitiveness, speed and impatience, and hostility; if it measures only some but not all of these components, then the instrument is said to lack content validity.

Traditionally, researchers have relied largely on subjective impressions in judging an instrument's content validity. It has been common practice for test developers simply to inspect a questionnaire's constituent items visually to evaluate the thoroughness of content coverage. Occasionally, researchers have obtained experts' judgments as to whether a measure taps a particular construct with sufficient breadth and depth.

More recently, researchers have evaluated content validity using multivariate statistical procedures, such as exploratory factor analysis, to determine the dimensions or content domains that an instrument assesses. Researchers can use principal components analysis (PCA), for example, to discover (a) the number of different facets of the underlying construct that the instrument taps, (b) the specific questions or items that constitute each facet, (c) how strongly each of the constituent items defines its relevant content domain, and (d) how strongly content domains relate to one another (see Bryant & Yarnold, 1995, for a detailed discussion of PCA). By revealing the dimensions that an instrument inadequately covers, PCA also indicates ways in which content validity can be improved. Thus, if PCA identifies domains that have only a few diagnostic items or domains that are missing entirely, then new items can be added to the instrument to make content coverage more thorough. The issue of which particular dimensions underlie a measurement instrument is also known as *factorial validity* or *structural validity*.

As an example, Klohnen (1996) used PCA to examine the components underlying the construct of ego resiliency, as measured by a new self-report inventory. A sample of 725 adults completed the 26-item instrument. Consistent with theory, the analysis disclosed four dominant factors—confident optimism, productive activity, interpersonal warmth and insight, and skilled expressiveness—that explained 61% of the total variance among the 26 items. These findings support the content validity of the new instrument as a measure of ego resiliency.

As another multivariate strategy for assessing content validity, researchers can also use confirmatory factor analysis (CFA) to test alternative hypotheses about the content domains that underlie an instrument. With CFA, the researcher imposes a measurement model on the data (i.e., the researcher stipulates which items reflect which content domains and whether these content domains interrelate) and evaluates how well this model reproduces the observed pattern of relationships among the items. An acceptable goodness-of-fit provides evidence in support of the instrument's content validity. A major advantage of CFA over exploratory procedures is that it enables researchers to compare the goodness-of-fit of competing alternative measurement models systematically and to test hypotheses about which model or models best fit the data (see Bryant & Yarnold, 1995, for a detailed discussion of CFA).

As with exploratory factor analysis, content domains that have few diagnostic items or domains that fail to emerge altogether in CFA represent areas in need of improved measurement.

Besides using exploratory PCA, for example, Klohnen (1996) also used CFA to compare the goodness-of-fit of three alternative models for a new measure of ego resiliency: (a) four uncorrelated factors reflecting the four facets found in the exploratory analysis (Model 1), (b) four correlated latent factors (Model 2), and (c) four underlying factors that are, in turn, influenced by a single, general, higher order, ego-resiliency factor (Model 3). Confirming hypotheses, Models 2 and 3 fit the data equally well, and both of these models fit the data significantly better than Model 1. Consistent with the results of the earlier exploratory analysis, these CFA findings support the content validity of the new measure of ego resiliency, and they suggest that ego resiliency is best conceptualized as a unitary construct consisting of four interrelated personality domains.

Criterion Validity

Criterion validity concerns how accurately an instrument predicts a well-accepted indicator of a given concept, or a *criterion*. If a test measures what it is supposed to measure, then it should be useful in predicting future, present, and past outcomes of interest. For example, if a measure of criminal behavior is valid, then it should be possible to use it to predict whether an individual (a) will be arrested in the future for a criminal violation, (b) is currently breaking the law, and (c) has a previous criminal record. In criterion-based validation, bivariate statistical procedures, such as the Pearson product-moment correlation coefficient, are often used to relate test scores to the *criterion measure*. When a correlation coefficient is used to evaluate criterion validity in this way, the correlation is called a *validity coefficient*.

Predictive

Researchers can examine at least three different types of criterion validity, depending on when the test score and the criterion measure are obtained. First, the researcher might acquire the test score before measuring the criterion to assess how well the former predicts the latter. This form of criterion validity is known as *predictive validity*, or *prospective validity*. To assess the predictive validity of a test of predisposition to midlife depression, for example, a researcher could use a longitudinal design to follow the same individuals over time to see if people who score higher on the test during their 20s or 30s actually are more likely to become depressed during their 40s (i.e., in mid life), relative to people who score lower on the test at the earlier age. Prospective validation is strongest when the process of testing could not reasonably be expected to influence the criterion measure (Anastasi, 1988).

When the criterion is measured on a continuous scale, predictive validity is often assessed using bivariate statistical procedures such as the Pearson correlation coefficient. However, researchers also have used multiple regression analysis, a multivariable technique, to control for the effects of demographic variables (e.g., sex, age, race), while using test scores to predict the criterion measure (for a detailed discussion of the assumptions and mechanics of multiple regression, see Licht, 1995.) For example, Watson and Walker (1996) examined the long-term predictive validity of measures of dispositional positive and negative affect, by first administering these measures to a sample of undergraduates and then assessing depression and anxiety in these same respondents 6 years later. Multiple regression and correlation analyses demonstrated that initial scores on measures of both positive and negative affect significantly predicted later levels of affect and symptomology, controlling for the effects of marital status, employment status, and education level. These results suggest that the measures of dispositional affect have predictive validity.

Another multivariate statistical technique often used to assess predictive validity is logistic regression analysis. Logistic regression is appropriate when the criterion measure is dichotomous, although it can also be extended to criteria that have been measured using three or more categories. As with traditional linear regression, researchers can use logistic regression to control for demographic variables and to examine the independent effect of test score as a predictor of criterion measures. Logistic regression expresses the predictive power of a test in terms of the increase in odds of membership in one or the other of the levels of the criterion measure, for every one-unit increase in test score (for a detailed discussion of the assumptions and mechanics of logistic regression analysis, see Wright, 1995.)

For example, Pierce, Choi, Gilpin, Farkas, and Merritt (1996) used logistic regression in a prospective validation study evaluating the predictive validity of a measure of smoking susceptibility (i.e., the absence of a firm decision not to smoke). Pierce et al. (1996) first administered the susceptibility measure to a large, nationally representative sample of adolescents who had never experimented with smoking and then 4 years later assessed whether or not these same respondents had experimented with smoking. Logistic regression analyses demonstrated that initial susceptibility to smoking was a stronger independent predictor of experimentation than the presence of smokers among either family or best friends. These findings support the predictive validity of the measure of smoking susceptibility.

An alternative multivariate approach to assessing predictive validity that is more powerful than regression is structural equation modeling (SEM). With SEM, the researcher uses multiple measures as indicators of both the underlying construct to be validated and of the criterion construct, and then estimates the causal influence between the two latent constructs. The principal advantage of SEM is that it controls for the attenuating effects of unreliability by removing measurement error from the latent constructs. SEM also enables the researcher to partial out systematic sources of measurement error by allowing for correlated error between measured variables. Because this type of causal modeling requires multiple measures of each latent construct and sufficient sample size, however, it may be inappropriate in many cases (for a detailed discussion of the assumptions and logic of SEM, see chapter 7 in this volume).

Prospective validation is popular within the field of industrial–organizational psychology as a means of developing tests of job-related ability. With this approach, for example, a researcher first administers a battery of pre-employment screening tests to a group of job applicants and then later assesses their performance on the job. The researcher can then use multiple regression analysis to examine how well the set of tests as a whole predicts job performance and how much each of the tests independently contributes to the prediction of performance (see Anastasi, 1988).

Typically, industrial–organizational psychologists are also interested in determining whether tests of job-related ability are equally accurate in predicting the job performance of people from different racial or ethnic groups. Does a test of management skills, for example, predict the job performance of White and African American applicants equally well, or does it predict job performance better for one group than for the other? This issue of whether a test predicts scores on a criterion measure differently for different subgroups is known as *differential validity*. The case in which a test shows validity for one particular group but not for another is known as *single-group validity* (Anastasi, 1988).

Statisticians have developed multivariate analyses to detect differential validity. For example, Cleary (1968) pioneered a series of statistical tests, based on multiple regression analysis, that evaluates differences in the regression lines relating test scores to the criterion measure for each group. Evidence for differential validity exists when there are between-group differences in either the slope, y intercept, or standard error of the estimate for the regression lines. Differences in slope and in the standard error of the estimate indicate that the test predicts performance more accurately in one group than in the other. Intercept differences indicate that the test systematically underpredicts or overpredicts job performance for a particular group. Because the Uniform Guidelines of the Equal Employment Opportunity Commission outlaw tests in the workplace that display differential validity across racial groups, personnel psychologists routinely conduct analyses of differential validity in the course of test development (see Miner & Miner, 1978).

Although a relationship between initial test scores and later criterion measures supports a test's predictive validity, it is critical to rule out alternative methodological explanations for the observed predictive relationship. In validating personnel selection tests, for example, supervisors may become aware of employees' test scores, and this knowledge may influence the supervisors' ratings of the employees' job performance. Supervisors may expect employees with low test scores to perform poorly on the job and may rate them accordingly, and supervisors may give high-scoring employees the benefit of the doubt when preparing performance appraisals. This potential source of bias in predictive validation is known as *criterion contamination* (Anastasi, 1988). To prevent this type of methodological artifact, it is essential to keep test scores and criterion measures independent, so that knowledge of the former does not influence the latter.

Concurrent

As an alternative to prospective validation, the researcher might obtain the test score and the criterion measure at the same point in time and see how strongly the two correlate. This form of criterion validity is known as *concurrent validity*. To assess the concurrent validity of a test of the strength of people's sex drive, for example, a researcher could use a cross-sectional survey to measure sex drive and reported sexual activity in a sample of respondents at a single point in time to see if people's test scores relate to their reported level of current sexual activity. Because it is quicker and easier to collect test scores and criterion measure at the same time, researchers use concurrent validation more often than prospective validation. Often, however, concurrent validation provides weaker evidence for criterion validity than does predictive validation. When both the test and the criterion are obtained from self-reports, for example, respondents' desire to answer consistently may inflate the apparent relationship between the test and the criterion (Cook & Campbell, 1979). Thus, concurrent validation is strongest when the test score and the criterion measure are collected independently for the same individuals.

As with predictive validity, when the criterion is measured on a continuous scale, researchers often use simple bivariate correlations, as well as multiple regression analyses, to assess concurrent validity. For example, Clark, Bormann, Cropanzano, and James (1995) examined the concurrent validity of three self-report measures of coping by administering each of these questionnaires to a sample of 306 undergraduates, along with a set of criterion measures assessing daily hassles and uplifts, physical symptoms, life satisfaction, and positive and negative affect. Supporting concurrent validity, multiple regression analyses indicated that all three coping measures were significantly related to the outcome variables.

Investigators also use canonical correlation analysis to assess concurrent validity (see chapter 9 in this volume). For example, Edwards and Baglioni (1991) used a multivariate form of multiple regression, equivalent to canonical correlation, to examine the relationships between (a) 3 global measures and 15 subscales assessing Type A behavior and (b) 10 measures of mental and physical symptoms in a cross-sectional survey of 240 business executives. Results indicated that subscale scores were superior to the global measures in terms of the number of relationships detected, interpretability, and total explanatory power. These findings support the concurrent validity of the Type A subscales.

Retrospective

As a another form of criterion validity, the researcher might acquire the test score after measuring the criterion. This form of criterion validity is known as *retrospective validity*, or *postdictive validity*. To evaluate the retrospective validity of a creativity test, for example, a researcher could administer the test to a sample of adults and then ask the adults' former high school teachers to rate how creative these adults were as students —a significant correlation between test score and teacher ratings of creativity would be evidence of retrospective validity. The retrospective measurement of criteria is particularly weak methodologically, however, because knowledge about the present may distort respondents' recollection of the past. Nevertheless, retrospective measurement is common practice in some research areas.

For example, psychiatric researchers have used retrospective validation procedures to examine the relationship between early parental loss and depression. With this approach, psychiatric in-patients first complete measures of depression and then answer questions about whether or not they experienced parental loss during their childhood (see Roy, 1988). Although researchers in this area have not typically used multivariate statistical techniques, adopting such strategies (e.g., multiple regression, logistic regression) would enhance statistical conclusion validity. Whenever the respondents' recollections cannot be verified, however, retrospective validation remains methodologically weak.

As an alternative to using respondents' recall to assess criteria from the past, researchers can rely on archival records, when available, to quantify criterion measures obtained at an earlier point in time. An advantage of this approach over retrospection is that it avoids the possible distortion of recall. For example, Wilson and Mather (1974) assessed the postdictive validity of using palm "lifelines" to measure longevity. Using a sample of corpses, these researchers first measured the length of the lifeline and then examined medical records to determine date of birth. Results failed to support the postdictive validity of lifelines as a measure of lifespan.

Construct Validity

Within the context of test validation, *construct validity* concerns whether a given measure, or *operational definition*, actually assesses the underlying conceptual variable, or construct, that the measure is intended to represent. Does a questionnaire that is intended to assess participants' levels of aggression, for example, truly tap aggression, or does it in fact tap another concept, such as assertiveness or social dominance? Does an instrument that is designed to evaluate willingness to help others really measure helpfulness, or does it actually measure empathy and need for approval? Previous researchers have used a variety of terms to denote the conceptual accuracy of measurement, including *trait validity* (Campbell, 1960) and *referent validity* (Snow, 1974). Construct validation —that is, determining whether an instrument actually measures what it is supposed to measure—remains among the most difficult challenges in the social sciences (cf. Fiske, 1973; and Webb, Campbell, Schwartz, Sechrest, & Grove, 1981).

The crucial first step in establishing construct validity is what Cook and Campbell (1979) have termed *preoperational explication*. This process involves the initial formulation of a clear and explicit definition of the underlying characteristic of interest, carefully specifying the necessary components or key ingredients that constitute the construct, and what distinguishes it from related but separate constructs. Investigators should formulate conceptual definitions before choosing operational definitions (i.e., before deciding how to measure research variables)—hence the term. Without an adequate conceptual definition at the outset, it is impossible to establish construct validity. To know whether you have validly measured a particular construct, you must first understand its meaning. Otherwise, there is no clear standard to use in determining whether the conditions necessary for inferring the construct's presence exist.

Imagine, for example, that you created an instrument to measure the extent to which an individual is a "nerd." To demonstrate construct validity, you would need a clear initial definition of what a nerd is to show that the instrument in fact measures "nerdiness." Furthermore, without a precise definition of nerd, you would have no way of distinguishing your measure of the nerdiness construct from measures of potentially related constructs, such as shyness, introversion, or nonconformity.

Related to construct validity is the concept of *face validity*. Whereas construct validity concerns the degree to which an instrument actually measures the underlying concept or behavior it is intended to measure, face validity concerns the degree to which the particular instrument appears to measure what it is intended to measure. Face validity is a purely subjective judgment on the part of those who either administer or complete the instrument, concerning whether the instrument seems superficially (i.e., on its face) to measure what it is supposed to measure. Face validity may have nothing to do with construct validity: An instrument may appear on its surface to tap a particular construct, regardless of whether or not it actually does. Asking someone whether a glass of water is half empty or half full, for example, has face validity as a measure of optimism, but has unknown construct validity. Thus, face validity is really nothing more than a superficial impression.

Some instruments are transparent with respect to the underlying construct they are designed to tap. Such measures are said to have face validity. Measures of subjective well-being, for example, typically consistent of questions whose intended focus is relatively obvious. Questions such as "During the past week, what percentage of the time did you feel happy?" and "Would you say you're very happy, pretty happy, or not too happy these days?" are clearly recognizable as measures of happiness.

The intended focus of other instruments, in contrast, is not at all apparent. Such measures are said to lack face validity. Projective measures, such as the Rorschach Test (Rorschach, 1945) in which respondents describe what each of a series of inkblots looks like to them, for example, typically have little or no face validity, although such measures may or may not have construct validity.

Convergent

There are two main forms of construct validity in the social sciences. The first of these is *convergent validity*, or the degree to which multiple measures of the same construct demonstrate agreement or convergence. Measures that supposedly assess the same thing should correlate highly if they are in fact valid measures of the same underlying concept. For example, if pupil dilation is a valid measure of attraction to a visual stimulus, then the amount of pupil dilation should correlate significantly with self-reported liking of the stimulus.

As with the assessment of other forms of test validity, convergent validity is often evaluated simply in terms of bivariate correlation coefficients relating multiple measures that are expected to intercorrelate. In developing a multidimensional measure of life satisfaction for children, for example, Huebner (1994) explored the convergent validity of the measure by administering it to a sample of 413 children in Grades 3–5, along with instruments assessing loneliness and social dissatisfaction, perceived quality of school life, and self-concept. Supporting convergent validity, each subscale of the new satisfaction instrument—family, friends, school, living environment, and self—correlated significantly with a related subscale from the other measures.

As a multivariate strategy for assessing convergent validity, however, CFA is far superior to the visual inspection of bivariate correlations (Judd, Jessor, & Donovan, 1986). With CFA, researchers can test the hypothesis that all of the measures under investigation converge (i.e., share a single underlying construct), and they can contrast the goodness-of-fit of this particular model with one that stipulates the absence of convergent validity. Although PCA is also used to determine whether a set of measures converge to form a single factor, CFA provides a more systematic means of evaluating convergent validity.

In actual practice, convergent validity is usually assessed in conjunction with its relational opposite, divergent validity. Researchers typically use multivariate statistical procedures to assess both forms of construct validity at the same time.

Discriminant

Conversely related to convergent validity is *discriminant validity* (the degree to which multiple measures of different concepts are distinct). Measures that supposedly assess different things should not correlate highly if they are in fact valid measures of these different concepts. For example, if happiness is not merely the absence of depression but is instead a distinct phenomenon, then measures of happiness should not be so negatively correlated with measures of depression as to be indistinguishable.

Paralleling the statistical approaches to convergent validation, discriminant validation often consists of nothing more than an inspection of bivariate correlation coefficients. These correlational analyses are sometimes accompanied by multiple regression analyses, however, when the instrument being validated has multiple subscales. For example, Swinkels and Giuliano (1995) used multiple regression analyses to examine the relationship of two measures of mood awareness—mood labeling and mood monitoring—to measures of personality and emotion. The results revealed that these two forms of mood awareness showed different patterns of association with the outcome measures, supporting the discriminant validity of the distinction between mood labeling and mood awareness.

Researchers have also relied on factor analysis as a multivariate tool for evaluating discriminant validity. If a particular measure truly is distinct from other measures, then it should form its own separate dimension when factor analyzed along with the other measures. For example, Harder and Zalma (1990) used PCA to examine the discriminant validity of scales designed to measure the separate constructs of shame and guilt. Including the individual items that composed the scales in the same analysis, these researchers found a factor structure consisting primarily of two dimensions, one of which was defined by the shame items and the other by the guilt items. These results support the discriminant validity of the distinction between shame and guilt.

As with convergent validity, however, CFA provides a much more powerful means of assessing discriminant validity than either multiple regression or principal components analysis. With CFA, researchers can test the hypothesis that the measures under investigation diverge (i.e., represent separate underlying constructs), and they can contrast the goodness-of-fit of this particular model with one that stipulates that the measures converge to form a single latent construct. (For a detailed discussion of the use of CFA in construct validation, see Judd et al., 1986.)

For example, Bryant and Baxter (1997) used CFA to examine the discriminant validity of separate measures of positive (Ingram & Wisnicki, 1988) and negative (Hollon & Kendall, 1980) automatic cognition. Their results revealed that negative automatic thoughts were more strongly related to negative affect than to positive affect, whereas positive automatic thoughts were more strongly related to positive affect than to negative affect; and that negative automatic thoughts were also more strongly related to negative affect, anxiety, and neuroticism than were positive automatic thoughts, whereas positive automatic thoughts were more strongly related to positive affect than were negative automatic thoughts. These findings support the discriminant validity of separate measures of positive and negative automatic thinking.

As another means of assessing discriminant validity, researchers sometimes evaluate the accuracy with which scores on an instrument can be used to classify respondents into groups known to differ on a well-accepted criterion measure (*criterion groups*). This approach requires that the researcher (a) establish mutually exclusive criterion groups by first categorizing participants who meet predetermined guidelines for membership in respective groups, (b) administer the instrument to be validated to each of the criterion groups, and (c) examine classification accuracy using the scores on the instrument to discriminate the criterion groups. When the criterion groups consist of individuals exhibiting different types of psychological disorders, this type of work is known as *clinical validation.*

Although some researchers use univariate or multivariate analyses of variance to test hypotheses about mean differences between criterion groups on the instrument, other researchers use discriminant analysis to evaluate classification accuracy, and logistic regression might similarly be used (for a discussion of the assumptions and mechanics of multivariate analysis of variance, see Weinfurt, 1995; for a discussion of the requirements and logic of discriminant analysis, see Silva & Stam, 1995;

for a discussion of the requirements and logic of logistic regression analysis, see Wright, 1995.)

For example, Harrell and Ryon (1983) used discriminant analysis in a clinical validation study aimed at assessing the validity of the negative Automatic Thoughts Questionnaire (ATQ; Hollon & Kendall, 1980) as a tool for discriminating depressed versus nondepressed clinical populations. Harrell and Ryon first used self-report measures of depression and therapist ratings to categorize research participants into one of three criterion groups: (a) clinically depressed clients at a mental health center, (b) nondepressed mental health center clients with psychopathology, (c) or nondepressed hospital patients seeking treatment for medical problems. All participants then completed the ATQ, which taps the frequency of 30 negative self statements associated with depression. Discriminant analyses using all 30 ATQ items correctly classified 100% of the clinically depressed group, 84% of the nondepressed psychopathology group, and 95% of the nondepressed medical group. These findings support the clinical use of the ATQ as a tool for diagnosing depression.

Incremental

Closely related to discriminant validity is the concept of *incremental validity* (Sechrest, 1963), or the question of whether a particular measure (e.g., *Z*) provides explanatory power over and above another measure (e.g., *X*), in predicting a relevant criterion (e.g., *Y*). If measure *Z* is truly distinct from measure *X*, then *Z* should contribute to the prediction of *Y* after the effects of *X* have been accounted for. For example, if Graduate Record Examinations (GREs) scores reflect cognitive abilities that are distinct from academic achievement, then they should predict the quality of students' performance in graduate school after the effects of college grade point average have been statistically removed. Otherwise, GRE scores lack incremental validity in relation to grades.

Incremental validity is often assessed statistically using the multivariate procedure of hierarchical stepwise multiple regression (see Licht, 1995). With this procedure, one contrasts the proportion of variance explained in *Y* (i.e., R^2) using a baseline regression model that includes measure *X* (Step 1) with the R^2 obtained using an expanded regression model that includes both measures *X* and *Z* (Step 2). One then examines the statistical significance of the change in R^2 between the baseline and the expanded models so as to test the incremental validity of measure *Z*.

Alternatively, researchers can evaluate one test's incremental validity in relation to another by computing a partial correlation, which quantifies the proportion of variance in a criterion measure explained by one test after removing variance shared with the other test from both the first test and the criterion measure. For example, Klohnen (1996) examined the incremental validity of her new self-report measure of ego resiliency over standard personality measures of self-acceptance, empathy, and communality in predicting trained observers' ratings of ego resiliency. In particular, she computed partial correlation coefficients between the new self-report measure and observer ratings, controlling for the effects of self-acceptance, empathy, and communality. The new measure of ego resiliency significantly predicted observer ratings even after controlling for the joint effects of the other personality measures. These results support the new instrument's incremental validity in relation to the other personality scales.

Multitrait–Multimethod Matrix

Perhaps the most sophisticated research design for assessing construct validity is the MTMM matrix, first proposed by Campbell and Fiske (1959). The MTMM matrix is an integrative multivariable framework within which information about convergent and discriminant validity is systematically gathered in a single study. With this approach, one assesses two or more constructs (or traits) using two or more measurement techniques (or methods) and then intercorrelates these various measurements. As Campbell and Fiske (1959) originally noted,

> each test or task employed for measurement purposes is a trait–method unit, a union of a particular trait content with measurement procedures not specific to that content. The systematic variance among test scores can be due to responses to the measurement features as well as responses to the trait content . . . in order to estimate the relative contributions of trait and method variance, more than one trait as well as more than one method must be employed in the validation process. (p. 81)

Conceptual Components

Table 4.1 presents the correlations that form the conceptual basis of the MTMM matrix. This illustration involves three different traits (A, B, and C) and three different methods (1, 2, and 3). Following Campbell

Table 4.1

MTMM Matrix Conceptual Components

Traits		Method 1			Method 2			Method 3		
		A1	B1	C1	A2	B2	C2	A3	B3	C3
Method 1	A1	*r*								
	B1	**HM**	*r*							
	C1	**HM**	**HM**	*r*						
Method 2	A2	*v*	HH	HH	*r*					
	B2	HH	*v*	HH	**HM**	*r*				
	C2	HH	HH	*v*	**HM**	**HM**	*r*			
Method 3	A3	*v*	HH	HH	*v*	HH	HH	*r*		
	B3	HH	*v*	HH	HH	*v*	HH	**HM**	*r*	
	C3	HH	HH	*v*	HH	HH	*v*	**HM**	**HM**	*r*

Note. The table entries denote Pearson product–moment correlation coefficients relating pairs of trait–method combinations. MTMM = multitrait–multimethod matrix; *r* = reliability coefficient, or monotrait–monomethod coefficient (i.e., test–retest correlation between the same trait measured using the same method); *v* = validity coefficient, or monotrait–heteromethod coefficient (i.e., the correlation between the same trait measured using different methods); **HM** = heterotrait–monomethod coefficient (i.e., the correlation between different traits measured using the same method); HH = heterotrait–heteromethod coefficient (i.e., the correlation between different traits measured using different methods). There are 45 unique correlations in a 3 (Traits A, B, C) × 3 (Methods 1, 2, 3) multitrait–multimethod matrix: 9 reliability (*r*) coefficients, 9 validity (*v*) coefficients, 9 heterotrait–monomethod (**HM**) coefficients, and 18 heterotrait–heteromethod (HH) coefficients. In practice, it is important to control experiment-wise Type I error with such designs (e.g., by using a Bonferroni-type alpha adjustment). Boldface entries represent heterotrait–monomethod triangles.

and Fiske's (1959) original approach, the MTMM matrix consists of 4 types of correlations, each of which has a different label.

First, there are correlations between the same trait measured using the same method, or test–retest reliability coefficients. Campbell and Fiske called these *monotrait–monomethod values.* Note that there are three *reliability diagonals* in the matrix, one for each method, or a total of 9 reliability coefficients. Although the original formulation of the MTMM matrix conceived reliability in terms of test–retest correlations, later researchers have often substituted reliability indices of internal consistency, such as Cronbach's alpha or the split-half coefficient.

Second, immediately below each reliability diagonal is a triangle of correlations among different traits measured using the same method. Campbell and Fiske called these *heterotrait–monomethod coefficients.* Note that there are nine heterotrait–monomethod coefficients in the matrix. Each reliability diagonal and its adjacent heterotrait–monomethod triangle compose what Campbell and Fiske termed a *monomethod block.* The matrix has three monomethod blocks, each of which contains a single heterotrait–monomethod triangle.

Third, there are correlations between the same trait measured using different methods, or validity coefficients. Campbell and Fiske called these *monotrait–heteromethod coefficients.* Note that there are three *validity diagonals* in the matrix, one for each pair of different methods, or a total of nine validity coefficients.

Fourth, immediately above and below each validity diagonal is a triangle of correlations between different traits measured using different methods. Campbell and Fiske called these *heterotrait–heteromethod coefficients.* Note that there are 6 heterotrait-heteromethod coefficients for each validity diagonal or a total of 18 heterotrait–heteromethod coefficients in the matrix. Each validity diagonal and its two adjacent heterotrait–heteromethod triangles compose what Campbell and Fiske termed a *heteromethod block.* The matrix has three heteromethod blocks, each of which contains two heterotrait–heteromethod triangles.

A Hypothetical Illustration

To better understand the composition and meaning of the MTMM matrix, consider the following hypothetical example. Imagine that an investigator wishes to use the MTMM matrix to evaluate the convergent and discriminant validity of a new self-report measure of "kissing ability." This new measure, entitled the Scale for Measuring the Ability to

Creatively Kiss (SMACK), consists of 20 statements that respondents rate on a 7-point scale, ranging from *strongly disagree* (1) to *strongly agree* (7), for example, "I've been told that I'm a good kisser" and "I am talented at the art of kissing."

To create the MTMM matrix (see Table 4.1), the researcher needs to choose at least two traits besides kissing ability (Trait A) and at least two methods besides self-report (Method 1). As distinct constructs for discriminant validation, the researcher selects two traits that are conceptually similar to kissing ability, but that should be distinct: facial expressiveness (Trait B) and body awareness (Trait C). As alternative methods of measurement, the researcher selects direct participant observation (Method 2) and informant report (Method 3).

Thus, in addition to using the self-report SMACK (A1), kissing ability is also measured in two other ways using (a) a trained participant observer who rates the quality of the respondent's kissing behavior on a 0–100 scale under standardized testing conditions, i.e., candlelight, soft romantic music, breath freshener (A2), and (b) a knowledgeable prior romantic partner who provides a rating on a 1–10 scale of the respondent's kissing ability (A3). Likewise, facial expressiveness is measured in three ways using (a) a self-report measure (B1), (b) an observer trained to use a standardized facial expressiveness coding system (B2), and (c) a knowledgeable informant who provides a rating on a 1–10 scale (B3). Finally, body awareness is also measured in three ways, using (a) a self-report measure (C1), (b) an observer trained to use a standardized body awareness coding system (C2), and (c) a knowledgeable informant who provides a rating on a 1–10 scale (C3).

Campbell and Fiske's Classic Criteria for Evaluating Multitrait–Multimethod Matrices

To evaluate convergent and discriminant validity, Campbell and Fiske (1959) proposed four criteria that should exist based on visual inspection of a MTMM matrix. The first of these criteria concerns convergent validity, which is said to exist when the correlations among multiple methods of measuring the same construct (i.e., monotrait–heteromethod coefficients or validity coefficients) are "significantly different from zero and sufficiently large" (Campbell & Fiske, 1959, p. 82). If different methods of measuring the same trait do not intercorrelate, then it is unreasonable to conclude that they tap the same concept. Thus, to return to the hypothetical example, the self-report, participant observation, and informant report methods of measuring kissing ability should substantially intercorrelate. Otherwise, it is difficult to argue that the SMACK truly taps kissing ability.

Campbell and Fiske's (1959) remaining three criteria concern the assessment of discriminant validity. First, the correlations among multiple methods of measuring the same concept (i.e., validity coefficients) should be higher than the correlations among multiple methods of measuring different concepts (i.e., heterotrait–heteromethod coefficients). Thus, the correlations among the self-report, participant observation, and informant report methods of measuring kissing ability should be higher than the correlations between kissing ability and facial expressiveness, kissing ability and body awareness, or facial expressiveness and body awareness measured in different ways.

Second, correlations among different methods of measuring the same concept (i.e., validity coefficients) should also be higher than the correlations among the same methods of measuring different concepts (i.e., heterotrait–monomethod coefficients). Thus, the correlations among the self-report, participant observation, and informant report methods of measuring kissing ability should be higher than the correlations between kissing ability and facial expressiveness, kissing ability and body awareness, or facial expressiveness and body awareness measured in the same way.

Finally, the pattern of correlations among the different concepts should be identical, regardless of whether the same or different methods of measuring have been used. Thus, kissing ability, facial expressiveness, and body awareness should show the same interrelationships, regardless of how the three traits are assessed. When this final criterion is met, correlations among measures of concepts are independent of methods; when it is violated, however, method variance operates differently across the correlations. Imagine, for example, that kissing ability, facial expressiveness, and body awareness are more highly interrelated when measured using self-report than when measured using behavioral observation. This pattern of results might reflect the respondents' desire to be consistent in their self-reports.

Although Campbell and Fiske's (1959) four classic criteria are easy to examine visually, there are some important limitations to their use in analyzing MTMM matrices. At least four critical assumptions underlie Campbell and Fiske's classic criteria: (a) concepts (traits) and measurement methods are uncorrelated, (b) measurement methods influence all traits equally, (c) measurement methods are independent of one another, and (d) measurement methods are equally reliable (cf. Schmitt & Stults, 1986). Unfortunately, these critical assumptions are often untenable. In personality research, for example, self-ratings frequently correlate with peer ratings because both share implicit theories about the origin of individual characteristics, and alternative methods of measuring peer ratings often show at least moderate correlations (Bagozzi, 1993).

Another problem with these classic criteria is that no precise standards have been established for determining how well the criteria are met. The criteria are merely rough "rules of thumb" that depend on a qualitative assessment of confirming and disconfirming evidence. Although methods exist for comparing correlated correlation coefficients, as well as hypothesized patterns of intercorrelation (Meng, Rosenthal, & Rubin, 1992), these more systematic decision rules generally have not been used. Furthermore, Campbell and Fiske's (1959) criteria do not explicitly capitalize on the multivariate nature of the MTMM matrix, but instead rely implicitly on separate sets of bivariate analyses.

Techniques for Analyzing Multitrait–Multimethod Matrices

Since Campbell and Fiske (1959) first introduced the MTMM matrix, investigators have proposed a host of strategies for analyzing such data. Each analytic approach, however, has certain shortcomings. Thus, after more than 40 years, there is still no consensus on how best to analyze MTMM matrices (Fiske & Campbell, 1992; Kenny & Kashy, 1992). Although researchers occasionally rely solely on visual inspection of Campbell and Fiske's four classic criteria to examine MTMM data (e.g., Massey, 1997), the dominant contemporary analytic approach is multivariate. The following sections briefly review the most commonly used procedures for analyzing MTMM matrices, focusing on three available techniques: (a) analysis of variance (ANOVA), (b) generalized proximity function, and (c) CFA. Because it is now used most often, the last technique will receive the most coverage.

Briefly, other multivariate techniques for analyzing MTMM data, besides those discussed here, include exploratory factor analysis (Golding & Seidman, 1974; Jackson, 1975); smallest space analysis (Levin, Montag, & Comrey, 1983); the direct product model (Browne, 1984); first-order CFA with separate factors for traits, methods, and measure specificity (Kumar & Dillon, 1990); and second-order CFA with measures loading indirectly on traits and methods (Marsh & Hocevar, 1988). However, because these approaches are less prevalent and have greater limitations, I have chosen not to cover them.

Analysis of Variance

The earliest, and at one time the most common, method of analyzing MTMM matrices was with ANOVA (Boruch & Wolins, 1970; Stanley, 1961). With this approach, each observed variable is considered to be the sum of four components: (a) a general factor that underlies all measures for individuals across traits and methods of measuring, (b) a trait factor on which all measures specify the individual as being either above or below his or her standing on the general factor, (c) a method factor that reflects the extent to which a particular measurement method gives higher or lower scores on all traits to a particular individual, and (d) random error.

With the ANOVA approach, one computes three average correlation coefficients from the MTMM matrix: the average of all correlations involving (a) different traits measured using different methods (i.e., heterotrait–heteromethod coefficients), (b) different traits measured using the same method (i.e., heterotrait–monomethod coefficients), and (c) the same trait measured using different methods (i.e., monotrait–heteromethod coefficients). These average correlations are then examined in a three-way (people × traits × methods) ANOVA, in which the general factor is analogous to the person effect, the method factor analogous to the Person × Method interaction, and the trait effect analogous to the Person × Trait interaction. The ANOVA model uses F tests to assess the statistical significance of these three terms.

As usually applied, the ANOVA model includes several assumptions that limit its utility. First, each trait and method factor is assumed to be independent of the general factor and independent of each other. Unfortunately, this assumption precludes the estimation of trait intercorrelations, method intercorrelations, and trait–method intercorrelations —a liability that may be critical in many practical applications. Second, because the ANOVA model involves averaging correlations of different types (i.e., monotrait–heteromethod coefficients, heterotrait–monomethod coefficients, monomethod–heterotrait coefficients, and heterotrait–heteromethod coefficients), it also assumes that the variances of different trait and method factors are equal—an assumption often violated in applied measurement situations.

Generalized Proximity Function

As an alternative procedure for analyzing MTMM matrices, Hubert and Baker (1978, 1979) pioneered a nonparametric equivalent of the ANOVA model known as the *generalized proximity function*. This approach begins, as does the ANOVA model, by computing three indices: (a) the average of the correlations for the same trait measured using different methods, (b) the difference between the average of the same trait measured using different methods and the average for different traits measured using different methods, and (c) the difference between the average of the same trait measured using different methods and the average for different traits measured using the same methods. The next step is to test the null hypothesis that the MTMM matrix shows no evidence of influence due to methods or traits, or that the correlations in the matrix are distributed randomly. If the null hypothesis cannot be rejected, then the first computed index above should equal the average of all of the other correlations in the MTMM matrix, and the second and third indices should equal zero. Hubert and Baker developed nonparametric inferential tests, for which exact probabilities can be computed or estimated using a Monte Carlo simulation, for statistical testing in this context.

Like the ANOVA approach, the generalized proximity approach is intuitively appealing because it provides a summary of the entire MTMM matrix. Unlike the ANOVA approach, however, this procedure makes no assumptions about normality and homogeneity of variance. Some problems with the generalized proximity approach are that it does not consider the possibility of differential reliability of measurement, address trait–method intercorrelations, or include a test of the patterning of trait interrelationships (Campbell & Fiske's, 1959, fourth criterion).

Confirmatory Factor Analysis

In recent times, the most popular statistical approach to the analysis of MTMM matrices has been CFA (Wothke, 1996). (For a review of the assumptions and logic of CFA, see Bryant & Yarnold, 1995.) This approach involves imposing different models of the underlying factors that influence the MTMM matrix and evaluating how well they explain observed correlations. At least four explanatory models of increasing complexity can be examined systematically using the CFA approach (cf. Jöreskog, 1971; Werts & Linn, 1970; Wothke, 1996). These models hypothesize that the correlations in the MTMM matrix are explained completely by (a) measurement error, that is, traits and methods have no influence (Model 1, the "null" model); (b) traits and error, that is, methods of measuring have no influence (Model 2, the "trait-only" model 2); (c) methods of measuring and error, that is, traits have no influence (Model 3, the "method-only" model); and (d) traits, methods, and error (Model 4, the "trait–method" model). Widaman (1985) presented a complete taxonomy of CFA models that systematically vary different characteristics of trait and method factors.

With the CFA approach, the researcher imposes each model one at a time on the MTMM matrix to determine how accurately each model reproduces the correlations in the matrix. Each confirmatory analysis provides a maximum-likelihood chi-square statistic as well as other measures of goodness-of-fit that indicate how well the hypothesized model explains the pattern of MTMM correlations. Also because the four CFA models above are nested (i.e., they are identical, except that some contain more constraints than others), their chi-square values and degrees of freedom can be directly contrasted to test improvement in the goodness-of-fit across models.

For example, two model contrasts provide tests of the significance of trait variance: (a) the contrast between the chi-square values obtained for Model 1 (measurement error only) versus Model 2 (traits plus error) and (b) the contrast between the chi-square values obtained for Model 3 (methods plus error) versus Model 4 (methods and traits plus error). Likewise, two model contrasts provide tests of the significance of method variance: (a) the contrast between the chi-square values obtained for Model 1 (measurement error only) versus Model 3 (methods plus error) and (b) the contrast between the chi-square values obtained for Model 2 (traits plus error) versus Model 4 (methods and traits plus error). When subtracting the chi-square value of the more restrictive model (i.e., the latter model in each of these contrasts) from the chi-square value for the less restrictive model (i.e., the former model in each of these contrasts), a significant reduction in chi-square values indicates that the more restrictive model provides an improvement in goodness-of-fit over the less restrictive model.

Researchers can also use CFA to partition the variance in respondents' scores on each observed measure into separate components that reflect the influence of trait, method, and error. For a given measure, (a) the squared standardized loading on its trait factor represents the variance from trait, (b) the squared standardized loading on its method factor represents the variance from method, and (c) the standardized unique-error term represents the variance from error. In addition, CFA provides estimates of correlations among traits and among methods, controlling for the attenuating effects of measurement error.

Compared with other statistical approaches, CFA has at least four advantages in analyzing MTMM matrices. First, the CFA approach allows the researcher to estimate the reliability of each measure, rather than assuming that all measures are equally reliable. Second, each confirmatory analysis yields various measures of goodness-of-fit for the overall model, whereas other procedures offer little or no way of judging the overall adequacy of a particular model. Third, CFA allows researchers to compare alternative explanatory models systematically to test hypotheses about which factors influence the observed correlations in the MTMM matrix and how these factors interrelate. Fourth, CFA provides a way of partitioning the variance of the measures into separate trait, method, and error components.

There are, however, also some distinct disadvantages to the CFA approach. One problem is that the approach requires a minimum of three traits and three methods, four traits and two methods, or two traits and four methods, to be conducted (Bagozzi, 1993). Also, when analyzing MTMM data, CFA often encounters technical problems, such as parameter estimates that fail to converge, improper solutions involving parameter estimates that lie outside of their permissible range (e.g., negative unique errors), or standard errors of parameter estimates that are excessively large (Marsh, 1989).

To overcome these technical obstacles, Marsh (1989) proposed a variant of the CFA approach known as the correlated uniqueness model (see also Kenny, 1976). Instead of including separate trait and method factors, this model includes trait factors but represents the effects of the methods as correlations among the error terms for the observed measures. With this approach, one models method effects by specifying that measures sharing the same method of assessment have unique errors that are correlated with one another. One may then contrast alternative models in the same fashion as in the traditional CFA approach. The primary advantage of the correlated uniqueness model over the traditional CFA approach is that it rarely produces ill-defined solutions (Marsh & Bailey, 1991).

There are, however, some limitations to the correlated uniqueness model. First, it is not always easy to interpret the correlated error terms, especially if both positive and negative correlations emerge for the same method. Furthermore, the model assumes that methods of measurement do not correlate with one another. Although this assumption seems reasonable when researchers measure traits in ways that are strikingly different (e.g., self-report and unobtrusive behavioral measurement), the assumption may be violated when researchers use related measurement methods (e.g., different forms of self-report).

An Empirical Example

To illustrate the use of CFA to analyze a MTMM matrix, I describe Bagozzi's (1993) reanalysis of data originally reported by Van Tuinen and Ramanaiah (1979). The latter two researchers used the MTMM matrix with a sample of 196 undergraduates to examine the construct validity of three traits: global self-esteem (measured by the Coopersmith Self-Esteem Inventory and the Tennessee Self-Concept Scale), social self-esteem (measured by the Jackson Personality Inventory and the Janis–Field Feelings of Inadequacy Scale), and the need for order (measured by the Personality Research Form and the Order Scale of the Comrey Personality Scale). Van Tuinen and Ramanaiah also used three methods of measurement: a self-report inventory with true–false response format, a self-report inventory with multiple 2-point and 5-point scales, and a self-assessment of the accuracy of self-descriptive traits using a 7-point scale. Because they collected measures at only one point in time, Van Tuinen and Ramanaiah used coefficients of internal consistency, instead of test–retest correlations, to assess reliability. Table 4.2 presents this MTMM matrix.

Applying Campbell and Fiske's Classic Criteria

Consider first how the matrix fares when evaluated using Campbell and Fiske's (1959) four classic criteria. To establish convergent validity, different methods of measuring the same trait should be highly correlated (i.e., the validity coefficients should be large). Note that all nine of the validity coefficients—which range from .56 to .75 (median = .68)—are relatively large and statistically significant ($ps < .0001$). This is strong evidence of convergent validity for the measures of global self-esteem, social self-esteem, and need for order.

Campbell and Fiske's (1959) first criterion for assessing discriminant validity is that correlations among different methods of measuring the same concept (i.e., the validity coefficients) be greater than the correlations among different methods of measuring different concepts (i.e., the heterotrait–heteromethod coefficients). Note that there

Table 4.2

MTMM Matrix for Measures of Global Self-Esteem, Social Self-Esteem, and Need for Order

	True–False			Multipoint			Self-rating		
Trait measures	**A1**	**B1**	**C1**	**A2**	**B2**	**C2**	**A3**	**B3**	**C3**
Method 1: True–false									
A1. Global SE	(83)								
B1. Social SE	**58**	(85)							
C1. Order	**17**	**14**	(74)						
Method 2: Multipoint									
A2. Global SE	75	45	23	(93)					
B2. Social SE	72	74	16	**65**	(91)				
C2. Order	09	06	68	**25**	**08**	(85)			
Method 3: Self-rating									
A3. Global SE	58	53	14	62	68	09	(63)		
B3. Social SE	47	74	10	40	69	07	**58**	(74)	
C3. Order	22	18	63	34	22	56	**30**	**23**	(82)

Note. MTMM = multitrait–multimethod matrix; global SE = global self-esteem (measured by the Coopersmith Self-Esteem Inventory and the Tennessee Self-Concept Scale); social SE = social self-esteem (measured by the Jackson Personality Inventory and the Janis–Field Feelings of Inadequacy Scale); order = need for order (measured by the Personality Research Form and the Order Scale of the Comrey Personality Scale); Method 1 (true–false) = self-report inventory with true–false response format; Method 2 (multipoint) = self-report inventory with 2-point and 5-point scales; Method 3 (self-rating) = self-assessment of the accuracy of self-descriptive traits using a 7-point scale. Tabled are Pearson product–moment correlation coefficients with decimal points omitted. Alpha reliability coefficients are in parentheses. Validity coefficients are underscored. Heterotrait–monomethod coefficients (i.e., correlations between different traits measured using the same method) are shown in boldface. $N = 196$. $rs \geq .11$ are statistically significant at $p < .05$, one tailed. From "A Multimethod Analysis of Selected Self-Esteem Measures," by M. Van Tuinen and N. Ramanaiah, 1979, *Journal of Research in Personality, 13*, p. 21 (Table 1). Copyright by Academic Press. Adapted with permission.

are three validity coefficients for each trait, and each validity coefficient must be compared with its four corresponding heterotrait–heteromethod correlations within each heteromethod block, yielding 12 comparisons for each trait (or 3 traits × 12 comparisons = 36 total comparisons). Thus, for example, the validity coefficient of global self-esteem assessed by true-false format and by multipoint scales (.75) is higher than the correlations between global self-esteem assessed by self-rating and (a) social self-esteem assessed by multipoint scales (.72), (b) need for order assessed by multipoint scales (.09), (c) social self-esteem assessed by true–false format (.45), and (d) need for order assessed by true–false format (.23). All 12 comparisons for social self-esteem, all 12 comparisons for need for order, and 11 of the 12 comparisons for global self-esteem meet this first criterion for discriminant validity. Overall, only one of these 36 total comparisons failed the first criterion (i.e., $p = 1/36 = .028$), a proportion of failures that is less than what one would expect at a chance level of .05. The data satisfy the first criterion for discriminant validity.

The second, and more difficult, discriminant validity criterion requires that correlations among different methods of measuring the same concept (i.e., validity coefficients) be greater than the correlations among the same methods of measuring different concepts (i.e., heterotrait–monomethod coefficients). Here, each validity coefficient must be compared with its four corresponding heterotrait–monomethod correlations, yielding a total of 12 comparisons for each trait. Thus, for example, the validity coefficient of global self-esteem assessed by true–false format and by multipoint scales (.75) is higher than the correlations between (a) global self-esteem and social self-esteem assessed by true–false format (.58), (b) global self-esteem and need for order assessed by true–false format (.17), (c) global self-esteem and social self-esteem assess by multipoint scales (.65), and (d) global self-esteem and need for order assessed by multipoint scales (.25). Nine of the 12 comparisons for global self-esteem, all 12 of the comparisons for social self-esteem, and all 12 comparisons for need for order meet this second criterion for discriminant validity. Overall, 3 of these 36 total comparisons failed the second criterion (i.e., $p = 3/36 = .083$), a proportion of failures that is greater than what one would expect at a chance level of .05. The data only marginally satisfy the second discriminant validity criterion.

The third and most stringent criterion for discriminant validity is that the pattern of correlations among the traits must be the same in all monomethod blocks as well as in all heteromethod blocks. This entails comparing the three trait intercorrelations with each other, within each of the three monotrait–heteromethod triangles and within each of the six heterotrait–monomethod triangles. The MTMM data fully meet this criterion. In each trait–method triangle, the correlation between global self-esteem and social self-esteem was the largest, whereas the correlation between social self-esteem and need for order was the smallest; and the measures of global and social self-esteem were more strongly correlated with each other than with the measures of need for order. To evaluate statistical significance, the rank order of correlations can be compared across all nine trait–method triangles to test the hypothesis that the patterns of correlations are the same in all triangles. This yields a coefficient of concordance of .222, which is nonsignificant, $\chi^2(2, N = 196) = 4.00$, $p > .13$. Thus, because this fails to reject the hypothesis that the pattern of correlations is identical across trait–method triangles, one can conclude that the MTMM data satisfy the third discriminant validity criterion (see Bagozzi, 1993).

Using Confirmatory Factor Analysis

Van Tuinen and Ramanaiah (1979) relied solely on Campbell and Fiske's (1959) four classic criteria to assess the convergent and discriminant validity of the self-esteem measures. Bagozzi (1993), in contrast, used CFA to test four competing models of trait and method effects. These four models are the following:

- *Model 1*: The null model assumes that the measures are uncorrelated with one another and that there is no trait or method variance, only error variance.
- *Model 2*: The trait-only model assumes that there is no variance due to methods and that the measures reflect only trait and error variance.
- *Model 3*: The method-only model assumes that there is no trait variance and that the measures reflect only method and error variance.
- *Model 4*: The trait–method model assumes that variation in the measures is explained entirely by traits, methods, and error. Figure 4.1 presents a schematic diagram of this trait–method CFA model.

Table 4.3 summarizes the results of Bagozzi's (1993) CFAs of this MTMM matrix. There are two tests of *trait* effects: (a) subtract the χ^2

Figure 4.1

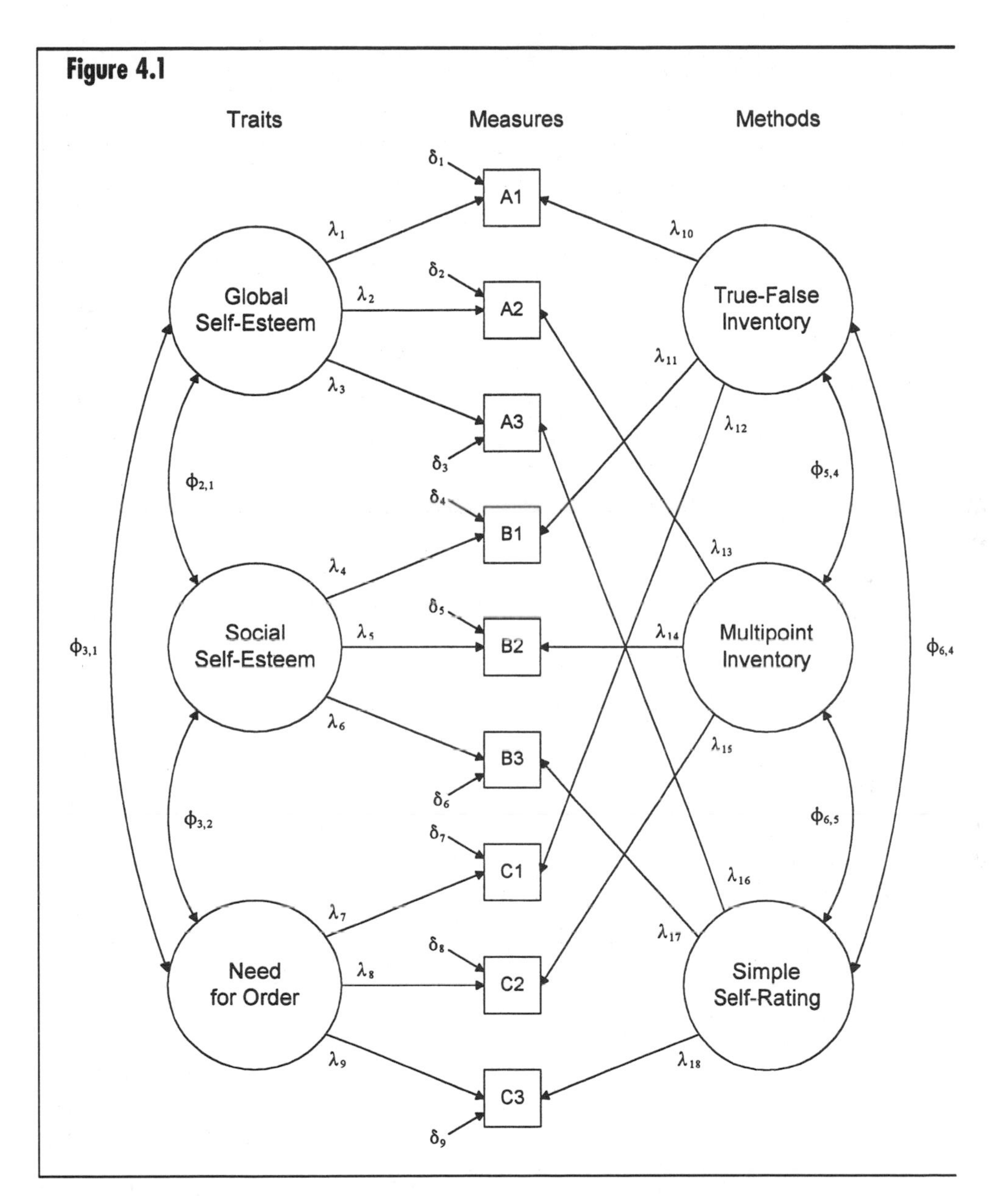

Schematic diagram illustrating the CFA trait–method model, containing three trait factors and three method factors, for Van Tuinen and Ramanaiah's (1979) nine measures. In the diagram, the observed measures are enclosed in boxes, corresponding to the nine observed measures (A1–A3, B1–B3, and C1–C3) in Van Tuinen and Ramanaiah's MTMM matrix (see Table 4.2). The three traits (global self-esteem, social self-esteem, and need for order) and the three methods (true–false inventory, multipoint inventory, and simple self-rating) are enclosed in circles, reflecting the hypothesis that they are underlying factors of influence on responses to the observed measures. The hypothesized effects of traits on observed measures are designated by $\lambda_1 - \lambda_9$ and are represented by the arrows connecting each trait to the three observed measures it is hypothesized to influence. The hypothesized effects of methods on observed measures are designated

Table 4.3

Results of Confirmatory Factor Analyses of an MTMM Matrix

Model	χ_2	df	*p*	NCNFI
1. Null	1,086.46	36	<.001	0.00
2. Trait only	111.27	24	<.001	0.92
3. Method only	358.04	24	<.001	0.68
4. Trait–method	11.27	24	>.50	1.00

Note. MTMM = multitrait–multimethod matrix; NCNFI = noncentralized normed fit index; this value, which ranges from 0 to 1, reflects the proportion of common variance in the set of measures that the given model explains relative to the null model. $N = 196$. Results are Figure 2 (p. 60), Table 2 (p. 63), and Table 3 (p. 64) of Bagozzi (1993).

value of (trait only) Model 2 from the χ^2 value of (null) Model 1 to obtain $\Delta\chi^2(12, N = 196) = 975.19$, $p < .001$, and (b) subtract the χ^2 value of (trait–method) Model 4 from the χ^2 value of (method-only) Model 3 to obtain $\Delta\chi^2(12, N = 196) = 346.77$, $p < .001$. These results indicate that a significant amount of trait variance is present. There are two tests of *method* effects: (a) subtract the χ^2 value of (method only) Model 3 from the χ^2 value of (null) Model 1 to obtain $\Delta\chi^2(12, N = 196) = 728.42$, $p < .001$, and (b) subtract the χ^2 value of (trait–method) Model 4 from the χ^2 value of (trait-only) Model 2 to obtain $\Delta\chi^2(12, N = 196) = 100.00$, $p < .001$. These results indicate that a significant amount of method variance also exists.

Table 4.3 further reveals that Model 4—specifying that variation in the observed measures is an additive function of trait, method, and error effects—fits the MTMM data well. Notice that Model 4 produces a nonsignificant χ^2 value ($p > .50$). This finding shows that overall the actual correlations in the MTMM matrix (see Table 4.2) are not significantly different from what Model 4 predicts they should be. Model 4 also yields a relative-fit index (i.e., the noncentralized normed fit index

---

by $\lambda_{10} - \lambda_{18}$ and are represented by the arrows connecting each method to the three observed measures it is hypothesized to influence. The curved arrows connecting trait factors to one another ($\phi_{3,1}$, $\phi_{2,1}$, and $\phi_{3,2}$) and connecting method factors to one another ($\phi_{6,4}$, $\phi_{5,4}$, and $\phi_{6,5}$) represent correlations between underlying factors. The absence of curved arrows connecting trait factors to method factors reflects the fact that this trait–method model assumes traits and methods are uncorrelated. Finally, the effect of unexplained error on each observed measure is designated by $\delta_1 - \delta_9$ and is represented by a small arrow connected to each measure.

[NCNFI]) of 1.00, which means that it explains 100% of the variance that the observed measures have in common, relative to the null model.

Inspecting the CFA solution for Model 4, however, Bagozzi (1993) found a technical problem. Namely, three of the observed measures had error variances that were negative. This type of improper solution is often encountered when using CFA to estimate the trait–method model (Marsh, 1989). Although one solution to the problem is to fix these negative variances to zero, Bagozzi (1993) found that fixing only the highest value to zero was sufficient to correct the problem. Results showed that Model 4 still fit the data well with this constraint, $\chi^2(13, N = 196) = 13.99$, $p > .37$, NCNFI = 0.999, and that no negative error estimates occurred.

Table 4.4 presents the factor loadings and factor intercorrelations from Bagozzi's (1993) CFA solution for Model 4. Note that the matrix of factor loadings in the top of this table corresponds to the pattern of trait influences (indicated by the symbols $\lambda_1 - \lambda_9$) and method influences (indicated using the symbols $\lambda_{10} - \lambda_{18}$) modeled in Figure 4.1. Also the matrix of factor intercorrelations in the bottom of this table corresponds to the 6 factor interrelationships (each indicated by the symbol ϕ) in Figure 4.1. Here is how Bagozzi (1993) summarized the findings reported in Table 4.4:

> The factor loadings for the true–false inventory and self-rating of global self-esteem and the multipoint inventory of social self-esteem are quite low. These results suggest that the measures are poor indicators of their respective factors. Notice also that the factor loadings of methods for all measures of global and social self-esteem are quite high. This indicates that methods biases strongly influence the measures of self-esteem. . . . As can be seen in the bottom of Table [4.4], all traits are distinct (i.e., each is correlated with the others at a level significantly less than 1.00) . . . all methods correlate at very high levels. (pp. 63–64)

Bagozzi (1993) also partitioned the variance of each measure into trait, method, and error components. To do this, he squared each measure's standardized factor loading on the trait and method factors and examined each measure's error term. Table 4.5 presents the results of these computations. The right column of this table corresponds to the effects of error on each observed measure ($\delta_1 - \delta_9$) modeled in Figure 4.1. Notice that (a) only the measures of need for order exhibit acceptably high levels of trait variance, (b) all measures of global and

Table 4.4

Standardized Factor Loadings and Factor Intercorrelations From a CFA Solution for the Trait–Method Model

	Traits			Methods		
Measures	Global SE	Social SE	Order	True–false	Multipoint	Self-rating
Global SE by						
True–false	21			83*		
Multipoint	64*				77*	
Self-rating	5					82*
Social SE by						
True–false		51*		74*		
Multipoint		17*			90*	
Self-rating		55*				71*
Order by						
True–false			84*	20*		
Multipoint			79*		12	
Self-rating			69*			33*

	Traits			Methods	
Traits	Global SE	Social SE	Order	True–false	Multipoint
Social SE	−32	—			
Order	26*	00			
Methods					
Multipoint				98*	
Self-rating				85*	92*

Note. CFA = confirmatory factor analysis; global SE = global self-esteem (measured by the Coopersmith Self-Esteem Inventory and the Tennessee Self-Concept Scale); social SE = social self-esteem (measured by the Jackson Personality Inventory and the Janis–Field Feelings of Inadequacy Scale); order = need for order (measured by the Personality Research Form and the Order Scale of the Comrey Personality Scale). Blank loadings and intercorrelations were fixed at zero. Decimals are omitted. $N = 196$. Results are Figure 2 (p. 60), Table 2 (p. 63), and Table 3 (p. 64) of Bagozzi (1993). $^*p < .05$.

Table 4.5

Partitioning Variance Into Trait, Method, and Error Components for the Trait–Method CFA Model

Measure	Variance components		
	Trait	Method	Error
Global Self-Esteem by			
True–false	.04	.69	.27
Multipoint	.41	.60	.00[a]
Self-rating	.02	.68	.32
Social Self-Esteem by			
True–false	.26	.55	.19
Multipoint	.03	.80	.17
Self-rating	.31	.51	.18
Need for Order by			
True–false	.71	.04	.26
Multipoint	.62	.02	.37
Self-rating	.48	.11	.42

Note. CFA = confirmatory factor analysis; global SE = global self-esteem (measured by the Coopersmith Self-Esteem Inventory and the Tennessee Self-Concept Scale); social SE = social self-esteem (measured by the Jackson Personality Inventory and the Janis–Field Feelings of Inadequacy Scale); order = need for order (measured by the Personality Research Form and the Order Scale of the Comrey Personality Scale); true–false = the method of measuring that used a self-report inventory with a true–false response format; multipoint = the method of measuring that used a self-report inventory with 2-point and 5-point scales; self-rating = the method of measuring that used a self-assessment of the accuracy of self-descriptive traits using a 7-point scale. Results are from Figure 2 (p. 60), Table 2 (p. 63), and Table 3 (p. 64) of Bagozzi (1993). [a]Constrained value, needed in this analysis to obtain a solution having no error-terms with negative variances. Of course, it is theoretically untenable that the true error variance for the multipoint measure of Global Self-Esteem is actually zero.

social self-esteem show high levels of method bias, and (c) error variance is generally low.

Considered as a whole, the results of Bagozzi's (1993) CFAs provide the strongest support for the convergent validity of measures of need for order and the weakest support for the convergent validity of measures of global self-esteem. Bagozzi's (1993) analyses also provide evidence for the discriminant validity of all three constructs in Van Tuinen and Ramanaiah's (1979) MTMM matrix. In sum, using CFA to analyze the MTMM matrix provides much more precise, detailed, and comprehensive information about validity than applying Campbell and Fiske's (1959) classic criteria does.

Summary

Validity concerns the soundness or truthfulness of inferences that researchers make. In this chapter, I primarily reviewed various types of validity in relation to measurement and discussed ways that researchers typically use multivariate statistics to assess them. Each type of validity involves a different kind of inference about a measurement. Is it thorough (content validity)? Is it useful in predicting relevant outcomes (criterion validity) that have occurred either after (predictive or prospective validity), while (concurrent validity), or before (postdictive or retrospective validity) the measurement was taken? If it really measures what it is supposed to measure (construct validity), then are its readings similar to those from other measures of the same thing (convergent validity)? Are its readings dissimilar to those from measures of different things (discriminant validity)? Furthermore, does it add anything to what they already know from measures of other things (incremental validity)? To help them answer these thorny questions about measurement, researchers use a broad array of multivariate statistical tools. Table 4.6 summarizes these types of measurement validity and appropriate multivariate procedures for assessing them.

As a strategy for construct validation, Campbell and Fiske's (1959) MTMM matrix provides a systematic way to gauge both convergent and discriminant validity at the same time. Over the past four decades, techniques for analyzing MTMM data have evolved from the four rules of thumb originally proposed to multivariate CFA. Yet there remains little agreement as to how best to extract critical information from the matrix. Although CFA is the modern-day method of choice in analyzing MTMM data, it is plagued by frequent technical problems in obtaining proper solutions. An optimal statistical means of harnessing the full power of the MTMM matrix awaits discovery.

Suggestions for Further Reading

For introductory-level treatments of validity in the context of empirical research, see Crano and Brewer (1973), Dooley (1984), and Shaughnessy and Zechmeister (1997). For more advanced discussions of validity in relation to experimental and quasi-experimental methodologies, see Campbell and Stanley (1966) and Cook and Campbell (1979).

For introductory-level treatments of the process of test validation,

Table 4.6

Forms of Measurement Validity and Appropriate Means of Multivariate Statistical Analysis

Primary type of measurement validity	Central measurement issue	Form of the primary type	Appropriate means of multivariate statistical assessment
Content	Thoroughness	Factorial/structural	Principal components analysis, confirmatory factor analysis
Criterion	Prediction	Predictive/prospective	Multiple regression, discriminant analysis, logistic regression, structural equation modeling
		Concurrent	Multiple regression, canonical correlation analysis
		Postdictive/retrospective	Multiple regression, discriminant analysis, logistic regression
Construct	Meaning	Convergent	Principal components analysis, path analysis, confirmatory factor analysis
		Discriminant	Multiple regression, principal components analysis, confirmatory factor analysis, discriminant analysis
		Incremental	Multiple regression, partial correlations
Face	Appearance	None	None

see Aiken (1976), Anastasi (1988), and Cronbach (1990). Readers interested in learning more about construct validation should consult Cronbach and Meehl (1955) and Messick (1989, 1995). For more advanced discussion of test validation and psychometrics, see Nunnally (1978), Wainer and Braum (1988), and Wiersma and Jurs (1990).

Details of how to construct and analyze MTMM matrices appear in Bagozzi (1993, 1994), Byrne (1989), Fiske (1982), Marsh (1989), Marsh and Bailey (1991), Marsh and Grayson (1995), Schmitt and Stults (1986), Widaman (1985), and Wothke (1996).

Glossary

Clinical Validation The empirical process of assessing or establishing the accuracy of using test scores to identify individuals who are experiencing some form of clinical disorder.

Concurrent Validity The issue of whether test scores correlate with a criterion measure that is assessed at the same point in time.

Construct An abstract, hypothetical concept that cannot be directly observed, but that is represented by concrete operational definitions, such as a particular experimental manipulation or a particular measure. For example, love is a construct that cannot be directly observed but that can be operationally defined or measured in terms of self-report questions or compassionate behaviors. The degree of match between the underlying construct and its operational definition is the essence of construct validity.

Construct Validity The issue of whether an instrument actually measures what it is supposed to measure, or the confidence one can have in labeling operational definitions in theory-relevant, conceptual terms. Also known as *trait validity* or *referent validity*.

Content Validity The issue of whether an instrument covers all relevant aspects of the conceptual or behavioral domain that it is intended to measure.

Convergent Validity The issue of whether test scores correlate with other measures of the same underlying construct. A form of construct validity.

Criterion A relevant, external outcome that is a well-accepted indicator of a particular concept and to which test scores should relate, if the test actually measures what it is intended to measure.

CRITERION CONTAMINATION A methodological problem in validation studies that occurs when persons providing criterion ratings are aware of respondents' initial test scores, and this knowledge potentially influences their ratings of respondents on the criterion.

CRITERION GROUPS Research participants who have been classified into separate groups on the basis of differences on a well-accepted criterion measure. Researchers often assess discriminant validity by determining whether test scores can be used accurately to identify which criterion groups respondents belong to. Often used in clinical validation.

CRITERION MEASURE An empirical assessment of a relevant, external outcome that is a well-accepted indicator of a given concept and to which test scores should relate, if the test actually measures what it is intended to measure.

CRITERION VALIDITY The issue of whether scores on an instrument can be used to predict a relevant, external outcome, or criterion. If a test measures what it is supposed to measure, then it should be useful in predicting future, present, and past outcomes of interest.

CROSS-VALIDATION The empirical process of assessing or establishing external validity, or the generalizability of research findings across different populations, settings, or time periods.

DIFFERENTIAL VALIDITY The issue of whether a test predicts scores on a criterion measure differently for different subgroups.

DISCRIMINANT VALIDITY The issue of whether test scores are uncorrelated with measures of different underlying constructs. A form of construct validity.

ECOLOGICAL VALIDITY The issue of whether the environment, or situational context, of a research study is representative of the natural, "real-world" social environment.

EXTERNAL VALIDITY The issue of whether research results are generalizable to or across different populations, settings, and time periods.

FACE VALIDITY The issue of whether a test appears "on its face" to measure what it is supposed to measure, regardless of whether it actually does.

FACTORIAL VALIDITY The issue of whether the dimensions or conceptual components hypothesized to make up an instrument actually underlie people's responses to the instrument. Usually evaluated by means of exploratory or confirmatory factor analysis. Also known as *structural validity.*

HETEROMETHOD BLOCK A validity diagonal and its two adjacent heterotrait–heteromethod triangles in a multitrait–multimethod matrix.

HETEROTRAIT–HETEROMETHOD COEFFICIENTS Pearson product–moment correlation coefficients within a multitrait–multimethod matrix that represent relationships between different traits, or constructs, measured using different methods. These correlations are located in a triangle immediately above and below each validity diagonal in a multitrait–multimethod matrix.

HETEROTRAIT–MONOMETHOD COEFFICIENTS Pearson product–moment correlation coefficients within a multitrait–multimethod matrix that represent relationships between different traits, or constructs, measured using the same method. These correlations are located in a triangle immediately below each reliability diagonal in a multitrait–multimethod matrix.

INCREMENTAL VALIDITY The issue of whether a particular measure provides explanatory power over and above another measure, in predicting a relevant criterion.

INTERNAL VALIDITY The issue of whether a particular experimental manipulation or treatment caused the effects observed in a particular outcome measure.

MONOMETHOD BLOCK A reliability diagonal and its adjacent heterotrait–monomethod triangle in a multitrait–multimethod matrix.

MONOTRAIT–HETEROMETHOD COEFFICIENTS Pearson product–moment correlation coefficients within a multitrait–multimethod matrix that represent relationships between the same trait, or construct, measured using different methods. Also known as *validity coefficients.*

MONOTRAIT–MONOMETHOD COEFFICIENTS Pearson product–moment correlation coefficients within a multitrait–multimethod matrix that represent relationships between the same trait, or construct, measured using the same method at different points in time. Also known as *test–retest reliabilities.* Researchers sometimes substitute reliability indices of internal consistency in place of test–retest reliabilities.

MULTITRAIT–MULTIMETHOD MATRIX (MTMM) An integrative multivariable framework for systematically gathering information about convergent and discriminant validity in a single study. With this approach, one assesses two or more constructs (or traits) using two or more measurement techniques (or methods), and then intercorrelates these various measurements.

NOMOLOGICAL NET A broadly integrative theoretical framework that identifies the key constructs associated with a phenomena of interest and the interrelationships among these key constructs.

NOMOLOGICAL VALIDITY The issue of whether obtained data patterns fit theoretical predictions about such data patterns.

OPERATIONAL DEFINITION A specific, concrete means of representing an abstract, hypothetical concept (or construct) in a research study. For example, heart rate might be used as operational definition of the construct "fear." The degree of match between an operational definition and its intended "target" construct is the essence of construct validity. An operational definition is also known as a *research operation* or an *operationalization.*

POSTDICTIVE VALIDITY The issue of whether test scores can be used to predict a criterion measure that has been measured earlier in time, such as archived records or past events. Also known as *retrospective validity.*

PREDICTIVE VALIDITY The issue of whether test scores obtained earlier can be used to predict a criterion measure that is assessed later. Also known as *prospective validity.*

PROSPECTIVE VALIDITY The issue of whether test scores obtained earlier can be used to predict a criterion measure that is assessed later. Also known as *predictive validity.*

PSYCHOMETRICS A specialized research area concerned with assessing and establishing the validity and reliability of measurement instruments.

REFERENT VALIDITY The issue of whether research operationalization correspond to the referent terms in the original research proposition. Also known as *trait validity* or *construct validity.*

Reliability Diagonal A line of Pearson product–moment correlation coefficients within a multitrait–multimethod matrix that represent relationships between the same trait, or construct, measured using the same method at different points in time (i.e., test–retest reliabilities). Researchers sometimes substitute reliability indices of internal consistency in place of test–retest reliabilities. Also known as *monotrait–monomethod coefficients.*

Retrospective Validity The issue of whether test scores correlate with a criterion measure that focuses on the past, such as respondents' recollection of their experiences at an earlier time. Also known as *postdictive validity.*

Single–Group Validity The issue of whether a test predicts scores on a criterion measure for one particular subgroup but not for another.

Statistical Conclusion Validity The issue of whether the inferences drawn from statistical analyses are accurate and appropriate.

Structural Validity The issue of whether the dimensions or conceptual components hypothesized to make up an instrument actually underlie people's responses to the instrument. Usually evaluated by means of exploratory or confirmatory factor analysis. Also known as *factorial validity.*

Trait Validity The issue of whether the conceptual definitions underlying the research variables match the actual operationalizations used in a study. Also known as *referent validity* or *construct validity.*

Validity In general, an overall evaluative judgment of the degree to which empirical evidence and theoretical rationales support the adequacy and appropriateness of conclusions drawn from some form of assessment.

Validation The process of assessing or establishing validity through systematic investigation.

Validity Coefficient A Pearson product–moment correlation coefficient relating test scores to a criterion measure. Also *monotrait–heteromethod coefficients* in a *multitrait–multimethod matrix.*

Validity Diagonal A line of Pearson product–moment correlation coefficients within a multitrait–multimethod matrix that represent relationships between the same trait, or construct, measured using different methods. Also known as *monotrait–heteromethod values.*

References

Aiken, L. R., Jr. (1976). *Psychological testing and assessment* (2nd ed.). Boston: Allyn & Bacon.

Anastasi, A. (1988). *Psychological testing.* Upper Saddle River, NJ: Prentice-Hall.

Bagozzi, R. P. (1993). Assessing construct validity in personality research: Applications to measures of self-esteem. *Journal of Research in Personality, 27,* 49–87.

Bagozzi, R. P. (1994). Structural equation models in marketing research: Basic principles. In R. P. Bagozzi (Ed.), *Principles of marketing research* (pp. 317–385). Oxford, England: Blackwell.

Boruch, R. F., & Wolins, L. (1970). A procedure for estimation of trait, method, and error variance attributable to a measure. *Educational and Psychological Measurement, 30,* 547–574.

Browne, M. W. (1984). The decomposition of multitrait–multimethod matrices. *British Journal of Mathematical and Statistical Psychology, 37,* 1–21.

Bryant, F. B., & Baxter, W. J. (1997). The structure of positive and negative automatic cognition. *Cognition and Emotion, 11,* 225–258.

Bryant, F. B., & Yarnold, P. R. (1995). Principal-components analysis and exploratory and confirmatory factor analysis. In L. G. Grimm & P. R. Yarnold (Eds.), *Reading and understanding multivariate statistics* (pp. 99–136). Washington, DC: American Psychological Association.

Byrne, B. M. (1989). *A primer of LISREL: Basic applications and programming for confirmatory factor analysis models.* New York: Springer-Verlag.

Campbell, D. T. (1960). Recommendations for APA test standards regarding construct, trait, and discriminant validity. *American Psychologist, 15,* 546–553.

Campbell, D. T., & Fiske, D. W. (1959). Convergent and discriminant validity by the multitrait–multimethod matrix. *Psychological Bulletin, 56,* 81–105.

Campbell, D. T., & Stanley, J. C. (1966). *Experimental and quasi-experimental designs for research.* Chicago: Rand McNally.

Clark, K .K., Bormann, C. A., Cropanzano, R. S., & James, K. (1995). Validation evidence for three coping measures. *Journal of Personality Assessment, 65,* 434–455.

Cleary, T. A. (1968). Test bias: Prediction of grades of Negro and white students in integrated colleges. *Journal of Educational Measurement, 5,* 115–124.

Cook, T. D., & Campbell, D. T. (1979). *Quasi-experimentation: Design and analysis issues for field settings.* Chicago: Rand McNally.

Crano, W. D., & Brewer, M. B. (1973). *Principles of research in social psychology.* New York: McGraw-Hill.

Cronbach, L. J. (1990). *Essentials of psychological testing.* New York: Harper & Row.

Cronbach, L. J., & Meehl, P. E. (1955). Construct validity in psychological tests. *Psychological Bulletin, 52,* 281–302.

Dooley, D. (1984). *Social research methods.* Englewood Cliffs, NJ: Prentice-Hall.

Durlak, J. A. (1995). Understanding meta-analysis. In L. G. Grimm & P. R. Yarnold (Eds.), *Reading and understanding multivariate statistics* (pp. 319–352). Washington, DC: American Psychological Association.

Edwards, J. R., & Baglioni, A. J. (1991). Relationship between Type A behavior pattern and mental and physical symptoms: A comparison of global and component measures. *Journal of Applied Psychology, 75,* 276–290.

Fiske, D. W. (1973). Can a personality construct be validated empirically? *Psychological Bulletin, 80,* 89–92.

Fiske, D. W. (1982). Convergent–discriminant validation in measurements and research strategies. In D. Brinberg & L. Kidder (Eds.), *New directions for methodology of social and behavioral science: Forms of validity in research* (Vol. 12, pp. 72–92). San Francisco: Jossey-Bass.

Fiske, D. W., & Campbell, D. T. (1992). Citations do not solve problems. *Psychological Bulletin, 112,* 393–395.

Golding, S. L., & Seidman, E. (1974). Analysis of multitrait–multimethod matrices: A two-step principal components procedure. *Multivariate Behavioral Research, 9,* 479–496.

Harder, D. H., & Zalma, A. (1990). Two promising shame and guilt scales: A construct validity comparison. *Journal of Personality Assessment, 55,* 729–745.

Harrell, T. H., & Ryon, N. B. (1983). Cognitive–behavioral assessment of depression: Clinical validation of the Automatic Thoughts Questionnaire. *Journal of Consulting and Clinical Psychology, 51,* 721–725.

Hollon, S. D., & Kendall, P. C. (1980). Cognitive self-statements in depression: Development of an automatic thoughts questionnaire. *Cognitive Therapy and Research, 4,* 383–395.

Hubert, L. J., & Baker, F. B. (1978). Analyzing the multitrait–multimethod matrix. *Multivariate Behavioral Research, 13,* 163–179.

Hubert, L. J., & Baker, F. B. (1979). A note on analyzing the multitrait–multimethod matrix: An application of a generalized proximity function comparison. *British Journal of Mathematical and Statistical Psychology, 32,* 179–184.

Huebner, E. S. (1994). Preliminary development and validation of a multidimensional life satisfaction scale for children. *Psychological Assessment, 6,* 149–158.

Ingram, R. E., & Wisnicki, K. S. (1988). Assessment of positive automatic cognition. *Journal of Consulting and Clinical Psychology, 56,* 898–902.

Jackson, D. N. (1975). Multimethod factor analysis: A reformulation. *Multivariate Behavioral Research, 10,* 259–275.

Jöreskog, K. G. (1971). Statistical analysis of sets of congeneric tests. *Psychometrika, 36,* 109–133.

Judd, C. M., Jessor, R., & Donovan, J. E. (1986). Structural equation models and personality research. *Journal of Personality, 54,* 149–198.

Kenny, D. (1976). An empirical application of confirmatory factor analysis to the multitrait–multimethod matrix. *Journal of Experimental Social Psychology, 12,* 247–252.

Kenny, D., & Kashy, D. A. (1992). Analysis of the multitrait–multimethod matrix by confirmatory factor analysis. *Psychological Bulletin, 112,* 165–172.

Klohnen, E. C. (1996). Conceptual analysis and measurement of the construct of ego-resiliency. *Journal of Personality and Social Psychology, 70,* 1067–1079.

Kumar, A., & Dillon, W. R. (1990). On the use of confirmatory measurement models in the analysis of multiple-informant reports. *Journal of Marketing Research, 27,* 102–111.

Levin, J., Montag, I., & Comrey, A. L. (1983). Comparison of multitrait–multimethod, factor, and smallest space analysis on personality scale data. *Psychological Reports, 53,* 591–596.

Licht, M. H. (1995). Multiple regression and correlation. In L.G. Grimm & P. R. Yarnold (Eds.), *Reading and understanding multivariate statistics* (pp. 19–64). Washington, DC: American Psychological Association.

Marsh, H. W. (1989). Confirmatory factor analyses of multitrait–multimethod data: Many problems and a few solutions. *Applied Psychological Measurement, 13,* 335–361.

Marsh, H. W., & Bailey, M. (1991). Confirmatory factor analyses of multitrait–

multimethod data: A comparison of the behavior of alternative models. *Applied Psychological Measurement, 15,* 47–70.

Marsh, H. W., & Grayson, D. (1995). Latent variable models of multitrait–multimethod data. In R. H. Hoyle (Ed.), *Structural equation modeling: Concepts, issues, and applications* (pp. 177–198). Thousand Oaks, CA: Sage.

Marsh, H. W., & Hocevar, D. (1988). A new, more powerful approach to multitrait–multimethod analyses: Application of second-order confirmatory factor analysis. *Journal of Applied Psychology, 73,* 107–117.

Massey, A. J. (1997). Multitrait–multimethod/multiform evidence for the validity of reporting units in national assessments in science at age 14 in England and Wales. *Educational and Psychological Measurement, 57,* 108–117.

Meng, X., Rosenthal, R., & Rubin, D. B. (1992). Comparing correlated correlation coefficients. *Psychological Bulletin, 111,* 172–175.

Messick, S. (1989). Validity. In R. L. Linn (Ed.), *Educational measurement* (3rd ed., pp. 13–103). New York: Macmillan.

Messick, S. (1995). Validity of psychological assessment: Validation of inferences from persons' responses and performance as scientific inquiry into score meaning. *American Psychologist, 50,* 741–749.

Miner, M. G., & Miner, J. B. (1978). *Employee selection within the law.* Washington, DC: Bureau of National Affairs.

Nunnally, J. C. (1978). *Psychometric theory.* New York: McGraw-Hill.

Pierce, J. P., Choi, W. S., Gilpin, E. A., Farkas, A. J., & Merritt, R. K. (1996). Validation of susceptibility as a predictor of which adolescents take up smoking in the United States. *Health Psychology, 15,* 355–361.

Popper, K. R. (1959). *The logic of scientific discovery.* New York: Basic Books.

Rorschach, (1945). *Rorschach technique.* San Antonio, TX: Psychological Corporation.

Roy, A. (1988). Early parental loss and depression. In F. Flach (Ed.), *Affective disorders* (pp. 19–28). New York: Norton.

Schmitt, N., & Stults, D. N. (1986). Methodology review: Analysis of multitrait–multimethod matrices. *Applied Psychological Measurement, 10,* 1–22.

Sechrest, L. (1963). Incremental validity: A recommendation. *Educational and Psychological Measurement, 23,* 153–158.

Shaughnessy, J. J., & Zechmeister, E. B. (1997). *Research methods in psychology* (5th ed.). New York: McGraw-Hill.

Silva, A. P. D., & Stam, A. (1995). Discriminant analysis. In L. G. Grimm & P. R. Yarnold (Eds.), *Reading and understanding multivariate statistics* (pp. 277–318). Washington, DC: American Psychological Association.

Snow, R. E. (1974). Representative and quasi-representative designs for research on teaching. *Review of Educational Research, 44,* 265–291.

Stanley, J. C. (1961). Analysis of unreplicated three-way classifications, with applications to rater bias and trait independence. *Psychometrika, 26,* 205–219.

Swinkels, A., & Giuliano, T. A. (1995). The measurement and conceptualization of mood awareness: Monitoring and labeling one's mood states. *Personality and Social Psychology Bulletin, 21,* 934–949.

Van Tuinen, M., & Ramanaiah, N. (1979). A multimethod analysis of selected self-esteem measures. *Journal of Research in Personality, 13,* 16–24.

Wainer, H., & Braum, H. I. (Eds.). (1988). *Test validity.* Hillsdale, NJ: Erlbaum.

Watson, D., & Walker, L. M. (1996). The long-term stability and predictive validity of trait measures of affect. *Journal of Personality and Social Psychology, 70,* 567–577.

Webb, E. J., Campbell, D. T., Schwartz, R. D., Sechrest, L., & Grove, J. B. (1981). *Nonreactive measures in the social sciences* (2nd ed.). Boston: Houghton-Mifflin.

Weinfurt, K. (1995). Multivariate analysis of variance. In L.G. Grimm & P. R. Yarnold (Eds.), *Reading and understanding multivariate statistics* (pp. 245–276). Washington, DC: American Psychological Association.

Werts, C. E., & Linn, R. L. (1970). Path analysis: Psychological examples. *Psychological Bulletin, 74,* 193–212.

Widaman, K. F. (1985). Hierarchically nested covariance structure models for multitrait–multimethod data. *Applied Psychological Measurement, 9,* 1–26.

Wiersma, W., & Jurs, S. G. (1990). *Educational measurement and testing* (2nd ed.). Boston: Allyn & Bacon.

Wilson, M., & Mather, L. (1974). Life expectancy. *Journal of the American Medical Association, 229,* 1421–1422.

Wothke, W. (1996). Models for multitrait–multimethod matrix analysis. In G. A. Marcoulides & R. E. Schumacker (Eds.), *Advanced structural equation modeling: Issues and applications* (pp. 7–56). Mahwah, NJ: Erlbaum.

Wright, R. E. (1995). Logistic regression. In L. G. Grimm & P. R. Yarnold (Eds.), *Reading and understanding multivariate statistics* (pp. 217–244). Washington, DC: American Psychological Association.

5 Cluster Analysis

Joseph F. Hair, Jr., and William C. Black

Cluster analysis is a group of multivariate techniques whose primary purpose is to assemble objects based on the characteristics that they possess. Cluster analysis classifies *objects* (e.g., respondents, products, or other entities), so that each object is similar to others in the cluster with respect to a predetermined selection criterion. The resulting clusters of objects should then exhibit high internal (within-cluster) homogeneity and high external (between-cluster) heterogeneity. Thus, if the classification is successful, the objects within the clusters will be close together when plotted geometrically and different clusters will be farther apart. A common use of cluster analysis in clinical psychology is the identification of types (clusters) of disorders. For example, a researcher may want to know if it is important to identify different types of attention deficit hyperactivity disorder (ADHD). A cluster analysis may reveal that a syndrome that specifies attention deficits should, perhaps, be separated from a syndrome that emphasizes hyperactivity. The cluster analysis can reveal what symptoms discriminate the two categories. Subsequent validation, as described later in this chapter, is needed to justify the use of the proposed ADHD clusters.

In cluster analysis, the concept of the variate is again a central issue but in a different way from other multivariate techniques. The *cluster variate* is the set of variables representing the characteristics used to compare objects in the cluster analysis. Because the cluster variate in-

This chapter is from *Multivariate Data Analysis*, by J. R. Hair, Jr., R. E. Anderson, R. L. Tatham, and W. C. Black, 1998, pp. 469–518. Copyright 1998 by Prentice-Hall, Inc. Adapted with permission of Prentice-Hall Inc., Upper Saddle River, NJ.

cludes only the variable used to compare objects, it determines the "character" of the objects. Cluster analysis is the only multivariate technique that does not estimate the variate empirically but uses the variate as specified by the researcher. The focus of cluster analysis is on the comparison of objects based on the variate, not on the estimation of the variate itself. This makes the researcher's definition of the variate a critical step in cluster analysis.

Cluster analysis has been referred to as "Q analysis," "typology," "classification analysis," and "numerical taxonomy." This variety of names is due in part to the use of clustering methods in such diverse disciplines as psychology, biology, sociology, economics, engineering, and business. Although the names differ across disciplines, the methods all have a common dimension: classification according to natural relationships (Aldenderfer & Blashfield, 1984; Anderberg, 1973; Bailey, 1994; Green & Carroll, 1978; Punj & Stewart, 1983; Sneath & Sokal, 1973). This common dimension represents the essence of all clustering approaches. As such, the primary value of cluster analysis lies in the classification of data, as suggested by "natural" groupings of the data themselves. Cluster analysis is comparable with factor analysis in its objective of assessing structure; cluster analysis differs from factor analysis in that cluster analysis groups objects, whereas factor analysis is primarily concerned with grouping variables.

Cluster analysis is useful in many situations. For example, a researcher who has collected data with a questionnaire may be faced with a large number of observations that are meaningless unless classified into manageable groups. Using cluster analysis, the researcher can perform data reduction objectively by decreasing the information from an entire population or sample to information about specific, smaller subgroups. For example, if we can understand the attitudes of a population by identifying the major groups within the population, then we have reduced the data for the entire population into profiles of a number of groups. In this fashion, the researcher has a more concise, understandable description of the observations, with minimal loss of information.

Cluster analysis is also useful when a researcher wishes to develop hypotheses concerning the nature of the data or to examine already stated hypotheses. For example, a researcher may believe that attitudes toward the consumption of diet versus regular soft drinks could be used to separate soft drink consumers into logical segments or groups. Cluster analysis can classify soft drink consumers by their attitudes about diet versus regular soft drinks, and the resulting clusters, if any, can be profiled for demographic similarities and differences. Alternatively, a clinical psychologist could use cluster analysis to form symptom clusters to identify potential subtypes of a disorder.

These examples are just a small fraction of the types of applications of cluster analysis. Ranging from the derivation of taxonomies in biology for grouping all living organisms to psychological classifications that are based on personality and other personal traits, to the segmentation analyses of marketers, cluster analysis has always had a strong tradition of grouping individuals. The result has been an explosion of applications in almost every area of inquiry, creating not only a wealth of knowledge on the use of cluster analysis but also the need for a better understanding of the technique to minimize its misuse.

Yet with the benefits of cluster analysis come some caveats. Cluster analysis can be characterized as descriptive, atheoretical, and noninferential. Cluster analysis has no statistical basis on which to draw statistical inferences from a sample to a population and is primarily used as an exploratory technique. The solutions are not unique because the cluster membership for any number of solutions is dependent on many elements of the procedure; many different solutions can be obtained by varying one or more elements. Moreover, cluster analysis always creates clusters, regardless of the "true" existence of any structure in the data. Finally, the cluster solution is totally dependent on the variables used as the basis for the similarity measure. The addition or deletion of relevant variables can have a substantial effect on the resulting solution. Thus, the researcher must take care in assessing the effect of each decision involved in performing a cluster analysis.

Brief Descriptions

Before continuing with the details of cluster analysis, two brief descriptions of published studies that use cluster analysis are provided.

A Typology of Customer Complaints

Singh (1990) performed a cluster analysis to develop a categorization system of styles of consumer complaint behavior (CCB). Three dimensions of complaint intentions or behaviors that differ on the basis of the type of response (actions directed at the seller, negative word-of-mouth, or complaints to third parties) were recorded from a random sample of store customers seen in a 2-year period. Based on these three behaviors, cluster analysis was used to identify groups of similar individuals. Four clusters of consumer groups were identified: (a) no action; (b) voice actions only; (c) voice and private actions; and (d) voice, private, and third-party actions. The results of testing whether the response styles would reproduce differences in actual behavior were offered as support for the validity of the cluster solution. Finally, a number of demographic, personality–attitudinal, and situational variables—identified in the research literature as important to understanding consumer complaints—were used to profile the CCB styles. In addition, the author used multiple discriminant analysis to determine the relative importance of demographic variables for each cluster. This study extends previous research and demonstrates the multifaceted nature of complaint styles. Such findings should be of interest to retail managers by increasing knowledge about customers and improving the handling of customer complaints.

Police Officers' Perceptions of Rape

Campbell and Johnson (1997) examined how police officers defined rape, irrespective of the legal definition in their state. Officers' narrative responses to the following question were content analyzed:

> As you know, it's the legislators that make the laws and decide how to define crimes and what punishments will be. But you are actually in the community, dealing with victims and criminals. Based on your work as a police officer, how do you define rape–sexual assault?

Three clusters of definitions were retained. The smallest cluster (19% of officers) was characterized as "force definition of rape" and corresponded closely to the state's legal definition of rape. In contrast to the most salient feature of rape (force), as defined by the state in which the study was conducted, a second cluster was comprised of 31% of the sample and emphasized penetration and consent ("consent definition of rape"). The third cluster included 50% of the sample and was labeled "mixed definition of rape." Officers provided definitions that departed from the state's legal definition and included such stereotypic attitudes as "Sometimes a guy can't stop himself; he gets egged on by the girl."

Campbell and Johnson (1997) then provided validity data by comparing the groups on measures not included in the cluster analysis. Included among the variables that significantly differed across groups were job title, training in sexual assault law, attitudes toward women, and attitudes toward interpersonal violence. Because only one cluster represented officer definitions that closely corresponded with state law and this was the smallest cluster, the authors recommended that law enforcement agencies pay closer attention to formal training in the legal definition of rape and sexual assault.

How Does Cluster Analysis Work?

The nature of cluster analysis can be illustrated by a simple bivariate example. Suppose a marketing researcher wishes to determine market segments (clusters) in a small community on the basis of patterns of loyalty to brands and stores. A small sample of seven of the respondents is selected as a pilot test of how cluster analysis is applied. Two measures of loyalty, V_1 (store loyalty) and V_2 (brand loyalty), were measured for each respondent on a scale of 0 to 10 (0 = *not loyal at all*; 10 = *highly loyal*). The values for each respondent are shown in Figure 5.1, along with a scatter diagram depicting each observation on the variables.

The primary objective of cluster analysis is to define the structure of the data by placing the most similar observations into groups. To accomplish this task, however, the research must address three questions. First, how does one measure similarity? It requires a method for simultaneously comparing observations on the two clustering variables (V_1 and V_2).

Several methods are possible, including the correlation between objects; a measure of association used in other multivariate techniques; or perhaps a measure of their proximity in two-dimensional space, such that the distance between observations indicates similarity. Second, how does one form clusters? No matter how similarity is measured, the procedure must group those observations that are most similar into a cluster. This procedure must determine for each observation which other observations it will be grouped with. Third, how many groups does one form? Any number of "rules" might be used, but the fundamental task is to assess the "average" similarity across clusters, such that as the average increases, the clusters become less similar. The researcher then faces a trade-off: fewer clusters versus less homogeneity. Simple structure, in striving toward parsimony, is reflected in as few clusters as possible. Yet as the number of clusters decreases, the homogeneity within the clusters necessarily decreases. Thus, a balance must be found be-

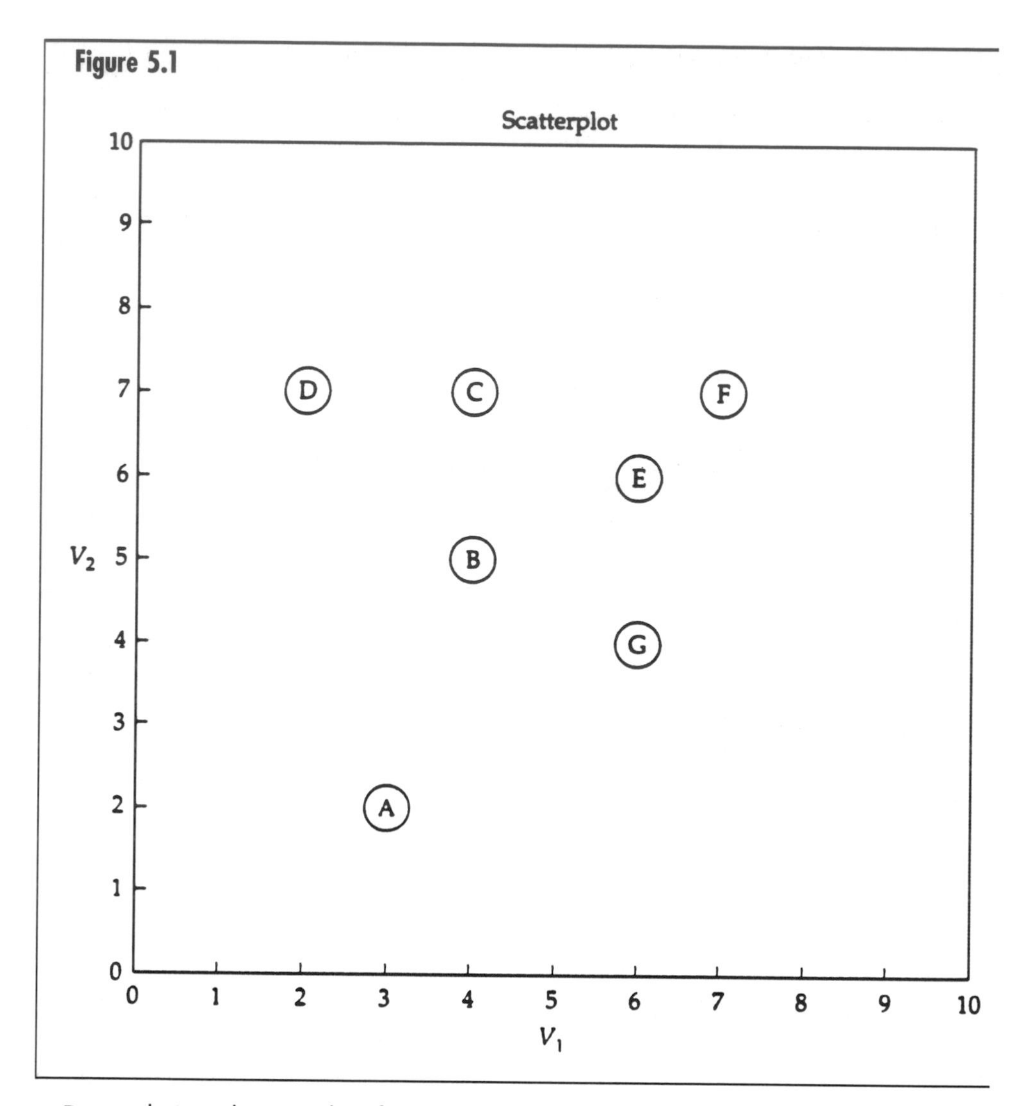

Data values and scatterplot of seven observations based on two variables.

tween defining the most basic structure (fewer clusters) that still achieves the necessary level of similarity within the clusters. Once we have procedures for addressing each issue, we can perform a cluster analysis.

Measuring Similarity

We illustrate a cluster analysis for the seven observations (respondents) using simple procedures for each of the issues. First, similarity is measured as the Euclidean distance (straight line) between each pair of

Table 5.1

Proximity Matrix of Euclidean Distances Between Observations

	Observation						
Observation	A	B	C	D	E	F	G
A	—						
B	3.162	—					
C	5.099	2.000	—				
D	5.099	2.828	2.000	—			
E	5.000	2.236	2.236	4.123	—		
F	6.403	3.606	3.000	5.000	1.414	—	
G	3.606	2.236	3.606	5.000	2.000	3.162	—

observations. Table 5.1 contains measures of proximity between each respondent. In using distance as the measure of proximity, we must remember that smaller distances indicate greater similarity, such that Observations E and F are the most similar (1.414), whereas A and F are the most dissimilar (6.403).

Forming Clusters

We must next develop a procedure for forming clusters. As we see later in this chapter, many methods have been proposed; but for our purposes here, we use this simple rule: identify the two most similar (closest) observations not already in the same cluster, and combine their clusters. We apply this rule repeatedly, starting with each observation in its own "cluster," and combine two clusters at a time until all observations are in a single cluster. This is termed a *hierarchical procedure* because it moves in a stepwise fashion to form an entire range of cluster solutions. It is also an *agglomerative method* because clusters are formed by the combination of existing clusters.

Table 5.2 details the steps of the hierarchical process, first depicting the initial state with all seven observations in single-member clusters. Then clusters are joined in the agglomerative process until only one cluster remains. Step 1 identifies the two closest observations (E and F) and combines them into a cluster, moving from seven to six clusters. Next, Step 2 finds the next closest pairs of observations. In this case, three pairs have the same distance of 2.000 (E and G, C and D, and B and C). Let us start with E and G. Although G is a single member cluster, E was combined in the prior step with F, so the cluster formed

Table 5.2

Agglomerative Hierarchical Clustering Process

	Agglomeration process		Cluster solution		
Step	**Minimum distance between unclustered observations**[a]	**Observation pair**	**Cluster membership**	**No. of clusters**	**Overall similarity measure (average within-cluster distance)**
	Initial solution		**(A) (B) (C) (D) (E) (F) (G)**	**7**	**0**
1	1.414	E–F	(A) (B) (C) (D) (E–F) (G)	6	1.414
2	2.000	E–G	(A) (B) (C) (D) (E–F–G)	5	2.192
3	2.000	C–D	(A) (B) (C–D) (E–F–G)	4	2.144
4	2.000	B–C	(A) (B–C–D) (E–F–G)	3	2.234
5	2.236	B–E	(A) (B–C–D–E–F–G)	2	2.896
6	3.162	A–B	(A–B–C–D–E–F–G)	1	3.420

[a]Euclidean distance between observations.

at this stage now has three members: G, E, and F. Step 3 combines the single member clusters of C and D, whereas Step 4 combines B with the two-member cluster C–D that was formed in Step 3. At this point, there are Cluster 1 (A), Cluster 2 (B, C, and D), and Cluster 3 (E, F, and G).

The next smallest distance is 2.236 for three pairs of observations (E and B, B and G, C and E). We use only one of these distances, however, as each observation pair contains a member from each of the two existing clusters (B, C, D vs. E, F, G). Thus, Step 5 combines the two three-member clusters into a single six-member cluster. The final step (6) is to combine Observation A with the remaining cluster (six observations) into a single cluster at a distance of 3.162. Note that there are distances smaller than or equal to 3.162, but they are not used because they are between members of the same cluster.

The hierarchical clustering process can be portrayed graphically in several ways. Figurc 5.2 illustrates two such methods. First, because the process is hierarchical, the clustering process can be shown as a series of nested groupings (see Figure 5.2A). This process, however, can only represent the proximity of the observations for two or three clustering variables in the scatter plot or three-dimensional graph. A more common approach is the *dendrogram*, which represents the clustering process in a treelike graph. The horizontal axis represents the agglomeration coefficient, in this instance the distance used in joining clusters. This approach is useful in identifying outliers, such as Observation A. It also depicts the relative size of varying clusters, although it becomes unwieldy when the number of observations increases.

Determining the Number of Clusters in the Final Solution

A hierarchical method results in a number of cluster solutions, in this case ranging from a one-cluster to a six-cluster solution. But which one should we choose? We know that as we move from single-member clusters, homogeneity decreases. So why not stay at seven clusters, the most homogeneous possible? The problem is that we did not define any structure with seven clusters. So the researcher must view each cluster solution for its description of structure balanced against the homogeneity of the clusters.

In this example, we use a simple measure of homogeneity, namely, the average distances of all observations within clusters (refer to the right-most column of Table 5.2). In the initial solution with seven clus-

Figure 5.2

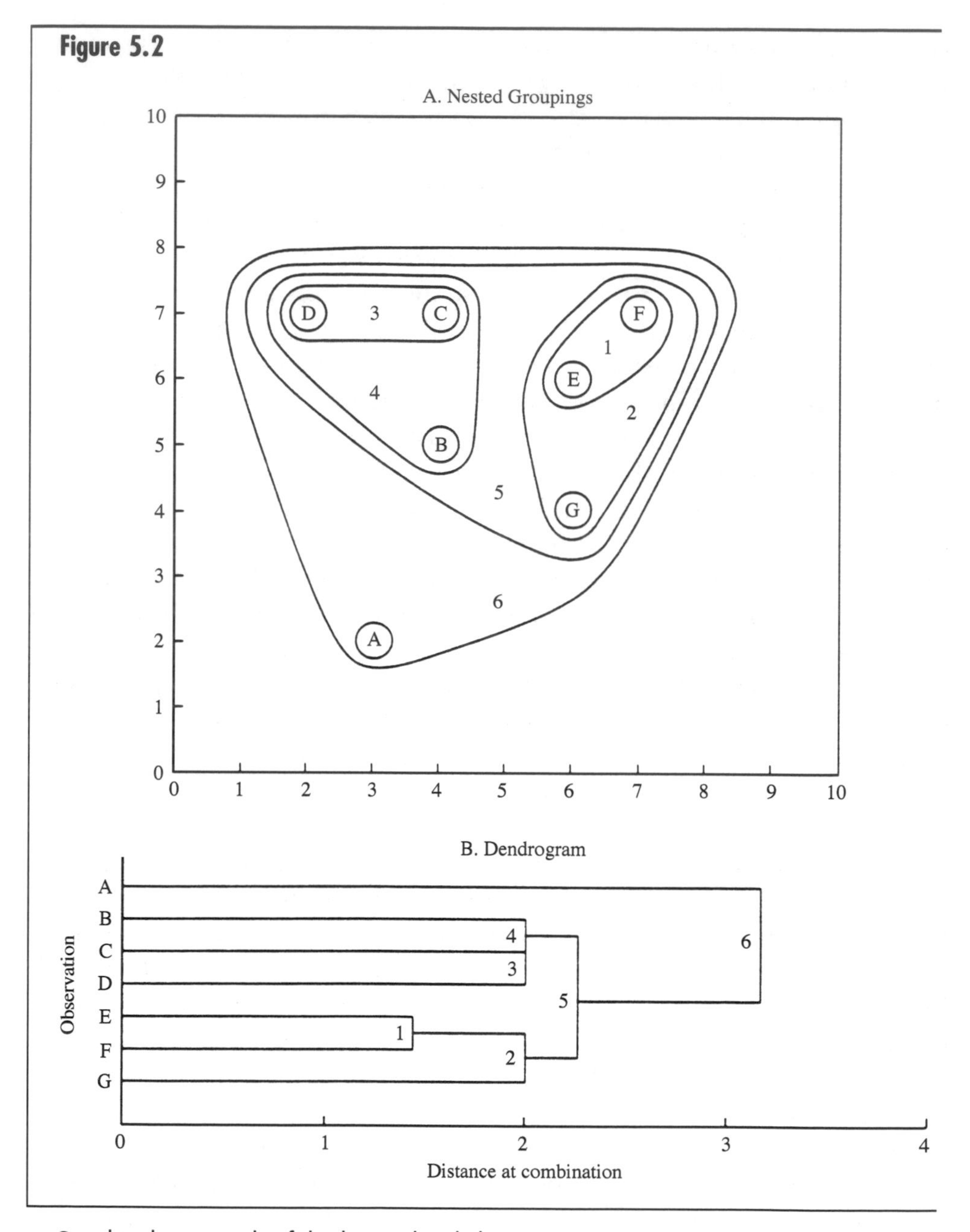

Graphical portrayals of the hierarchical clustering process.

ters, our overall similarity measure is 0—no observation is paired with another. For the six-cluster solution, the overall similarity is the distance between the two observations (1.414) joined at Step 1. Step 2 forms a three-member cluster (E, F, and G), so that the overall similarity measure is the mean of the distances between E and F (1.414), E and G (2.000), and F and G (3.162), for an average of 2.192. In Step 3, a new two-member cluster is formed with a distance of 2.000, which causes the overall average to fall slightly to 2.144. We can proceed to form new clusters this manner until a single-cluster solution is formed (Step 6), where the average of all distances in the distance matrix is 3.412.

Now, how do we use this overall measure of similarity to select a cluster solution? Remember that we are trying for the simplest structure possible that still represents homogeneous groupings. If we monitor the overall similarity measure as the number of clusters decreases, large increases in the overall measure indicates that two clusters were not that similar. In our example, the overall measure increases when we first join two observations (Step 1) and then again when we make our first three-member cluster (Step 2). But in the next two steps (3 and 4), the overall measure does not change substantially. This indicates that we are forming other clusters with essentially the same homogeneity of the existing clusters. But when we get to Step 5, which combines the two three-member clusters, we see a large increase.

This indicates that joining these two clusters resulted in a single cluster that was markedly less homogeneous. We would consider the cluster solution of Step 4 much better than that found in Step 5. We can also see that in Step 6 the overall measure actually decreases slightly, indicating that even though the last observation remained separate until the last step, when it was joined it did not change the cluster homogeneity markedly. However, given the rather unique profile of Observation A as compared with the others, it might best be designated as a member of the *entropy group*, those observations that are outliers and independent of the existing clusters. Thus, when reviewing the range of cluster solutions, the three-cluster solution of Step 4 seems the most appropriate for a final-cluster solution, with two equally sized clusters and the single outlying observation.

As one would probably realize by now, the selection of the final-cluster solution requires substantial researcher judgment, and it is considered by many as too subjective. Even though more sophisticated methods have been developed to assist in evaluating the cluster solutions, it still falls to the researcher to decide as to the number of clusters to accept as the final solution. Cluster analysis is rather simple in this bivariate case because the data are two dimensional. In most marketing and social sciences research studies, however, more than two variables are measured on each object, and the situation is much more complex with many more observations. We discuss how the researcher can use more sophisticated procedures to deal with the increased complexity of "real-world" applications in the remainder of this chapter.

Cluster Analysis Decision Process

Cluster analysis can be viewed from a six-stage model-building approach. Starting with research objectives that can be either exploratory or confirmatory, the design of a cluster analysis deals with the partitioning of the data set to form clusters, the interpretation of the clusters, and the validation of the results. The partitioning process determines how clusters may be developed. The interpretation process involves understanding the characteristics of each cluster and developing a name or label that appropriately defines its nature. The third process involves assessing the validity of the cluster solution (i.e., determining its stability and generalizability), along with describing the characteristics of each cluster to explain how they may differ on relevant dimensions such as demographics. The following sections detail all these issues through the six stages of the model-building process. Figure 5.3 depicts the first three stages of the cluster analysis decision tree.

Stage 1: Objectives

The primary goal of cluster analysis is to partition a set of objects into two or more groups based on the similarity of the objects for a set of specified characteristics (cluster variate). In forming homogeneous groups, the researcher can achieve any of three objectives:

1. *Taxonomy description.* The most traditional use of cluster analysis has been for exploratory purposes and the formation of a *taxonomy*—an empirically based classification of objects. As described earlier, cluster analysis has been used in a range of applications for its partitioning ability. However, cluster analysis can also generate hypotheses related to the structure of the objects. Although viewed principally as an exploratory technique, cluster analysis can be used for confirmatory purposes. If a proposed structure can be defined for a set of objects, cluster analysis can be applied and a proposed *typology* (theoretically based classification) can be compared with that derived from the cluster analysis.

Figure 5.3

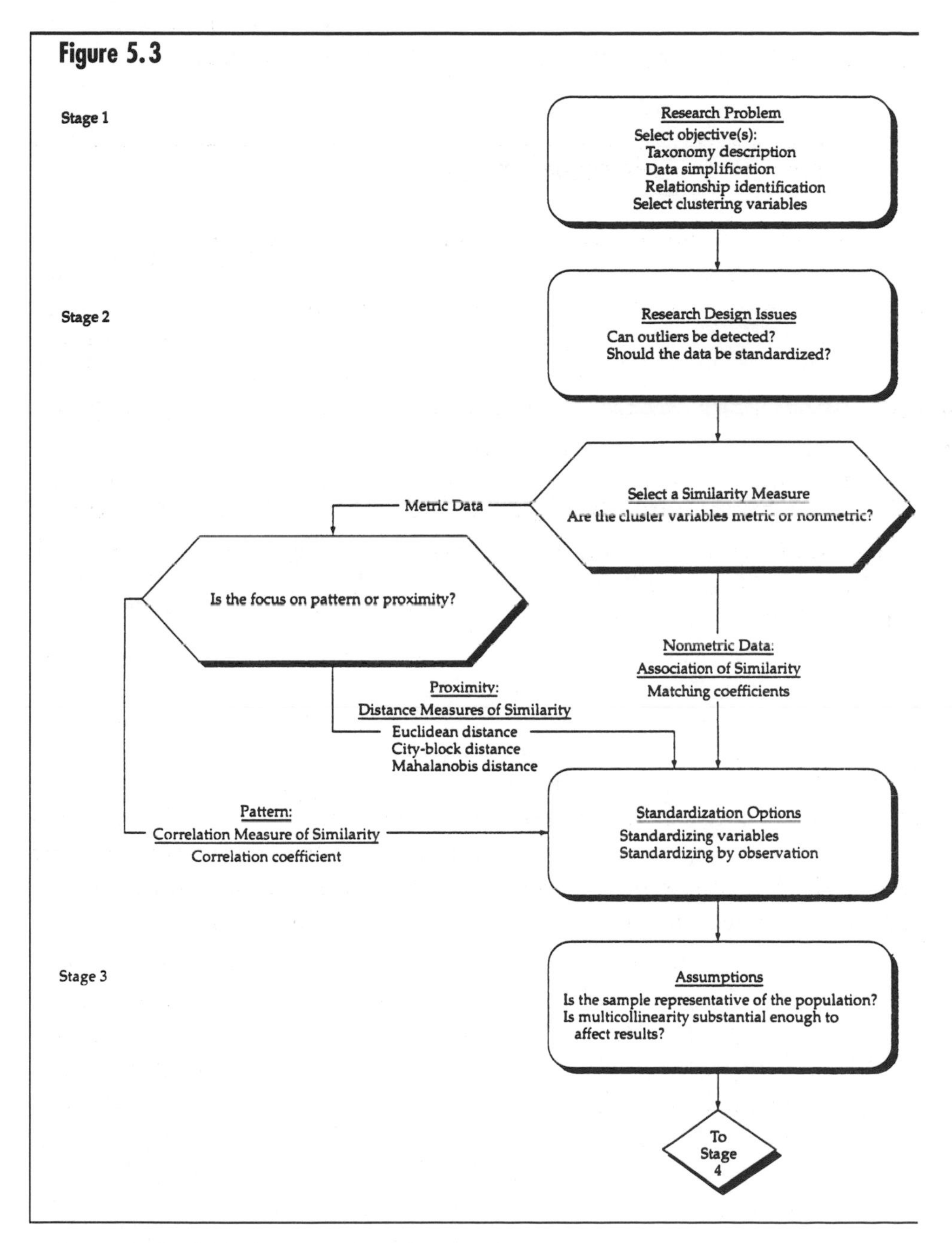

Stages 1–3 of the cluster analysis decision diagram.

2. *Data simplification.* In the course of deriving a taxonomy, cluster analysis also achieves a simplified perspective on the observations. With a defined structure, the observations can be grouped for further analysis. Whereas factor analysis attempts to provide "dimensions" or structure to variables, cluster analysis performs the same task for observations. Thus, instead of viewing all of the observations as unique, they can be viewed as members of a cluster and profiled by its general characteristics.
3. *Relationship identification.* With the clusters defined and the underlying structure of the data represented in the clusters, the researcher has a means of revealing relationships among the observations that was perhaps not possible with the individual observations. For example, one could use discriminant analysis to discover which variables, not used in the cluster analysis, discriminate among the cluster groups.

In any application, the objectives of cluster analysis cannot be separated from the selection of variables used to characterize the objects to be clustered. Whether the objective is exploratory or confirmatory, the researcher effectively limits the possible results by the variables selected for use. The derived clusters can only reflect the inherent structure of the data as defined by the variables.

Selecting the variables to be included in the cluster variate must be done with regard to both theoretical–conceptual and practical considerations. Any application of cluster analysis must have some rationale on which variables are selected. Whether the rationale is based on an explicit theory, past research, or supposition, the researcher must realize the importance of including only those variables that (a) characterize the objects being clustered and (b) relate specifically to the objectives of the cluster analysis. The cluster analysis technique has no means of differentiating the relevant from irrelevant variables. It only derives the most consistent, yet distinct, groups of objects across all variables. The inclusion of an irrelevant variable increases the chance that outliers will be created on these variables, which can have a substantive effect on the results. Thus, one should never include variables indiscriminately, but instead should carefully choose the variables with the research objective as the criterion for selection.

In a practical vein, cluster analysis can be dramatically affected by the inclusion of only one or two inappropriate or undifferentiated variables (Milligan, 1980). The researcher is always encouraged to examine the results and to eliminate the variables that are not distinctive (i.e., that do not differ significantly) across the derived clusters. This procedure allows the cluster techniques to maximally define clusters based on only those variables exhibiting differences across the objects.

Stage 2: Research Design

With the objectives defined and variables selected, the researcher must address three questions before starting the partitioning process: (a) Can outliers be detected and, if so, should they be deleted? (b) How should object similarity be measured? and (c) Should the data be standardized? Many approaches can be used to answer these questions. However, none has been evaluated sufficiently to provide a definitive answer to any of these questions, and unfortunately, many of the approaches provide different results for the same data set. Thus, cluster analysis, along with factor analysis, is much more of an art than a science. For this reason, our discussion reviews these issues in a general way by providing examples of the most commonly used approaches and an assessment of the practical limitations where possible.

The importance of these issues and the decisions made in later stages becomes apparent when we realize that although cluster analysis is seeking structure in the data, it must actually impose a structure through a selected methodology. Cluster analysis cannot evaluate all of the possible partitions because for even the relatively small problem of partitioning 25 objects into 5 nonoverlapping clusters there are 2.4×10^{15} possible partitions (Anderberg, 1973). Instead, based on the decisions of the researcher, the technique identifies one of the possible solutions as "correct." From this viewpoint, the research design issues and the choice of methodologies made by the researcher have greater effect than perhaps with any other multivariate technique.

Irrelevant Variables and Outliers

In its search for structure, cluster analysis is sensitive to the inclusion of irrelevant variables. Such variables essentially constitute "noise," and thus, the use of irrelevant variables to define typologies serves to degrade the reliability (or reproducibility) of the group assignments and, accordingly, inhibits the performance of group status assignment in the validity component of the study. Cluster analysis is also sensitive to outliers (objects that are very different from all others). Outliers can represent either (a) truly aberrant observations that are not representative of the general population or (b) an undersampling of actual groups in the population that causes an underrepresentation of the groups in the sample. In both cases, the outliers distort the true structure and make the derived clusters unrepresentative. For this reason, a preliminary screening for outliers is always necessary. Probably the easiest way to conduct this screening is to prepare a graphic profile diagram, such as that shown in Figure 5.4. The *profile diagram* lists the variables along the horizontal axis and the variable values along the vertical axis. Each point on the graph represents the value of the corresponding variable, and the points are connected to facilitate visual interpretation. Profiles for all objects are then plotted on the graph, a line for each object. Outliers are those objects with very different profiles, most often characterized by extreme values on one or more variables.

Obviously, such a procedure becomes cumbersome with a large number of objects (observations) or variables. For the observations shown in Figure 5.4, there is no obvious outlier that has all extremely high or low values. Just as in detecting multivariate outliers in other multivariate techniques, however, outliers may also be defined as having unique profiles that distinguish them from all of the other observations. Also, they may emerge in the calculation of similarity. By whatever

Figure 5.4

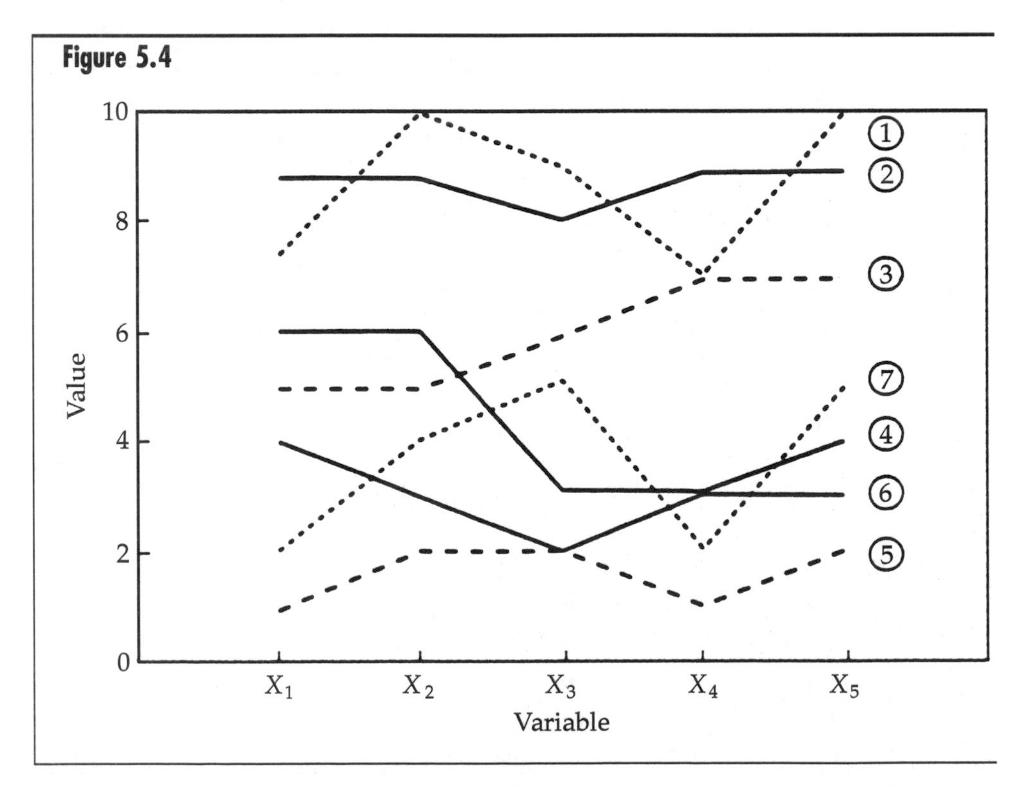

Profile diagram for screening outliers (1–7).

means used, observations identified as outliers must be assessed for their representativeness of the population and deleted from the analysis if deemed unrepresentative. As in other instances of outlier detection, however, the researcher should exhibit caution in deleting observations from the sample, as they may distort the actual structure of the data.

Similarity Measures

The concept of similarity is fundamental to cluster analysis. *Interobject similarity* is a measure of correspondence or resemblance between objects to be clustered. In factor analysis, we create a correlation matrix between variables that is then used to group variables into factors. A comparable process occurs in cluster analysis. Here, the characteristics defining similarity are first specified. Then, the characteristics are combined into a similarity measure calculated for all pairs of objects, just as we used correlations between variables in factor analysis. In this way, any object can be compared with any other object through the similarity measure. The cluster analysis procedure then proceeds to group similar objects together into clusters.

Interobject similarity can be measured in a variety of ways, but three methods dominate the applications of cluster analysis: (a) correlational measures, (b) distance measures, and (c) association measures. Each method represents a particular perspective on similarity, dependent on both its objectives and the type of data. Both the correlational and distance measures require metric (continuous) data, whereas the association measures are for nonmetric (categorical) data.

Correlational measures. The interobject measure of similarity that probably comes to mind first is the correlation coefficient between a pair of objects measured on several variables. In effect, instead of correlating two sets of variables, we invert the objects' X variables matrix so that the columns represent the objects and the rows represent the variables. Thus, the correlation coefficient between the two columns of numbers is the correlation (or similarity) between the profiles of the two objects. High correlations indicate similarity, and low correlations denote a lack of it. This procedure is followed in the application of Q-type factor analysis (see chapter 11).

Correlational measures represent similarity by the correspondence of patterns across the characteristics (X variables). This is illustrated by examining the example of seven observations (cases) shown in Figure 5.4. A correlational measure of similarity looks not at the magnitude of the values but instead at the patterns of the values. In Table 5.3, which contains the correlations among these seven observations, we can see two distinct groups. First, Cases 1, 5, and 7 all have similar patterns and corresponding high positive intercorrelations (i.e., for Cases 1 and 5, $r = .963$; Cases 1 and 7, $r = .891$; and Cases 5 and 7, $r = .963$). Likewise, Cases 2, 4, and 6 also have high positive correlations among themselves but low or negative correlations with the other observations. Case 3 has low or negative correlations with all other cases, thereby perhaps forming a group by itself. Thus, correlations represent patterns across the variables much more than the magnitudes. Correlational measures, however, are rarely used because the emphasis in most applications of cluster analysis is on the magnitudes of the objects, not the patterns of values.

Distance measures. Although correlational measures have an intuitive appeal and are used in many other multivariate techniques, they are not the most commonly used measure of similarity in cluster analysis. Distance measures of similarity, which represent similarity as the proximity of observations to one another across the variables in the cluster variate, are the similarity measure most often used. Distance measures are actually a measure of dissimilarity, with larger values denoting lesser similarity. Distance is converted into a similarity measure by using an inverse relationship. A simple illustration of this was shown in the hypothetical example in which clusters of observations were defined based on the proximity of observations to one another when each observation's scores on two variables were plotted graphically (see Figure 5.2).

Comparison to correlational measures. The difference between correlational and distance measures can be seen by referring again to Figure 5.4. Distance measures focus on the magnitude of the values and portray as similar cases that are close together but that may have very different patterns across the variables. Table 5.3 also contains distance measures of similarity for the seven cases, and we see a very different clustering of cases emerging than that found when using the correlational measures. With smaller distances representing greater similarity, Cases 1 and 2 form one group, while Cases 4, 5, 6, and 7 make up another group. These groups represent those with higher versus lower values. A third group, consisting of only Case 3, can be seen as differing from the other two groups as it has values that are both low and high. Whereas the two clusters using distance measures have different members than those using correlations, Case 3 is unique in either measure

Table 5.3

Calculating Correlational and Distance Measures of Similarity

Original Data

Case	Variables X_1	X_2	X_3	X_4	X_5
1	7	10	9	7	10
2	9	9	8	9	9
3	5	5	6	7	7
4	6	6	3	3	4
5	1	2	2	1	2
6	4	3	2	3	3
7	2	4	5	2	5

Similarity Measure: Correlation

Case	Case 1	2	3	4	5	6	7
1	1.000						
2	−0.147	1.000					
3	0.000	0.000	1.000				
4	0.087	0.516	−0.824	1.000			
5	0.963	−0.408	0.000	−0.060	1.000		
6	−0.466	0.791	−0.354	0.699	−0.645	1.000	
7	0.891	−0.516	0.165	−0.239	0.963	−0.699	1.000

Similarity Measure: Euclidean Distance

Case	Case 1	2	3	4	5	6	7
1	nc						
2	3.32	nc					
3	6.86	6.63	nc				
4	10.24	10.20	6.00	nc			
5	15.78	16.19	10.10	7.07	nc		
6	13.11	13.00	7.28	3.87	3.87	nc	
7	11.27	12.16	6.32	5.10	4.90	4.36	nc

Note. nc = distances not calculated. From *Multivariate Data Analysis*, by J. R. Hair, Jr., R. E. Anderson, R. L. Tatham, and W. C. Black, 1998, p. 485. Copyright 1998 by Prentice-Hall, Inc. Adapted with permission of Prentice-Hall, Inc., Upper Saddle River, NJ.

of similarity. The choice of a correlational measure, rather than the more traditional distance measure, requires a different interpretation of the results by the researcher. Clusters based on correlational measures may not have similar values but may instead have similar patterns. Distance-based clusters have more similar values across the set of variables, but the patterns can be quite different.

Types of distance measures. Several distance measures are available. The most commonly used is Euclidean distance. An example of how Euclidean distance is obtained is shown geometrically in Figure 5.5.

Suppose that two points in two dimensions have coordinates (X_1, Y_1) and (X_2, Y_2), respectively. The Euclidean distance between the points is the length of the hypotenuse of a right triangle, as calculated by the formula under the figure. This concept is easily generalized to more than two variables. The Euclidean distance is used to calculate several specific measures, one being the simple Euclidean distance (calculated as described above) and the other is the squared, or absolute, Euclidean distance, in which the distance value is the sum of the squared differ-

Figure 5.5

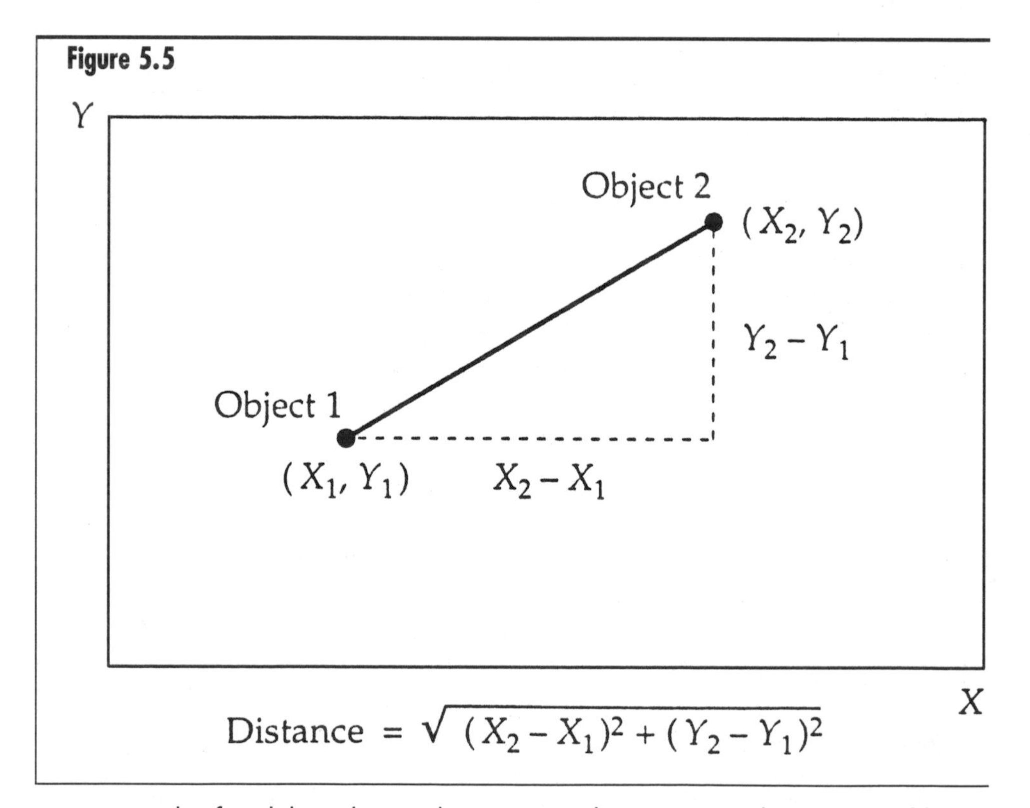

An example of Euclidean distance between two objects measured on two variables.

ences without taking the square root. The squared Euclidean distance has the advantage of not taking the square root, which speeds computations markedly, and is the recommended distance measure for the centroid and Ward's methods of clustering (see below).

Several options not based on the Euclidean distance are also available. One of the most widely used alternative measures involves replacing the squared differences by the sum of the absolute differences of the variables. This procedure is the absolute, or city-block, distance function. The *city-block approach* to calculating distances may be appropriate under certain circumstances, but it causes several problems (Shephard, 1966). One such problem is the assumption that the variables are not correlated with one another; if they are correlated, the clusters are not valid. Other measures that use variations of the absolute differences or the powers applied to the differences (other than just squaring the differences) are also available in most cluster programs.

Effect of unstandardized data values. A problem faced by all of the distance measures that use unstandardized data is the inconsistencies between cluster solutions when the scale of the variables is changed. For example, suppose three objects, A, B, and C, are measured on two variables, probability of purchasing brand X (in percentages) and amount of time spent viewing commercials for brand X (in minutes or seconds). The values for each observations are shown in Table 5.4(A).

From this information, distance measures can be calculated. In our example, we calculate three distance measures for each object pair: (a) simple Euclidean distance, (b) squared or absolute Euclidean distance, and (c) city-block distance. First, we calculate the distance values based on purchase probability and the viewing time in minutes. These distances, with smaller values indicating greater proximity and similarity, and their rank order are shown in Table 5.4B. As we can see, the most similar objects (with the smallest distance) are B and C, followed by A and C, with A and B the least similar (or least proximal). This ordering holds for all three distance measures, but the relative similarity or dispersion between objects is the most pronounced in the squared Euclidean distance measure.

The ordering of similarities can change markedly with only a change in the scaling of one of the variables. If we measured the viewing time in seconds instead of minutes, then the rank orders would change (see Table 5.4C). Although B and C are still the most similar, the pair A–B is now next most similar and is almost identical to the similarity of B–C. Yet when we used minutes of viewing time, pair A–B was the

Table 5.4

Variations in Distance Measures Based on Alternative Data Scales

A. Original Data

Object	Purchase probability	Commercial viewing time	
		Minutes	Seconds
A	60	3.0	180
B	65	3.5	210
C	63	4.0	240

B. Distance Measures Based on Minutes

Object pair	Simple Euclidean distance		Squared or absolute Euclidean distance		City-block distance	
	Value	Rank	Value	Rank	Value	Rank
A–B	5.025	3	25.25	3	5.5	3
A–C	3.162	2	10.00	2	4.0	2
B–C	2.062	1	4.25	1	2.5	1

C. Distance Measures Based on Seconds

Object pair	Simple Euclidean distance		Squared or absolute Euclidean distance		City-block distance	
	Value	Rank	Value	Rank	Value	Rank
A–B	30.41	2	925	2	35	3
A–C	60.07	3	3,609	3	63	2
B–C	30.06	1	904	1	32	1

D. Distance Measures Based on Standardized Values

Object pair	Standardized values		Simple Euclidean distance		Squared or absolute Euclidean distance		City-block distance	
	Purchase probability	Minutes/seconds of viewing time	Value	Rank	Value	Rank	Value	Rank
A–B	−1.06	−1.0	2.22	2	4.95	2	2.99	2
A–C	0.93	0.0	2.33	3	5.42	3	3.19	3
B–C	0.13	1.0	1.28	1	1.63	1	1.79	1

least similar by a substantial margin. What has occurred is that the scale of the viewing time variable has dominated the calculations, making purchase probability less significant in the calculations. The reverse was true, however, when we measured viewing time in minutes because purchase probability was dominant in the calculations. The researcher should thus note the tremendous effect that variable scaling can have on the final solution. Standardization of the clustering variables, whenever possible conceptually, should be used to avoid such instances as found in our example. The issue of standardization is discussed in the following section.

A commonly used measure of Euclidean distance that directly incorporates a standardization procedure is the *Mahalanobis distance.* The Mahalanobis approach not only performs a standardization process on the data by scaling in terms of the standard deviations but also sums the pooled within-group variance–covariance, which adjusts for intercorrelations among the variables. Highly intercorrelated sets of variables in cluster analysis can implicitly overweight one set of variables in the clustering procedures, as is discussed in the next section. In short, the Mahalanobis generalized distance procedure computes a distance measure between objects comparable to the R^2 in regression analysis. Although many situations are appropriate for use of the Mahalanobis distance, many computer software programs do not include it as a measure of similarity. In such cases, the researcher usually selects the squared Euclidean distance.

In attempting to select a distance measure, the researcher should remember the following caveats. Different distance measures or changing the scales of the variables may lead to different cluster solutions. Thus, it is advisable to use several measures and compare the results with theoretical or known patterns. Also when the variables are intercorrelated (either positively or negatively), the Mahalanobis distance measure is likely to be the most appropriate because it adjusts for intercorrelations and weights all variables equally. Of course, if the researcher wishes to weight the variables unequally, other procedures are available (Overall, 1964).

Association measures. Association measures of similarity are used to compare objects whose characteristics are measured only in nonmetric terms (nominal or ordinal measurement). As an example, respondents could answer yes or no on a number of statements. An association measure could assess the degree of agreement or matching between each pair of respondents. The simplest form of association measure would be the percentage of times there was agreement (both respondents answered yes or both said no to a question) across the set of questions. Extensions of this simple matching coefficient have been developed to accommodate multicategory nominal variables and even ordinal measures. Many computer programs, however, have limited support for association measures, and the researcher is often forced to first calculate the similarity measures and then input the similarity matrix into the cluster program. Reviews of the various types of association measures can be found in several sources (Everitt, 1980; Sneath & Sokal, 1973).

Data Standardization

With the similarity measure selected, the researcher must address only one more question: Should the data be standardized before similarities are calculated? In answering this question, the researcher must address several issues. First, most distance measures are sensitive to differing scales or magnitude among the variables. We saw this effect earlier when we changed from minutes to seconds on one of our variables. In general, variables with larger dispersion (i.e., larger standard deviations) have a greater effect on the final similarity value. Consider another example to illustrate this point. Assume that we want to cluster individuals on three variables: an attitude toward a product, age, and income. Now assume that we measure attitude on a 7-point scale of *liking* (7) to *disliking* (1), whereas age is measured in years, and income in dollars. If we plotted this on a three-dimensional graph, the distance between points (and their similarity) would be almost totally based on the income differences. The possible differences in attitude range from one to seven, whereas income may have a range perhaps a thousand times greater. Thus, graphically we would not be able to see any difference on the dimension associated with attitude. For this reason, the researcher must be aware of the implicit weighting of variables based on their relative dispersion that occurs with distance measures.

Standardization by variables. The most common form of standardization is the conversion of each variable to standard scores (also known as "*z* scores") by subtracting the mean and dividing by the standard deviation for each variable. This is an option in all computer programs, and many times it is even directly included in the cluster analysis procedure. This is the general form of a *normalized distance function*, which uses a Euclidean distance measure amenable to a normalizing transformation of the raw data. This process converts each raw data score into a standardized value with a zero mean and a unit standard deviation.

This transformation, in turn, eliminates the bias introduced by the differences in the scales of the several attributes or variables used in the analysis.

The benefits of standardization can be seen in Table 5.4D, where two variables (purchase probability and viewing time) have been standardized before computing the three distance measures. First, it is much easier to compare between the variables because they are on the same scale (a mean of 0 and standard deviation of 1). Positive values are above the mean and negative values below, with the magnitude representing the number of standard deviations the original value was from the mean. Second, there is no difference in the standardized values when only the scale changes. For example, when viewing time in minutes and then seconds is standardized, the values are the same. Thus, using standardized variables truly eliminates the effects due to scale differences not only across variables but also for the same variable. The researcher should not always apply standardization without consideration for its consequences, however. There is no reason to absolutely accept the cluster solution using standardized variables versus unstandardized variables. If there is some natural relationships reflected in the scaling of the variables, then standardization may not be appropriate. One example of a natural relationship would be a study that uses both overall and specific measures where the researcher wants the overall measure to be attributed a greater weight. For example, the study may use 5-point measures of job satisfaction for specific attributes and a 10-point measure for life satisfaction. In this study, due to the natural relationship, the researcher would not want to standardize the variables. The decision to standardize has both empirical and conceptual implications and should always be made with careful consideration.

Standardization by observation. What about standardizing respondents or cases? Why would one ever do this? For example, suppose that we had collected a number of ratings on a 10-point scale from respondents on the importance of several attributes in their purchase decision for a product. We could apply cluster analysis and obtain clusters, but one distinct possibility is that we would get clusters of people who said everything was important, some who said everything had little importance, and perhaps some clusters in between. This is in the clusters are *response-style effects*, which are the systematic patterns of responding to a set of questions, such as yea-sayers (those who answer very favorably to all questions) or nay-sayers (those who answer unfavorably to all questions).

If we want to identify groups according to their response style, then standardization is not appropriate. But in most instances, what is desired is the relative importance of one variable to another. In other words, is attribute 1 more or less important than the other attributes? Can clusters of respondents be found with similar patterns of importance? In this instance, standardizing by respondent would standardize each question not to the sample's average but instead to that respondent's average score. This *within-case* or *row-centering standardization* can be effective in removing response effects, and it is especially suited to many forms of attitudinal data. This is similar to a correlational measure in highlighting the pattern across variables, but the proximity of cases still determines the similarity value.

Stage 3: Assumptions

Cluster analysis, like multidimensional scaling, is not a statistical inference technique in which parameters from a sample are assessed as possibly being representative of a population. Instead, cluster analysis is an objective methodology for quantifying the structural characteristics of a set of observations. As such, it has strong mathematical properties, not statistical foundations. The requirements of normality, linearity, and homoscedasticity that are so important in other techniques have little bearing on cluster analysis. The researcher must focus, however, on two other critical issues: (a) representativeness of the sample and (b) multicollinearity.

Representativeness of the Sample

In very few instances does the researcher have a census of the population to use in the cluster analysis. Instead, a sample of cases is obtained and the clusters derived in the hope that they represent the structure of the population. The researcher must therefore be confident that the obtained sample is truly representative of the population. As mentioned earlier, outliers may really be only an undersampling of divergent groups that when discarded, introduce bias in the estimation of structure. The researcher must realize that cluster analysis is only as good as the representativeness of the sample. Therefore, all efforts should be taken to ensure that the sample is representative and that the results are generalizable to the population of interest.

Multicollinearity

Multicollinearity is an issue in other multivariate techniques (e.g., multiple regression) because it makes it difficult to discern the true effect of multicollinear variables. In cluster analysis, however, the effect is different because those variables that are multicollinear are implicitly weighted more heavily. Let us start with an example that illustrates its effect. Suppose that respondents are being clustered on 10 variables, all attitudinal statements concerning a service. When multicollinearity is examined, we see that there are really two sets of variables: the first made up of eight statements and the second consisting of the remaining two statements. If our intent is to really cluster the respondents on the dimensions of the product (in this case represented by the two groups of variables), then use of the original 10 variables would be misleading. Because each variable is weighted equally in cluster analysis, the first dimension has four times as many chances (eight items to two items) to affect the similarity measure as does the second dimension.

Multicollinearity acts as a weighting process not apparent to the observer but affecting the analysis nonetheless. For this reason, the researcher is encouraged to examine the variables used in cluster analysis for substantial multicollinearity and, if found, to either reduce the variables to equal numbers in each set or use one of the distance measures, such as Mahalanobis distance, that compensates for this correlation. There is debate over the use of factor scores in cluster analysis, as some research has shown that the variables that truly discriminate among the underlying groups are not well represented in most factor solutions. Thus, when factor scores are used, it is possible that a poor representation of the true structure of the data is obtained (Schaninger & Bass, 1986). The researcher must deal with both multicollinearity and discriminability of the variables to arrive at the best representation of structure.

Stage 4: Derivation of Clusters and Assessment of Overall Fit

With the variables selected and the similarity matrix calculated, the partitioning process begins. The researcher must first select the clustering algorithm used for forming clusters and then make the decision on the number of clusters to be formed. Both decisions have substantial implications not only on the results that are obtained but also on the interpretation that can be derived from the results. Each issue is discussed in the following sections. Figure 5.6 illustrates Stages 4–6 of the cluster analysis decision tree.

Figure 5.6

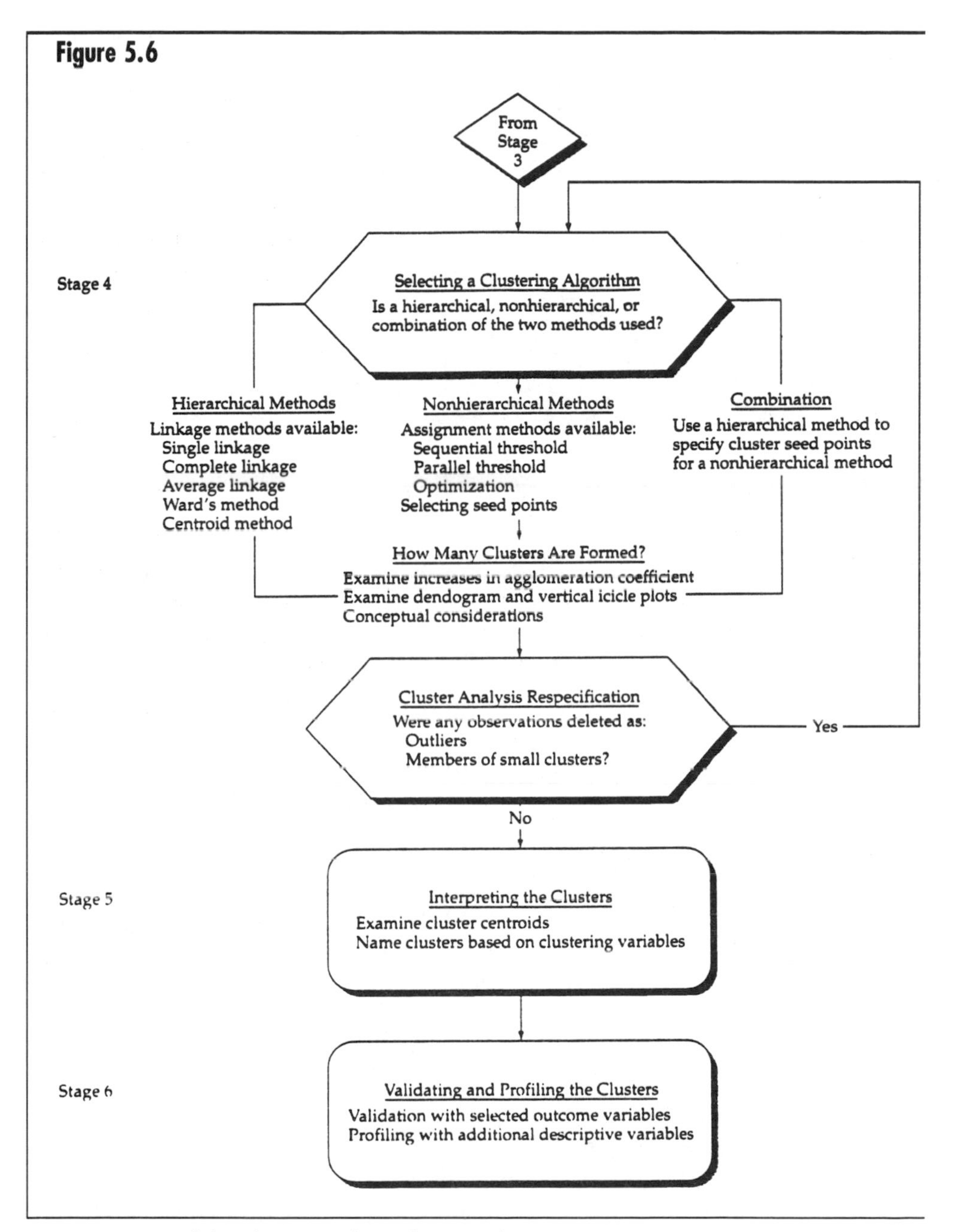

Stages 1–6 of the cluster analysis decision diagram.

Clustering Algorithms

The first major question to answer in the partitioning phase is, "What procedure should be used to place similar objects into groups or clusters?" That is, what clustering algorithm or set of rules is the most

Figure 5.7

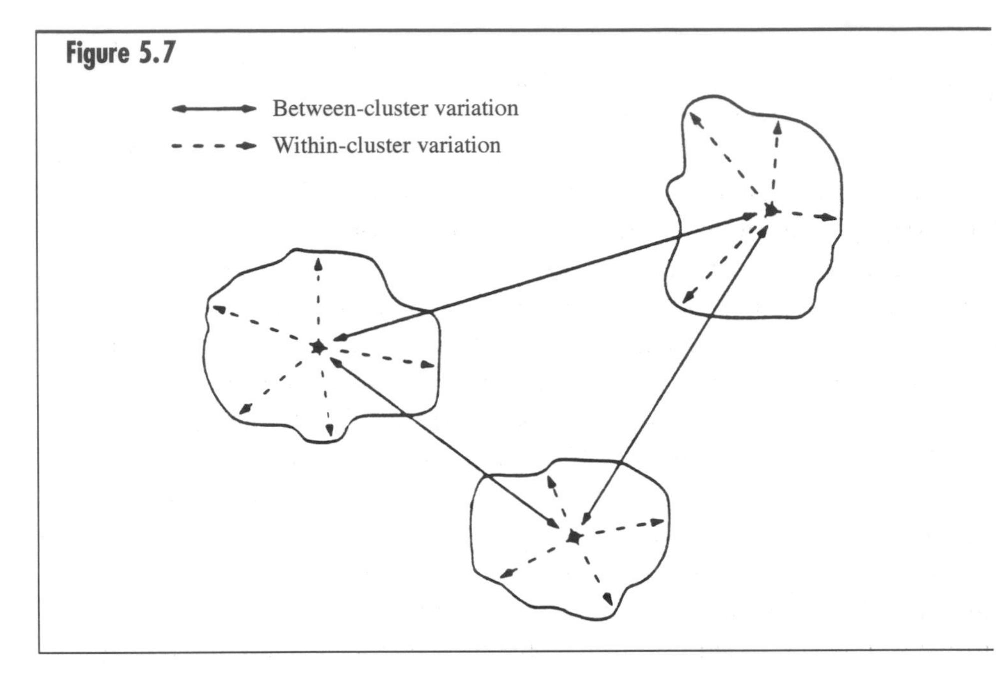

Cluster diagram showing between- and within-cluster variation.

appropriate? This is not a simple question because hundreds of computer programs using different algorithms are available, and more are always being developed. The essential criterion of all the algorithms, however, is that they attempt to maximize the differences between clusters relative to the variation within the clusters, as shown in Figure 5.7. The ratio of the between-cluster variation to the average within-cluster variation is then comparable to (but not identical to) the F ratio in analysis of variance. Most commonly used clustering algorithms can be classified into two general categories: (a) hierarchical and (b) nonhierarchical.

Hierarchical Cluster Procedures

Hierarchical procedures involve the construction of a hierarchy of a treelike structure. There are basically two types of hierarchical clustering procedures: (a) agglomerative and (b) divisive. In the agglomerative methods, each object or observation starts out as its own cluster. In subsequent steps, the two closest clusters (or individuals) are combined into a new aggregate cluster, thus reducing the number of clusters by one in each step. In some cases, a third individual joins the first two in a cluster. In others, two groups of individuals formed at an earlier stage may join together in a new cluster. Eventually, all individuals are grouped into one large cluster; for this reason, agglomerative procedures are sometimes referred to as "build-up methods."

An important characteristic of hierarchical procedures is that the results from an earlier stage are always nested within the results in a later stage, creating its similarity to a tree. For example, a six-cluster solution is obtained by joining two of the clusters found at the seven-cluster stage. Because clusters are formed only by joining existing clusters, any member of a cluster can trace its membership in an unbroken path to its beginning as a single observation. This process is shown in Figure 5.8; the representation is referred to as a *dendrogram* or *tree graph.*

When the clustering process proceeds in the opposite direction to agglomerative methods, it is referred to as a *divisive method.* In divisive methods, we begin with one large cluster containing all of the observations (objects). In succeeding steps, the observations that are most dissimilar are split off and made into smaller clusters. This process continues until each observation is a cluster in itself. In Figure 5.8, agglomerative methods would move from left to right, and divisive methods would move from right to left. Because most commonly used computer

Figure 5.8

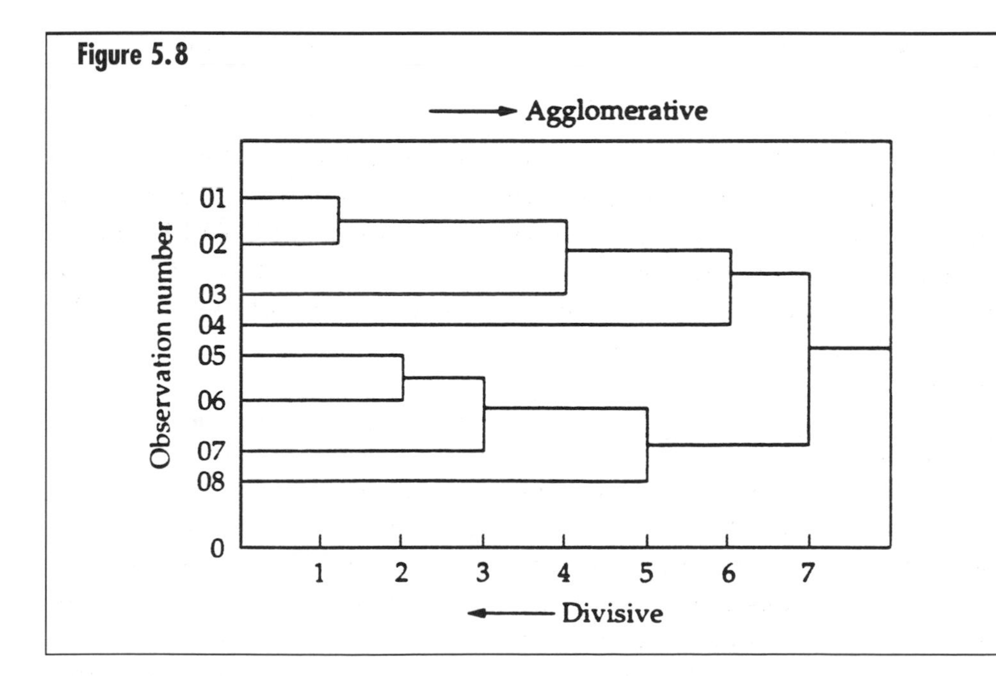

Dendogram illustrating hierarchical clustering.

packages use agglomerative methods, and divisive methods act almost as agglomerative methods in reverse, we focus on the agglomerative methods in our subsequent discussions.

Five popular agglomerative algorithms used to develop clusters are (a) single linkage, (b) complete linkage, (c) average linkage, (d) Ward's method, and (e) centroid method. These rules differ in how the distance between clusters is computed.

Single linkage. *Single linkage* is based on minimum distance. It finds the two objects separated by the shortest distance and places them in the first cluster. Then the next-shortest distance is found, and either a third object joins the first two to form a cluster or a new two-member cluster is formed. The process continues until all objects are in one cluster. This procedure has also been called the "nearest-neighbor approach."

The distance between any two clusters is the shortest distance from any point in one cluster to any point in the other. Two clusters are merged at any stage by the single shortest or strongest link between them. This was the rule applied in the example at the beginning of the chapter. Problems occur, however, when clusters are poorly delineated. In such cases, single linkage procedures can form long, snakelike chains, and eventually all individuals are placed in one chain. Individuals at opposite ends of a chain may be very dissimilar.

An example of this arrangement is shown in Figure 5.9. Three clusters (A, B, and C) are to be joined. The single-linkage algorithm, focusing on only the closest points in each cluster, would link Clusters A and B because of their short distance at the extreme ends of the clusters. Joining Clusters A and B creates a cluster that encircles Cluster C. Yet in striving for within-cluster homogeneity, it would be much better to join Cluster C with either A or B. This is the principal disadvantage of the single-linkage algorithm.

Complete linkage. *Complete linkage* is similar to single linkage except that the cluster criterion is based on maximum distance. For this reason, it is sometimes referred to as the "furthest-neighbor approach" or a "diameter method." The maximum distance between individuals in each cluster represents the smallest (minimum-diameter) sphere that can enclose all objects in both clusters. This method is complete because all of the objects in a cluster are linked to each other at some maximum distance or by minimum similarity. We can say that within-group similarity equals group diameter. This technique eliminates the snaking problem identified with single linkage.

Figure 5.9

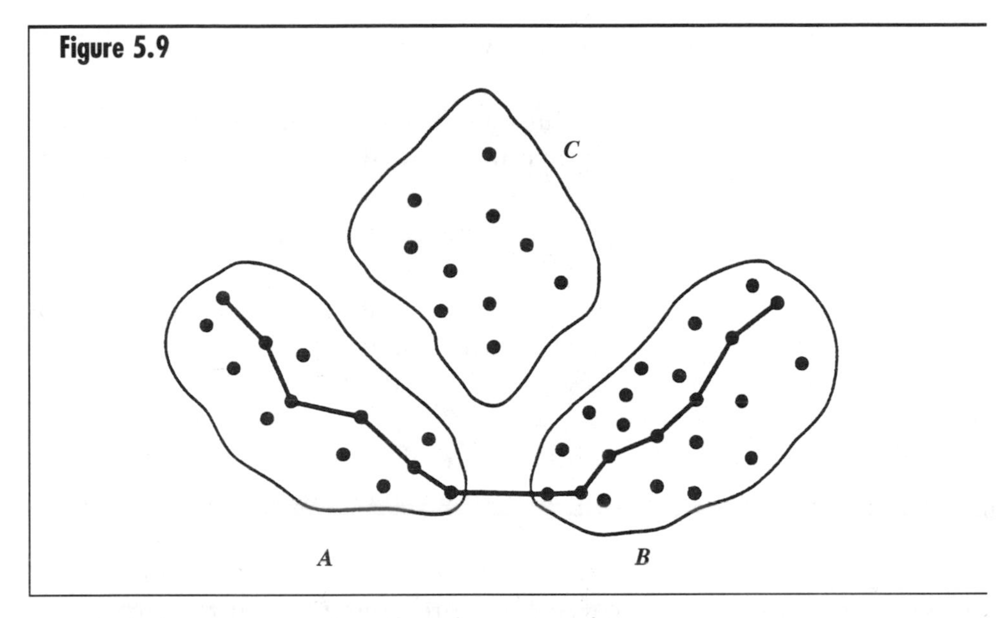

A single linkage joining dissimiliar Clusters A and B.

Figure 5.10 shows how the shortest (single linkage) and longest (complete linkage) distances represent similarity between groups. Both measures reflect only one aspect of the data. The use of the shortest distance reflects only a single pair of objects (the closest), while the complete linkage again reflects only a single pair, but this time it is the

Figure 5.10

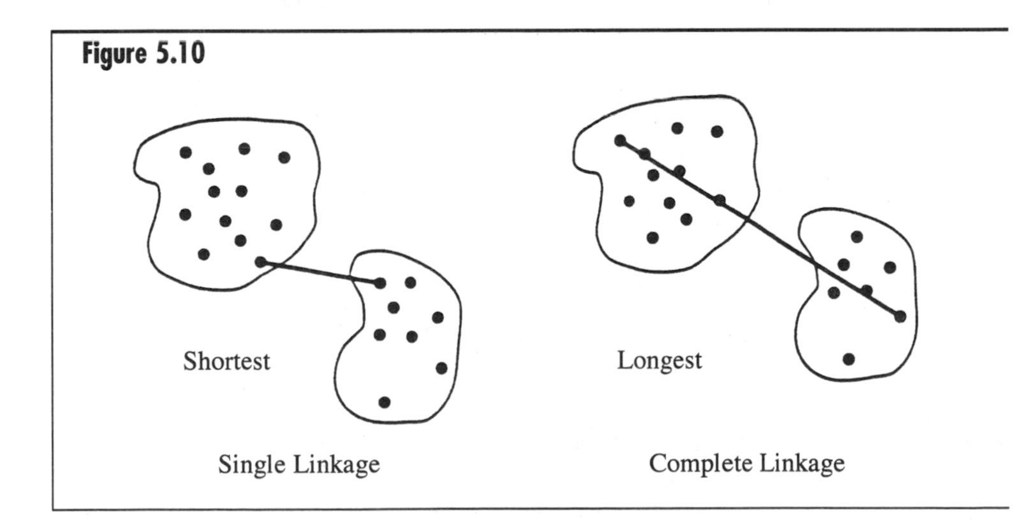

Comparison of distance measures for single linkage and complete linkage.

two most extreme. It is thus useful to visualize the measures as reflecting the similarity of most similar pair or least similar pair of objects.

Average linkage. *Average linkage* starts out the same as a single or a complete linkage, but the cluster criterion is the average distance from all individuals in one cluster to all individuals in another. Such techniques do not depend on extreme values, as do single linkage or complete linkage, and partitioning is based on all members of the clusters rather than on a single pair of extreme members. Average-linkage approaches tend to combine clusters with small within-cluster variation. They also tend to be biased toward the production of clusters with approximately the same variance.

Ward's method. In *Ward's method,* the distance between two clusters is the sum of squares between the two clusters summed over all variables. At each stage in the clustering procedure, the within-cluster sum of squares is minimized over all partitions (the complete set of disjoint or separate clusters) obtainable by combining two clusters from the previous stage. This procedure tends to combine clusters with a small number of observations. It is also biased toward the production of clusters with approximately the same number of observations.

Centroid method. In the *centroid method,* the distance between two clusters is the distance (typically squared Euclidean or simple Euclidean) between their centroids. *Cluster centroids* are the mean values of the observations on the variables in the cluster variate. In this method, every time individuals are grouped, a new centroid is computed. Cluster centroids migrate as cluster mergers take place. In other words, there is a change in a cluster centroid every time a new individual or group of individuals is added to an existing cluster. This method is the most popular with biologists, but it may produce messy and often confusing results. The confusion occurs because of reversals; that is, instances when the distance between the centroids of one pair may be less than the distance between the centroids of another pair merged at an earlier combination. The advantage of this method is that it is less affected by outliers than are other hierarchical methods.

Nonhierarchical Clustering Procedures

In contrast to hierarchical methods, *nonhierarchical procedures* do not involve the treelike construction process. Instead, they assign objects into clusters once the number of clusters to be formed is specified. Thus, the six-cluster solution is not just a combination of two clusters from the seven-cluster solution but is based only on finding the best six-cluster solution. In a simple example, the process works this way. The first step is to select a cluster seed as the initial cluster center, and all objects (individuals) within a prespecified threshold distance are then included in the resulting cluster. Then, another cluster seed is chosen and the assignment continues until all objects are assigned. Then, objects may be reassigned if they are closer to another cluster than the one originally assigned. There are several approaches for selecting cluster seeds and assigning objects, which we discuss in the next section. Nonhierarchical clustering procedures are frequently referred to as "K-means clustering," and they typically use one of the following three approaches for assigning individual observations to one of the clusters (Green, 1978, p. 428).

Sequential threshold. The *sequential threshold method* starts by selecting one cluster seed and including all objects within a prespecified distance. When all objects within the distance are included, a second cluster seed is selected, and all objects within the prespecified distance are included in that cluster. Then a third seed is selected, and the process continues as before. When an object is clustered with a seed, it is no longer considered for subsequent seeds.

Parallel threshold. In contrast, the *parallel threshold method* selects several cluster seeds simultaneously in the beginning and assigns objects within the threshold distance to the nearest seed. As the process evolves, threshold distances can be adjusted to include fewer or more objects in the clusters. Also, in some methods, objects remain unclustered if they are outside the prespecified threshold distance from any cluster seed.

Optimization. *Optimizing procedure* is similar to the other two except that it allows for the reassignment of objects. If in the course of assigning objects, an object becomes closer to another cluster that is not the cluster to which it was originally assigned, then an optimizing procedure switches the object to the more similar (closer) cluster.

Seed point selection. Nonhierarchical procedures are available in a number of computer programs, including all of the major statistical packages. The sequential threshold procedure (e.g., FASTCLUS program in SAS) is an example of a nonhierarchical clustering program designed for large data sets. After the researcher specifies the maximum number of clusters allowed, the procedure begins by selecting cluster seeds, which are used as initial guesses of the means of the clusters. The first seed is the first observation in the data set with no missing values. The second seed is the next complete observation (no missing data) that is separated from the first seed by a specified minimum distance. The default option is a zero minimum distance. After all seeds have been selected, the program assigns each observation to the cluster with the nearest seed. The researcher can specify that the cluster seeds be revised (updated) by calculating seed cluster means each time an observation is assigned. In contrast, the parallel threshold methods (e.g., QUICK CLUSTER in SPSS) establish the seed points as user-supplied points or select them randomly from all observations.

The major problem faced by all nonhierarchical clustering procedures is how to select the cluster seeds. For example, with a sequential threshold option, the initial and probably the final cluster results depend on the order of the observations in the data set, and shuffling the order of the data is likely to affect the results. Specifying the initial cluster seeds as in the sequential threshold procedure can reduce this problem. Even selecting the cluster seeds randomly, however, produces different results for each set of random seed points. Thus, the researcher must be aware of the effect of the cluster seed selection process on the final results.

Hierarchical Versus Nonhierarchical Methods

A definitive answer to the question of whether hierarchical or nonhierarchical methods should be used cannot be given for two reasons. First, the research problem at hand typically may suggest one method or the other. Second, what we learn with continued application of one method to a particular context may suggest one method over the other as more suitable for that context.

Pros and cons of hierarchical methods. In the past, hierarchical clustering techniques were more popular, with Ward's method and average linkage probably the best available (Milligan, 1980). Hierarchical procedures do have the advantage of being fast and, therefore, taking less computer time. Yet with the computing power of today, even personal computers can handle large data sets easily. Hierarchical procedures can be misleading, however, because undesirable early combinations may persist throughout the analysis and lead to artificial results. Of concern is the substantial effect of outliers on hierarchical methods, particularly with the complete linkage method. To reduce this possibility, the researcher may wish to cluster analyze the data several times, each time deleting problem observations or outliers. The deletion of cases, even those not found to be outliers, can many times distort the solution. Thus, the researcher must use extreme care in the deletion of observations for any reason.

Also, although computations of the clustering process are relatively fast, hierarchical methods are not amenable to analyzing very large samples. As sample size increases, the data storage requirements increase dramatically. For example, a sample of 400 cases requires storage of approximately 80,000 similarities, and this number increases to almost 125,000 for a sample of 500. Even given today's technological advances, problems of this size exceed the capacity of most personal computers, thus limiting the application of hierarchical methods in many instances. The researcher may take a random sample of the original observations to reduce its size but must now question the representativeness of the sample taken from the original sample.

Emergence of nonhierarchical methods. Nonhierarchical methods have gained acceptability and are applied increasingly. Their use, however, depends on the ability of the researcher to select the seed points according to some practical, objective, or theoretical basis. In these instances, nonhierarchical methods have several advantages over hierarchical techniques. The results are less susceptible to the outliers in the data, the distance measure used, and the inclusion of irrelevant or inappropriate variables. These benefits are realized, however, only with the use of nonrandom (i.e., specified) seed points; thus, the use of nonhierarchical techniques with random seed points is markedly inferior to the hierarchical techniques. Even a nonrandom starting solution does not guarantee an optimal clustering of observations. In fact, in many instances, the researcher gets a different final solution for each set of specified seed points. How is the researcher to select the "correct" answer? Only by analysis and validation can the researcher then select what is considered the "best" representation of structure, realizing that there are many alternatives that may be as acceptable.

A combination of both methods. Another approach is to use both methods (hierarchical and nonhierarchical) to gain the benefits of each (Milligan, 1980). First, a hierarchical technique can establish the number of clusters, profile the cluster centers, and identify any obvious outliers. After outliers are eliminated, the remaining observations are then clustered by a nonhierarchical method, with the cluster centers from the hierarchical results as the initial seed points. In this way, the advantages of the hierarchical methods are complemented by the ability of the nonhierarchical methods to "fine-tune" the results by allowing the switching of cluster membership.

Number of Clusters Formed

Perhaps the most perplexing issue for a researcher using cluster analysis is determining the final number of clusters to be formed (also known as the "stopping rule"). Unfortunately, no standard, objective selection procedure exists. Because there is no internal statistical criterion used for inference, such as the statistical significance tests of other multivariate methods, researchers have developed many criteria and guidelines for approaching the problem. The principal drawback is that these are ad hoc procedures that must be computed by the researcher, and many times they involve fairly complex procedures (Aldenderfer & Blashfield, 1984; Milligan & Cooper, 1985). One class of stopping rules that is relatively simple examines some measure of similarity or distance between clusters at each successive step, with the cluster solution defined when the similarity measure exceeds a specified value or when the successive values between steps makes a sudden jump. A simple example of this was used in the example at the beginning of the chapter, which looked for large increases in the average within-cluster distance. When a large increase occurs, the researcher selects the prior cluster solution on the logic that its combination caused a substantial decrease in similarity. This stopping rule has been shown to provide fairly accurate decisions in empirical studies (Milligan & Cooper, 1985). A second general class of stopping rules attempts to apply some form of statistical rule or to adapt a statistical test, such as the point-biserial/tau correlations or the likelihood ratio. Although some of these have been shown to have notable success, such as the cubic clustering criterion contained in SAS, many seem overly complex for the improvement that they provide over simpler measures. There are other specific procedures that have been proposed, but none have been found to be substantially better in all situations.

The researcher should also complement the strictly empirical judgment with any conceptualization of theoretical relationships that may suggest a natural number of clusters. Also one might start this process by specifying some criteria on the basis of practical considerations, such as saying "My findings will be more manageable and easier to communicate if I have between three and six clusters" and then solving for this number of clusters and selecting the best alternative after evaluating all of them. In the final analysis, however, it is probably best to compute a number of cluster solutions (e.g., two, three, and four) and then decide among the alternative solutions by using a priori criteria, practical judgment, common sense, or theoretical foundations. The cluster solutions are improved by restricting the solution according to conceptual aspects of the problem.

Respecification of the Cluster Analysis

When an acceptable cluster analysis solution is identified, the researcher should examine the fundamental structure represented in the defined clusters. Of note are widely disparate cluster sizes or clusters of only one or two observations. Researchers must examine widely varying cluster sizes from a conceptual perspective, comparing the actual results with the expectations formed in the research objectives. More troublesome are single-member clusters, which may be outliers not detected in earlier analyses. If a single-member cluster (or one of very small size compared with other clusters) appears, the researcher must decide if the cluster represents a valid structural component in the sample or if it should be deleted as unrepresentative. If any observations are deleted, especially when hierarchical solutions are used, the researcher should rerun the cluster analysis and start the process of defining clusters anew.

Stage 5: Interpretation of the Clusters

The interpretation stage involves examining each cluster in terms of the cluster variate to name or assign a label accurately describing the nature of the clusters. To clarify this process, let us refer to the example of diet versus regular soft drinks. Assume that an attitude scale was developed that consisted of statements regarding consumption of soft drinks. Examples of statements include "Diet soft drinks taste harsher," "Regular soft drinks have a fuller taste," "Diet drinks are healthier." Assume further that demographic and soft drink consumption data were also collected.

When starting the interpretation process, one measure frequently used is the cluster's centroid. If the clustering procedure were performed on the raw data, this would be a logical description. If the data were standardized or if the cluster analysis were performed using factor analysis (component factors), the researcher would have to go back to the raw scores for the original variables and compute average profiles using these data.

Continuing with the soft drink example, in this stage we examine the average score profiles on the attitude statements for each group and assign a descriptive label to each cluster. Many times discriminant analysis is applied to generate score profiles, but we must remember that statistically significant differences would not indicate an "optimal" solution because statistical differences are expected given the objective of cluster analysis. Examination of the profiles allows for a rich description of each cluster. For example, two of the clusters may have favorable attitudes about diet soft drinks, and the third cluster may have negative attitudes. Moreover, of the two favorable clusters, one may exhibit favorable attitudes toward only diet soft drinks, whereas the other may display favorable attitudes toward both diet and regular soft drinks. From this analytical procedure, one would evaluate each cluster's attitudes and develop substantive interpretations to facilitate labeling each. For example, one cluster might be labeled "Health and calorie conscious," whereas another might be "Get a sugar rush."

The profiling and interpretation of the clusters, however, achieve more than just description. First, they provide a means for assessing the correspondence of the derived clusters to those proposed by prior theory or practical experience. If used in a confirmatory mode, the cluster analysis profiles provide a direct means of assessing the correspondence. Second, the cluster profiles provide a route for making assessments of practical significance. The researcher may require that substantial differences exist on a set of clustering variables and the cluster solution is expanded until such differences arise. In either instance, the researcher has a plan for comparing the derived clusters to a preconceived typology.

Stage 6: Validation and Profile of the Clusters

Given the somewhat subjective nature of cluster analysis with regard to selecting an "optimal" cluster solution, the researcher should take great care in validating and ensuring the practical significance of the final cluster solution. Although no single method exists to ensure validity and practical significance, several approaches have been proposed to provide some basis for the researcher's assessment.

Validation of the Cluster Solution

Validation includes attempts by the researcher to ensure that the cluster solution is representative of the general population and, thus, is generalizable to other objects and stable over time. The most direct approach in this regard is to cluster analyze separate samples, comparing the cluster solutions and assessing the correspondence of the results. This approach, however, is often impractical because of time or cost constraints or the unavailability of research participants for multiple cluster analyses. In these instances, a common approach is to split the sample into two groups. Each is cluster analyzed separately, and the results are then compared. Other approaches include (a) a modified form of split sampling whereby cluster centers obtained from one cluster solution are used to define clusters from the other observations and the results are compared (McIntyre & Blashfield, 1980) and (b) a direct form of cross-validation (Punj & Stewart, 1983).

The researcher may also attempt to establish some form of criterion or predictive validity. To do so, the researcher selects variables not used to form the clusters but known to vary across the clusters. In our example, we may know from past research that attitudes toward diet soft drinks vary by age. Thus, we can statistically test for the differences in age between those clusters favorable to diet soft drinks and those that are not. The variables used to assess predictive validity should have strong theoretical or practical support because they become the benchmark for selecting among the cluster solutions.

Profile of the Cluster Solution

Profiling involves describing the characteristics of each cluster to explain how they may differ on relevant dimensions. This typically involves the use of discriminant analysis or some other appropriate statistic that tests for differences among means. The procedure begins after the clusters are identified. The researcher uses data not previously included in the cluster procedure to profile the characteristics of each cluster. These data typically are demographic characteristics, psychological profiles, consumption patterns, and so forth. Although there may not be a theoretical rationale for their difference across the clusters, such as is required for predictive validity assessment (see above), the data should at least have practical importance. Using discriminant analysis or analysis of variance, the researcher can then compare average score profiles for the clusters. The categorical dependent variable is the previously identified clusters, and the independent variables are the demographics, psychographics, and so on. From this analysis, assuming statistical significance, the researcher could conclude, for example, that the "Health and calorie conscious" cluster from our previous example consists of better educated, higher income professionals who are moderate consumers of soft drinks. In short, the profile analysis focuses on describing not what directly determines the clusters, but the characteristics of the clusters after they have been identified. Moreover, the emphasis is on the characteristics that differ significantly across the clusters and those that could predict membership in a particular attitude cluster.

An Example of Stages: Who Are the Homeless?

We provide another example of the use of cluster analysis. In the 1980s, citizens of the United States became increasingly aware of and sensitive to the plight of the homeless. It was not long before social scientists began to gather data in an attempt to identify the characteristics of the homeless population, embodied by the question "Who are the homeless?" This is the type of question that lends itself to cluster analysis because the analysis can identify naturally occurring groups, or types of homeless people. Within the context of the following example, we identify how the researchers addressed some of the issues that we present in our organizational schema, or stage model of cluster analysis.

Mowbray, Bybee, and Cohen (1993) identified 108 homeless people through various community agencies and shelters. Information about the research participants was obtained from agency referral forms, client interviews, ratings of outreach workers, and archival records. Because the researchers were interested in "dimensions thought to be important for community functioning," they were guided in the selection of measures; previous research had suggested good candidates: community living problems, depression, substance abuse, psychoticism, and aggression.

- *Stage 1: Objectives.* The authors used cluster analysis to establish a taxonomy of the homeless population. In addition, despite the exploratory nature of their study, variables were not selected arbitrarily; their selection was based on previous research that established a relationship between these variables and community functioning. Moreover, because the sample size was relatively small, a decision was made to limit the number of variables included in the analysis.
- *Stage 2: Research Design.* The researchers examined the data for outliers. Because, in their original format, the variables were scaled differently, a situation that can lead to undo weighting of scales with large ranges, the authors standardized the scales before submitting the data for analysis. In addition, they chose to use Euclidean distances to calculate interindividual similarities.
- *Stage 3: Assumptions.* The researchers did their best to obtain a representative sample. Although all participants came from one Midwestern state, about half the sample came from an urban

setting and half from a college town. About 50% of the clients were recruited by referrals from community mental health centers, 30% were recruited from psychiatric inpatient facilities, and 17% of the sample were living in shelters and had no records at the community health centers. The total sample had a large number of people who had current or past involvement with the mental health system. For this reason, the title of the researchers' article appropriately includes the wording "homeless mentally ill."

Another issue in cluster analysis is *multicollinearity,* the intercorrelation among variables used in clustering. Variables that are too highly correlated with one another distort the cluster solution by being overweighted. Of the 10 correlations in the matrix, there was a significant correlation (+.47) between community living problems and psychoticism. Because the correlation was moderate and the variables were conceptually distinct, the authors retained both variables for inclusion in the analysis.

The results of the analysis suggested the viability of three or four clusters. In examining the characteristics of the clusters, defined by only those variables used in the cluster analysis, the authors chose names for each cluster that best captured the meaning of the variables. The first cluster was "Hostile–psychotic" and included 35% of the sample. These clients scored the highest on measures of aggression and psychoticism. The second-largest cluster was "Best functioning" and included 28% of the sample. Clients in this group had low scores on all measures of dysfunction. In the identification of the third and fourth clusters, a decision had to be made. When viewed together as one cluster, the most apt name for this group was "Depressed." However, the results of the analysis suggested that a fourth cluster may have meaning. The Depressed group included individuals who abused substances and those who did not. In the authors' words, "Inspection of the clusters defined by each solution revealed significant and theoretically meaningful differences between the fourth cluster and the other three" (Mowbray et al., 1993, p. 86). Consequently, the Depressed group that was retained included 19% of the sample and was defined by high scores on depression and low scores on substance abuse. The fourth cluster, "Substance abusing," made up 19% of the total sample and was defined by high scores on substance abuse and the second-highest scores on depression.

- *Stage 4: Derivation of Clusters and Assessment of Overall Fit.* In deriving clusters, the researchers used a combination of hierarchical (Ward's method) and nonhierarchical (K means) clustering, with the centroids of the first method used as seed points in the second cluster analysis. Because the sample size was relatively small, the researchers selected only five variables for inclusion to guard against over fitting.
- *Stage 5: Interpretation of the Clusters.* Interpreting clusters involves examining the differences between the clusters with respect to the variables used in the cluster analysis. At first glance, one might think that because the analysis separated clusters on the basis of these variables, there would be a significant difference between clusters on all variables. This is rarely the case. The authors conducted univariate F tests on each variable and used the pattern of significant differences to name the clusters. For example, the Substance-abusing cluster reported significantly greater substance abuse than the other clusters, but there were no differences among the other clusters on this variable. Likewise, the Hostile–psychotic cluster showed high scores on hostility and psychoticism, which were significantly different than the other clusters, but no differences on these variables were observed among the remaining three clusters.

  The authors compared the groups on other variables not included in the cluster analysis. For example, the people in the Hostile–psychotic cluster scored higher on deviancy and showed the worst overall functioning; the Depressed group had the highest potential for suicide; women tended to be overrepresented in the Depressed group; and the Substance-abusing cluster tended to have younger people, while the Hostile–psychotic group tended to have more people over the age of 40.
- *Stage 6: Validation and Profile of the Clusters.* The authors stressed the profiling of clusters by comparing the groups on variables external to the cluster analysis. Moreover, they chose variables that would be expected to differ across the clusters. One outcome of the cluster analysis and profiling stage was that it may be possible to offer a narrative description of the characteristics of the members of the cluster.

Summary of the Decision Process

Cluster analysis provides researchers with an empirical and objective method for performing one of the most inherent tasks for humans: classification. Whether for purposes of simplification, exploration, or confirmation, cluster analysis is a potent analytical tool that has a wide range of applications. With this technique, however, comes a responsibility on the part of the researcher to apply the underlying principles appropriately. As mentioned in the beginning of this chapter, cluster analysis has many caveats that cause even the experienced researcher to apply it with caution. Yet when used appropriately, it has the potential to reveal structures within the data that could not be discovered by any other means. Thus, this powerful technique addresses a fundamental need of researchers in all fields with the knowledge that is can be as easily abused as used wisely.

An Example of An Application: The HATCO Database

To illustrate the application of cluster analysis techniques, I turn to the HATCO database.[1] The seven perceptions of HATCO provide a basis for illustrating one of the most common uses of cluster analysis: the formation of customer segments. In our example, we follow the sequence of stages (not enumerated) of the model-building process presented earlier in the chapter. The seven HATCO perceptions, or attributes, rated by each respondent included delivery speed, price, price flexibility, manufacturer's image, overall service, sales force image, and product quality.

We begin by cluster analyzing the ratings of HATCO customers as to the performance of HATCO on the seven attributes (X_1 to X_7). Our objective is to segment objects (customers) into groups with similar perceptions of HATCO. Once identified, HATCO can then formulate strat-

[1]HATCO stands for the Hair, Anderson, and Tatham Company, a large (although nonexistent) industrial supplier. The data set was obtained from a survey of 100 HATCO customers, assessed on 14 separate variables, collected through an established marketing research firm. Three types of information were collected. The first type is the perception of HATCO on seven attributes, identified in past studies as the most influential in the choice of suppliers. The respondents, purchasing managers of firms buying from HATCO, rated HATCO on each attribute. The second type of information relates to actual purchase outcomes, either the evaluations of each respondent's satisfaction with HATCO or the percentage of that respondent's product purchases from HATCO. The third type of information contains general characteristics of the purchasing companies, recorded as nonmetric data (e.g., firm size, or industry type).

egies with different appeals for the separate groups—the requisite basis for market segmentation. A primary concern is that the seven attributes used to form the clusters be adequate in scope and detail. From the examples in other chapters with the various multivariate techniques, we have found that these variables have sufficient predictive power to justify their use as the basis for segmentation. The sample of 100 observations was examined for outliers and was found to have no strong candidates for deletion (one should examine the cluster solutions in later stages and assess if outliers have emerged during the clustering process). Given that the set of seven variables is metric, squared Euclidean distance was chosen as the similarity measure. Standardization of variables was not undertaken because all variables were on the same scale, and within-case standardization was not appropriate because the magnitude of the perceptions was an important element of the segmentation objectives. The sample was considered a representative sample of HATCO customers. An analysis of multicollinearity identified only minimal levels that should not influence the cluster analysis in any substantial manner.

Ward's method was chosen to minimize the within-cluster differences and to avoid problems with "chaining" of the observations found in the single-linkage method. Because the data involve profiles of HATCO customers and our interest is in identifying types or profiles of these customers that may form the bases for differing strategies, a manageable number of clusters was deemed to be in the range of two to five. The clustering (agglomeration) coefficient showed rather large increases in going from four to three clusters, three to two clusters, and two to one cluster. The largest percentage of increase in the clustering coefficient occurred in going from two to one cluster, the next noticeable change in the percentage of increase occurred in combining four into three clusters. Thus, both the two- and four-cluster solutions were examined.

Table 5.5 contains the clustering variable profiles for both cluster solutions. An examination of the two-cluster profiles reveals two clusters that are almost mirror images of each other. Cluster 1 has high values on X_1, X_3, and X_5, whereas Cluster 2 has higher values for X_2, X_4, X_6, and X_7. Also shown in Table 5.5 are the profiles for the four-cluster solution, which reveal a number of patterns of high versus low values. Another aspect that varies from the two-cluster solution is that all of the clustering variables vary in a statistically significant manner across the four groups, versus only five of the variables in the two-cluster solution. Because all indicators (stopping rule, absence of outliers, and distinctive profiles) supported either the two- or four-cluster solutions, both solutions were carried forward into the nonhierarchical analysis to obtain final cluster solutions. The results (centroid values and cluster size) of the nonhierarchical are shown in Table 5.6 for both cluster solutions. As seen in the hierarchical methods, the two-cluster solution results in groups of almost equal size (52 vs. 48 observations) and the profiles correspond well with the cluster profiles from the hierarchical procedure. Again, only X_5 shows no significant differences between the clusters. For the four-group solution, the cluster sizes are similar to those from the hierarchical procedure, varying in size at the most by only four observations. Also the cluster profiles match well. The correspondence and stability of the two solutions between the nonhierarchical and hierarchical methods confirms the results subject to theoretical and practical acceptance.

Information essential to the interpretation and profiling stages is provided in Table 5.6. For each cluster, the centroid on each of the seven clustering variables is provided, along with the univariate *F* ratios and levels of significance comparing the differences between the cluster means. Factor analysis revealed that two factors underlie responses to those seven variables. The first factor contained X_1 (delivery speed), X_2 (price), X_3 (price flexibility), and X_7 (product quality; X_2 and X_7 were inversely related to X_1 and X_3); the second factor contained the two image items, X_4 (manufacturer image) and X_6 (salesforce image).

For the two-cluster solution, six of the seven variables produced statistically significantly effects (see Table 5.6). Only X_5, overall service, was not significantly different between the two clusters. In profiling customers, Cluster 1 is significantly higher on the first factor, whereas Cluster 2 has significantly higher perceptions of HATCO on the second factor—the two image variables (X_4 and X_6). The differences were much more distinctive on the first set of variables, and the image variables had substantially less delineation between the clusters, even though they were statistically significant. This should focus managerial attention on the four variables in the first set and relegate the image variables to a secondary role, although image variables are the key descriptors for the second cluster.

In the four-cluster solution, Cluster 1 split into Clusters 1 and 4, whereas Cluster 2 split into Clusters 2 and 3. Clusters 1 and 4 shared similar patterns on the first variable set (X_1, X_2, X_3, and X_7), maintaining their distinctiveness from Clusters 2 and 3. Their profiles varied, with Cluster 1 higher on X_2 and Cluster 4 higher on X_1, X_3, and X_7.

Table 5.5

Clustering Variable Profiles From the Hierarchical Cluster Analysis

Clustering Variable Profiles

Cluster	Clustering Variable Mean Values							
	X_1 Delivery speed	X_2 Price level	X_3 Price flexibility	X_4 Manufacturer image	X_5 Overall service	X_6 Salesforce image	X_7 Product quality	Cluster size
Two-cluster solution								
1	4.460	1.576	8.900	4.926	2.992	2.510	5.904	50
2	2.570	3.152	6.888	5.570	2.840	2.820	8.038	50
Four-cluster solution								
1	4.207	1.624	8.597	4.372	2.879	2.014	5.124	29
2	2.213	2.834	7.166	5.358	2.505	2.689	7.968	38
3	3.700	4.158	6.008	6.242	3.900	3.233	8.258	12
4	4.810	1.510	9.319	5.690	3.148	3.195	6.981	21

Significance Testing of Differences Between Cluster Centers

Variable	Cluster mean square	Degrees of freedom	Error mean square	Degrees of freedom	*F* value	Significance
Two-cluster solution						
X_2 Delivery speed	89.302	1	.851	98	104.95	.000
X_2 Price level	62.094	1	.811	98	76.61	.000
X_3 Price flexibility	101.204	1	.909	98	111.30	.000
X_4 Manufacturer image	10.368	1	1.187	98	8.73	.004
X_5 Overall service	.578	1	.564	98	1.02	.314
X_6 Salesforce image	2.402	1	.576	98	4.17	.044
X_7 Product quality	113.849	1	1.377	98	82.68	.000
Four-cluster solution						
X_1 Delivery speed	37.962	3	.613	96	61.98	.000
X_2 Price level	26.082	3	.659	96	39.56	.000
X_3 Price flexibility	39.927	3	.735	96	54.34	.000
X_4 Manufacturer image	12.884	3	.917	96	14.04	.000
X_5 Overall service	6.398	3	.382	96	16.75	.000
X_6 Salesforce image	7.367	3	.383	96	19.26	.000
X_7 Product quality	52.203	3	.960	96	54.37	.000

Note. From *Multivariate Data Analysis*, by J. R. Hair, Jr., R. E. Anderson, R. L. Tatham, and W. C. Black, 1998, p. 485 Copyright 1998 by Prentice-Hall, Inc. Adapted with permission of Prentice-Hall, Inc., Upper Saddle River, NJ.

Table 5.6

Solutions of the Nonhierarchical Cluster Analysis With Initial Seed Points From the Hierarchical Results

Two-Cluster Solution

	Mean values							
Cluster	**X_1 Delivery speed**	**X_2 Price level**	**X_3 Price flexibility**	**X_4 Manufacturer image**	**X_5 Overall service**	**X_6 Salesforce image**	**X_7 Product quality**	**Cluster size**
Final cluster centers								
1	4.383	1.581	8.900	4.925	2.958	2.525	5.904	52
2	2.575	3.212	6.804	5.598	2.871	2.817	8.127	48
Statistical significance of cluster differences								
F value	87.720	86.750	133.180	9.600	0.330	3.670	96.400	
Significance	0.000	0.000	0.000	0.003	0.566	0.058	0.000	

Profiling the clusters

	Cluster			
Predictive validity	**1**	**2**	***F* value**	**Significance**
X_9 Usage level	49.212	42.729	14.789	.000
X_{10} Satisfaction	5.133	4.379	23.826	.000

Four-Cluster Solution

Cluster	Mean values							
	X_1 Delivery speed	X_2 Price level	X_3 Price flexibility	X_4 Manufacturer image	X_5 Overall service	X_6 Salesforce image	X_7 Product quality	Cluster size
Final cluster centers								
1	4.094	1.621	8.630	4.415	2.830	2.079	5.273	33
2	2.171	2.846	7.123	5.403	2.489	2.683	8.194	35
3	3.662	4.200	5.946	6.123	3.900	3.177	7.946	13
4	4.884	1.511	9.368	5.811	3.179	3.300	7.000	19
Statistical significance of cluster differences								
F value	56.35	46.710	67.860	14.600	18.600	19.710	57.600	
Significance	0.000	0.000	0.000	0.000	0.000	0.000	0.000	

Profiling the clusters

Predictive validity	Cluster 1	Cluster 2	Cluster 3	Cluster 4	*F* value	Significance
X_9 Usage level	46.333	41.229	46.769	54.211	11.304	.000
X_{10} Satisfaction	4.839	4.134	5.038	5.642	22.212	.000

The marked difference comes on the image variables, where Cluster 1 follows the pattern seen in the two-cluster solution with lower scores, whereas Cluster 4 counters this trend by having high positive image scores. This represents the most marked difference between Clusters 1 and 4. Cluster 4 has relatively high perceptions of HATCO on all variables, whereas Cluster 1 is high only on the first set of variables.

The remaining clusters, 2 and 3, follow the pattern of having relatively high scores on the image variables and low scores elsewhere. They also differentiate themselves, however, as there is separation between them on the first set of variables, particularly X_1, X_2, and X_3. These findings reveal more in-depth profiles of types of consumers, and they can be used to develop marketing strategies directed toward each group's perceptions.

As a first validity check for stability of the cluster solution, a second nonhierarchical analysis was performed, this time allowing the procedure to randomly select the initial seed points for both cluster solutions. The results confirmed a consistency in the results for both cluster solutions. The cluster sizes were comparable for each solution, and the cluster profiles were very similar. Given the stability of the results between the specified seed points and random selection, management should feel confident that true differences do exist among customers in terms of their needs in a supplier and that the structure depicted in the cluster analysis is supported empirically.

To assess predictive validity, variables found in prior research—involving multiple regression and canonical correlation analyses—to have a relationship with clustering variables were assessed: X_9 (usage level) and X_{10} (satisfaction level). For the two-cluster solution, the univariate F ratios show that the cluster means for both variables are significantly different. The profiling process here shows that customers in Cluster 1, which rated HATCO higher on that first set of variables, had higher levels of usage and satisfaction with HATCO, as would be expected. Likewise, Cluster 2 customers had lower ratings on the first set of variables and lower ratings on the two additional variables. The four-cluster solution exhibits a similar pattern, with the clusters showing statistically significant differences on these two additional variables. Further predictive validity analyses involving additional variables not included in the clustering procedure [X_8 (firm size), X_{11} (specification buying), X_{12} (structure of procurement), X_{13} (industry type), and X_{14} (buying situation)] were undertaken and revealed that both the two-cluster and four-cluster solutions had distinctive profiles on this set of additional variables. Only X_{13} shows no difference across the two-cluster solution, with all other variables having significant differences for the two-cluster and four-cluster solutions. For example, Cluster 1 in the two-cluster solution was characterized as primarily small firms engaged in straight rebuys with HATCO that primarily use total-value analysis in a decentralized procurement system. Similar profiles could be developed for each cluster.

The cluster analysis of the 100 HATCO customers was successful in performing a market segmentation of HATCO customers. It not only created homogeneous groupings of customers based on their perceptions of HATCO but also found that these clusters met the tests of predictive validity and distinctiveness on additional sets of variables, all necessary for achieving practical significance. The segments represent different customer perspectives of HATCO, varying in both the types of variables that are viewed most positively and the magnitude of the perceptions.

One issue unresolved to this point is the selection of the final cluster solution between the two- and four-cluster solutions. Both provide viable market segments, as they are of substantial size and do not have any small cluster sizes caused by outliers. Moreover, they meet all of the criteria for a successful market segmentation. In either instance, the clusters (market segments) represent sets of consumers with homogeneous perceptions that can be uniquely identified, thus being prime candidates for differentiated marketing programs. The researcher can use the two-cluster solution to provide a basic delineation of customers that varying in perceptions and buying behavior or use the four-cluster solution for a more complex segmentation strategy that provides a highly differentiated mix of customer perceptions and well as targeting options.

Conclusion

Cluster analysis can be a useful data reduction technique. Because its application is more of an art than a science, however, it can easily be abused or misapplied by researchers. Different interobject measures and different algorithms can and do affect the results. The selection of the final cluster solution in most cases is based on both objective and subjective considerations. The prudent researcher, therefore, considers these issues and always assesses the effect of all decisions. If the researcher proceeds cautiously, however, cluster analysis can be an invaluable tool in identifying latent patterns suggesting useful groupings (clusters) of objects that are not discernible through other multivariate techniques.

Suggestions for Further Reading

Aldenderfer and Blashfield (1984) provided an excellent introduction to cluster analysis. For more in-depth coverage of the topic, Everitt's (1980) book is highly recommended.

Glossary

AGGLOMERATIVE METHODS *A hierarchical procedure* that begins with each object or observation in a separate cluster. In each subsequent step, the two *object* clusters that are most similar are combined to build a new aggregate cluster. The process is repeated until all objects are finally combined into a single cluster.

ALGORITHM A set of rules or procedures; similar to an equation.

AVERAGE LINKAGE The algorithm used in *agglomerative methods* that represents similarity as the average distance from all *objects* in one cluster to all objects in another. This approach tends to combine clusters with small variances.

CENTROID The average or mean value of the *objects* contained in the cluster on each variable, whether used in the *cluster variate* or in the validation process.

CENTROID METHOD The agglomerative *algorithm* in which similarity between clusters is measured as the distance between the two cluster *centroids.* When two clusters are combined, a new centroid is computed. Thus, cluster centroids migrate, or move, as the clusters are combined.

CITY-BLOCK APPROACH A method of calculating distances based on the sum of the absolute differences of the coordinates for the *objects.* This method assumes that the variables are uncorrelated and that unit scales are compatible.

CLUSTER CENTROID The average value of the *objects* contained in the cluster on all the variables in the *cluster variate.*

CLUSTER SEEDS The initial *centroids* or starting points for clusters. These values are selected to initiate *nonhierarchical procedures,* in which clusters are built around these prespecified points.

CLUSTER VARIATE A set of variables or characteristics representing the *objects* to be clustered and used to calculate the similarity between objects.

COMPLETE LINKAGE The agglomerative *algorithm* in which similarity is based on the maximum distance between *objects* in two clusters (the distance between the most dissimilar member of each cluster). At each stage of the agglomeration, the two clusters with the smallest maximum distance (most similar) are combined. Also referred to as *furthest-neighbor* or *diameter method.*

CRITERION VALIDITY The ability of clusters to show the expected differences on a variable not used to form the clusters. For example, if clusters were formed on performance ratings, standard marketing thought would suggest that clusters with higher performance ratings should also have higher satisfaction scores. If this were found to be so in empirical testing, then criterion validity is supported.

DENDROGRAM The graphical representation (tree graph) of the results of a *hierarchical procedure* in which each *object* is arrayed on the vertical axis, and the horizontal axis portrays the steps in the hierarchical procedure. Starting at the left side with each object represented as a separate cluster, the dendrogram shows graphically how the clusters are combined at each step of the procedure until all are contained in a single cluster.

DIVISIVE METHOD A clustering procedure, the opposite of the *agglomerative method,* that begins with all objects in a single, large cluster that is then divided at each step into two clusters that contain the most dissimilar objects.

ENTROPY GROUP A group of objects independent of any cluster (i.e., they do not fit into any cluster) that may be considered outliers and possibly eliminated from the cluster analysis.

EUCLIDEAN DISTANCE The most commonly used measure of the similarity between two *objects.* Essentially, it is a measure of the length of a straight line drawn between two objects.

HIERARCHICAL PROCEDURES The stepwise clustering procedures involving a combination (or division) of the *objects* into clusters. The two alternative procedures are the *agglomerative* and *divisive methods.* The result is the construction of a hierarchy or treelike structure (*dendrogram*) depicting the formation of the clusters, which produces $N - 1$ cluster solutions, where N is the number of objects. For example, if the agglomerative procedure starts with five objects in separate clusters, it shows how four clusters, then three, then two, and finally one cluster is formed.

INTEROBJECT SIMILARITY The correspondence or association of two *objects* based on the variables of the *cluster variate.* Similarity can be measured in two ways. First is a measure of association, such as higher positive correlation coefficients representing greater similarity. Second, "proximity" or "closeness" between each pair of objects can assess similarity, where measures of distance or difference are used, with smaller distances or differences representing greater similarity.

MAHALANOBIS DISTANCE A standardized form of *Euclidean distance.* Scaling responses in terms of standard deviations standardizes data, and adjustments are made for intercorrelations between the variables.

MULTICOLLINEARITY The extent to which a variable can be explained by the other variables in the analysis. As multicollinearity increases, it complicates the interpretation of the variate because it is more difficult to ascertain the effect of any single variable, owing to their interrelationships.

NONHIERARCHICAL PROCEDURES Procedures in which, instead of the treelike construction process found in the hierarchical procedures, *cluster seeds* are used to group objects within a prespecified distance of the seeds. The procedures produce only a single cluster solution for a set of cluster seeds. For example, if four cluster seeds are specified, only four clusters are formed. It does not produce results for all possible number of clusters as done with a *hierarchical procedure.*

NORMALIZED DISTANCE FUNCTION A process that converts each raw data score to a standardized variate with a mean of 0 and a standard deviation of 1, as a means to remove the bias introduced by differences in scales of several variables.

OBJECT A person, product–service, firm, or any other entity that can be evaluated on a number of attributes.

OUTLIER An observation that is substantially different from other observations (i.e., has an extreme value).

OPTIMIZING PROCEDURE A *nonhierarchical procedure* that allows for the reassignment of *objects* from the originally assigned cluster to another cluster on the basis of an overall optimizing criterion.

PARALLEL THRESHOLD METHOD A *nonhierarchical procedure* that selects the *cluster seeds* simultaneously in the beginning. *Objects* within the threshold distances are assigned to the nearest seed. Threshold distances can be adjusted to include fewer or more objects in the clusters. This method is the opposite of the *sequential threshold method.*

PREDICTIVE VALIDITY See *criterion validity.*

PROFILE DIAGRAM A graphical representation of data that aids in screening for *outliers* or the interpretation of the final cluster solution. Typically, the variables of the *cluster variate* or those used for validation are listed along the horizontal axis and the value scale on the vertical axis. Separate lines depict the scores (original or standardized) for individual *objects* or cluster *centroids* in a graphic plane.

RESPONSE-STYLE EFFECT A series of systematic responses by a respondent that reflect a "bias" or consistent pattern. Examples include always responding that an *object* is excellent or poor performing across all attributes with little or no variation.

ROW-CENTERING STANDARDIZATION See *within-case standardization.*

SEQUENTIAL THRESHOLD METHOD A *nonhierarchical procedure* that begins by selecting one *cluster seed.* All *objects* within a prespecified distance are then included in that cluster. Subsequent cluster seeds are selected until all objects are grouped in a cluster.

SINGLE LINKAGE A *hierarchical procedure* where similarity is defined as the minimum distance between any *object* in one cluster and any object in another. Simply, this means the distance between the closest objects in two clusters. This procedure has the potential for creating less compact, even chainlike clusters. This differs from the cluster *centroid method* that uses some measure of all objects in the cluster.

STOPPING RULE An *algorithm* for determining the final number of clusters to be formed. With no stopping rule inherent in cluster analysis, researchers have developed several criteria and guidelines for this problem. Two classes of rules exist that are applied post hoc and calculated by the researcher: (a) measures of similarity and (b) adapted statistical measures.

TAXONOMY An empirically derived classification of actual *objects* based on one or more characteristics. Typified by the application of cluster analysis or other grouping procedures. Can be contrasted to a *typology.*

TYPOLOGY A conceptually based classification of *objects* based on one or more characteristics. A typology does not usually attempt to group actual observations, but instead provides the theoretical foundation for the creation of a *taxonomy* that groups actual observations.

WARD'S METHOD A *hierarchical procedure* where the similarity used to join clusters is calculated as the sum of squares between the two clusters summed over all variables. It has the tendency to result in clusters of approximately equal size due to its minimization of within-group variation.

WITHIN-CASE STANDARDIZATION A method of standardization in which a respondent's responses are not compared with the overall sample but instead to their own responses. Also know as *ipsitizing,* each respondent's average response is used to standardize his or her own responses.

References

Aldenderfer, M. S., & Blashfield, R. K. (1984). *Cluster analysis.* Thousand Oaks, CA: Sage.

Anderberg, M. (1973). *Cluster analysis for applications.* New York: Academic Press.

Bailey, K. D. (1994). *Typologies and taxonomies: An introduction to classification techniques.* Thousand Oaks, CA: Sage.

Campbell, R., & Johnson, C. R. (1997). Police officer's perceptions of rape: Is there consistency between state law and individual beliefs? *Journal of Interpersonal Violence, 12*(2), 255–274.

Everitt, B. (1980). *Cluster analysis* (2nd ed.). New York: Halsted Press.

Green, P. E. (1978). *Analyzing multivariate data.* Hinsdale, IL: Holt, Rinehart & Winston.

Green, P. E., & Carroll, J. D. (1978). *Mathematical tools for applied multivariate analysis.* New York: Academic Press.

Hair, J. F., Jr., Anderson, R. E., Tatham, R. L., & Black, W. C. (1998). *Multivariate data analysis.* Upper Saddle River, NJ: Prentice-Hall.

McIntyre, R. M., & Blashfield, R. K. (1980). A nearest-centroid technique for evaluating the minimum-variance clustering procedure. *Multivariate Behavioral Research, 15,* 225–238.

Milligan, G. (1980). An examination of the effect of six types of error perturbation on fifteen clustering algorithms. *Psychometrika, 45,* 325–342.

Milligan, G. W., & Cooper, M. C. (1985). An examination of procedures for determining the number of clusters in a data set. *Psychometrika, 50*(2), 159–179.

Mowbray, C. T., Bybee, D., & Cohen, E. (1993). Describing the homeless mentally ill: Cluster analysis results. *American Journal of Community Psychology, 21*(2), 67–93.
Overall, J. (1964). Note on multivariate methods for profile analysis. *Psychological Bulletin, 61,* 195–198.
Punj, G., & Stewart, D. (1983). Cluster analysis in marketing research: Review and suggestions for application. *Journal of Marketing Research, 20,* 134–148.
Schaninger, C. M., & Bass, W. C. (1986). Removing response-style effects in attribute-determinance ratings to identify market segments. *Journal of Business Research, 14,* 237–252.
Shephard, R. (1966). Metric structures in ordinal data. *Journal of Mathematical Psychology, 3,* 287–315.
Singh, J. (1990). A typology of consumer dissatisfaction response styles. *Journal of Retailing, 66,* 57–99.
Sneath, P. H. A., & Sokal, R. R. (1973). *Numerical taxonomy.* San Francisco: Freeman Press.

Q-Technique Factor Analysis: One Variation on the Two-Mode Factor Analysis of Variables

Bruce Thompson

Factor analysis has been conceptually available to researchers since the turn of the 20th century (Spearman, 1904), but as a practical matter it has been widely used only with the more recent availability of both modern computers and user-friendly statistical software packages. Factor analysis examines patterns of relationships among factored entities (often variables) across replicates of the relationship patterns (usually people), with a view toward creating clusters or factors of the factored entities.

In this chapter, I explain how a particular kind of factor analysis, namely, *Q-technique factor analysis*, can be used to identify types or clusters of people with similar views. Q-technique factor analysis can be implemented with commonly available statistical software (e.g., SPSS) and addresses three questions:

1. How many types (factors) of people are there?
2. Are the expected people most associated with the expected person factors?
3. Which variables were and were not useful in differentiating the various person types or factors?

The Q-technique methods described here are well suited for studying person types, which is certainly a common endeavor within the field of psychology. Too frequently, psychologists inappropriately use factor analyses of variables in attempts to understand or identify person types

This chapter is a revised version of a paper presented at the annual Conference on Research Innovations in Early Intervention, Charleston, SC, May 2, 1998.

or factors. If one wishes to talk about types of people, the focus of one's analyses must be person factors.

For example, one often hears psychologists talk about "Type A personalities," "workaholics," and "introverts." Along these lines, Q-technique factor analysis is useful for (a) exploring data to identify new person types and thus developing typological theories or (b) collecting data to confirm or deny existing theories about person types. R technique is not directly useful for this purpose, even though many researchers incorrectly use R-technique methods to investigate questions about person types.

Excellent in-depth treatments of Q-technique factor analysis are available from Stephenson (1953), Kerlinger (1986, chap. 32), and Gorsuch (1983). Carr (1992) also provided an excellent short treatment. Campbell's (1996) treatment is more comprehensive and is equally readable.

Various Matrices of Association and Factor Techniques

All factor analyses, regardless of what entities are being factored (e.g., variables, people, occasions), begin with the calculation of some statistical measure of association of the factored entities with each other. These relationship statistics provide the basis for the actual clustering or factoring process. Numerous indices of association are available, including the variance–covariance matrix (Thompson & Borrello, 1987a), but many analysts use a matrix of bivariate correlation coefficients for this purpose.

The choice of an association matrix to analyze is important because the characteristics of the factors themselves are influenced, in turn, by the characteristics of the matrix of associations among the factored entities from which the factors are derived. A second important feature of the matrix of associations involves the organization of the data matrix on which the association matrix is based, because this organization determines what is being clustered or factor analyzed (e.g., variables or people).

Factorable Matrices of Association

Of course, even within the family of correlation coefficients, many choices are available, including Pearson's *r* and Spearman's rho. Dolenz-

Walsh (1996) provided a comprehensive and readable explanation of the data features that do and do not influence various correlation coefficients. A brief review may be helpful, however, to illustrate that different indices of association are sensitive to different data features.

The Pearson product–moment correlation coefficient asks two questions regarding data that are organized in a conventional manner (i.e., the rows of the data matrix correspond to people, whereas the columns delineate variables): (a) "do the two variables order the people in the same order?" and (b) "do the two variables have the same shape?" (i.e., skewness and kurtosis). Spearman's rho, however, asks only "do the two variables order the people in the same order?"

For example, consider the scores of the following four hypothetical people on two hypothetical variables, *X* and *Y*:

X	*Y*	Person
1	1	James Bowana
2	2	Becca Broker
3	3	Patricia Farragamo
4	4	Marcia Schumaker

Because the answers to both of the questions "do the two variables order the people in the same order?" and "do the two variables have the same shape?" are *yes*, the Pearson *r* of *X* and *Y* is +1. By the same token, because Spearman's rho only addresses the second question, but all four people are ordered exactly the same by both variables, for these data Spearman's rho also equals +1.

Additive constants do not affect statistics that measure distribution shape, such as the coefficients of skewness or of kurtosis (cf. Bump, 1991). Scores can be created on a new variable, *Y**, by adding 1 to every *Y* score from before, thus yielding

X	*Y**	Person
1	2	James Bowana
2	3	Becca Broker
3	4	Patricia Farragamo
4	5	Marcia Schumaker

Because both Pearson *r* questions are still answered with *yes*, *r* still equals +1. Because Spearman's rho addresses only the second question, which is answered *yes*, rho also still equals +1.

Multiplicative constants also do not affect coefficients of skewness or of kurtosis (cf. Bump, 1991). Scores can be created on a new variable, Y', by multiplying every Y score from earlier by 2, thus yielding

X	Y'	Person
1	2	James Bowana
2	4	Becca Broker
3	6	Patricia Farragamo
4	8	Marcia Schumaker

Because both Pearson r questions are still answered with *yes*, r still equals +1. Because Spearman's rho addresses only the second question, which is answered *yes*, rho also still equals +1.

The following data were created, however, without invoking either additive or multiplicative constants:

X	$Y\#$	Person
1	1	James Bowana
2	2	Becca Broker
3	3	Patricia Farragamo
4	999	Marcia Schumaker

Because both variables order the people constituting the rows in exactly the same order, Spearman's rho for these data still equals +1. Notwithstanding this fact, because the shapes of the two distributions are now unequal (i.e., the X data are symmetrical [coefficient of skewness = 0], whereas the $Y\#$ data are highly positively skewed [coefficient of skewness much greater than 0]), the Pearson r for these data does not equal +1.

Data Organization and the Six Two-Mode Techniques

Typically in factor analysis, the matrix of associations is computed from a two-dimensional raw data matrix (e.g., the rows represent scores of given people, while the columns represent the variables being measured). Analyses based on raw data matrices organized in this manner (i.e., matrices defined by two basic dimensions) are termed *two-mode factor analyses* (Gorsuch, 1983, chap. 15). Three-mode factor analyses that simultaneously consider variations across participants, variables, and occasions are mathematically possible (cf. Tucker, 1967) but are too complex and too infrequently reported to be considered here.

Although the most common two-mode analyses are based on data matrices with people defining rows and variables defining columns, there are a number of other two-mode analyses available to the researcher. Cattell (1966) conceptualized these possibilities as involving any combination of two dimensions (thus constituting a surface) from a "data box" defined by three dimensions: (a) variables, (b) participants (often people), and (c) occasions of measurement.

Table 6.1 presents the 6 two-mode "techniques" conceptualized and labeled by Cattell (1966) as well as illustrative applications of several of those techniques. Although all six techniques are available to researchers, R technique (cf. Thompson & Borrello, 1992b) and Q technique (cf. Thompson & Miller, 1984), respectively, are the most commonly used factor analysis techniques in contemporary practice.

Because R technique and Q technique both involve the same two elements (i.e., variables and people) and occasion is held constant, both R and Q are from the same surface of Cattell's data box. The raw data, however, are organized differently to differentiate these two factor analytic methods.

It is the organization of the raw data matrix that distinguishes the six techniques and not the mathematics of the factor-analytic process. For example, if people define the rows of the raw data matrix, with variables defining the columns, the analysis is R technique, regardless of how the factors are extracted (e.g., principal components, principal axis factoring, canonical factoring).

Thus, if the first person, Jennifer Loquacious, had scores of 1, 2,

Table 6.1

Six Variations of Two-Mode Factor Analysis

Technique label	Columns defining entities to be factored	Rows defining the patterns of associations	Example application
R	Variables	Participants	Thompson & Borrello (1987b)
Q	Participants	Variables	Thompson (1980b)
O	Occasions	Variables	Jones et al. (1980)
P	Variables	Occasions	Cattell (1953)
T	Occasions	Participants	Frankiewicz & Thompson (1979)
S	Participants	Occasions	[a]

[a] I am not aware of any study that uses S-technique factor analysis.

3, 4, and 5 on five variables, respectively, the first row of the raw data matrix for an R-technique factor analysis would present scores of 1, 2, 3, 4, and 5 in a horizontal fashion, and the raw data matrix would have n rows of data, each with $v = 5$ columns. So the raw data matrix would be an $n \times v$ matrix.

The same data could be transposed such that rows became columns and columns became rows. In this case, the first column of the raw data matrix would now present the scores of 1, 2, 3, 4, and 5, respectively. There would then be $v = 5$ rows in the raw data matrix, and the matrix would have n columns. Figure 6.1 illustrates the transposition process.

Obviously, any raw data matrix can be transposed in the manner illustrated in Figure 6.1. This may (albeit incorrectly) suggest that any single raw data matrix could be subjected to both R-technique (to factor variables) and Q-technique (to factor people) analyses.

The problem is that whichever technique is applied, it is generally preferable to have the number of row relationship–association pattern replications several times larger than the number of the column entities that are being factored. Thus, R technique should have several times more participants than factored variables, and Q technique should have several times more variables than factored people. This is to allow the patterns of relationships among the factored entities to be replicated over a number of rows in the raw data matrix, so that the stability of the estimated relationships can be ensured and, therefore, that the factors extracted from the matrix of associations are themselves also stable.

Figure 6.1

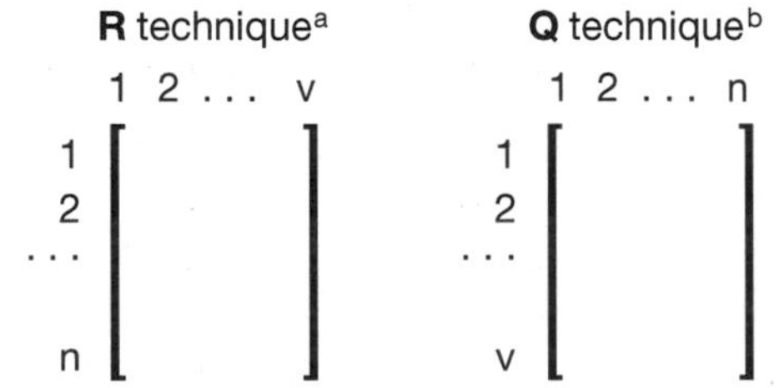

R technique[a]

$$\begin{array}{c} \\ 1 \\ 2 \\ \cdots \\ n \end{array}\begin{array}{c} 1\ 2\ \ldots\ v \\ \left[\begin{array}{c} \\ \\ \\ \\ \end{array}\right] \end{array}$$

Q technique[b]

$$\begin{array}{c} \\ 1 \\ 2 \\ \cdots \\ v \end{array}\begin{array}{c} 1\ 2\ \ldots\ n \\ \left[\begin{array}{c} \\ \\ \\ \\ \end{array}\right] \end{array}$$

Two two-mode (**R** and **Q**) variations from one of the three faces of Cattell's (1966) data box.
[a]In the raw data matrix used for **R**-technique factor analysis, each of the *n* persons defines a row of data, and each set of scores of the people on each of the *v* variables defines a column of data. [b]In the raw data matrix used for **Q**-technique factor analysis, each set of scores of the people on each of the *v* variables defines a row of data, and each of the *n* persons defines a column of data.

Q-Technique Factor Analysis Questions and Issues

Q-technique factor analysis is well suited to the relatively intensive study of a relatively small number of people. Q-technique isolates types (or prototypes) of people. In fact, researchers are often more interested in types of people than they are in clusters of variables, as emphasized previously.

Also as noted previously, Q-technique factor analysis can be used to address three primary questions:

1. How many types (factors) of people are there?
2. Which people are most associated with each type (e.g., are the expected people most associated with the expected person factors)?
3. Which variables were and were not useful in differentiating the various person types or factors?

Three sorts of methodological issues must be resolved in any Q-technique study to address these questions. First, who should be factored? Second, which variables should be measured to help define the person factors? Third, what response format should be used for data collection (i.e., a Q-technique study may or may not use a conventional *Q-sort* task)?

People

Q-technique factor analysis directly tests typological premises. As Kerlinger (1986) explained, in Q, "one tests theories on small sets of individuals carefully chosen for their 'known' or presumed possession of some significant characteristic or characteristics. One explores unknown and unfamiliar areas and variables for their identity, their interrelations, and their functioning" (p. 521). Thus, the people who are factored in Q-technique analysis must be carefully selected. The selection is all the more important because the Q-technique researcher has inherently elected to study (intensively) a small group of people (because even the most diligent participants cannot be expected to respond to more than 100 to 150 variables, and the number of factored people is limited by the number of variables, as noted previously).

Generally, distinct groups of people are factored. For example, Thompson and Miller (1984) sampled both school district administrators and program evaluators to determine whether job classification was associated with person types as regards perceptions of program evaluation. Similarly, Gillaspy, Campbell, and Thompson (1996) sampled two kinds of counselors to compare the therapist person factors defined by different perceptions of what love is.

Variables

The variables in a Q-technique analysis can be of many kinds, for example, statements responded to with respect to degree of agreement or disagreement or photographs responded to with regard to the portrayed physical attractiveness. There are two major choices regarding the selection of variables. One choice (e.g., Thompson, 1980b) is to use variables that are themselves implicitly *structured* (Kerlinger, 1986). For example, if the participants responded to the 42 items on the Love Attitudes Scale (Hendrick & Hendrick, 1990), the responses would be structured, because the scale includes seven items measuring each of the six types of love posited by Lee (1973/1976).

Alternatively, if the variables are presumed to be representative of a single population of items or variables, then the study would be considered *unstructured*. For example, if the participants responded to the 55 items on the Love Relationships Scale (Thompson & Borrello, 1987b), then the responses would be presumed to be unstructured, because the scale was developed inductively without premises regarding an underlying structure (Thompson & Borrello, 1992a).

Response Format

Quasinormal Q Sort

Although many response formats are candidates for the measurement protocols used to collect Q-study data (Daniel, 1989), most researchers use a Q-sort (Kerlinger, 1986, chap. 32) protocol in Q-technique studies. Q sorts require each participant to put stimuli (e.g., cards each listing a statement) into a predetermined number of categories, with exactly a predetermined number of items being placed in each category. Commonly, the predetermined numbers of stimuli that go into each category are created so as to yield a normal or a quasinormal, symmetrical distribution of scores. Kerlinger (1986, p. 509) provided an illustrative example for a Q sort involving 90 statements sorted as follows:

n items	3	4	7	10	13	16	13	10	7	4	3
Category	10	9	8	7	6	5	4	3	2	1	0

This response format yields data that are considered *ipsative* (Cattell, 1944), because the protocol invokes a forced-choice response format in which responses to one item inherently constrain the possible choices for subsequent items. Although ipsative data are not suitable for use in R-technique factor analysis (Thompson, Levitov, & Miederhoff, 1982), ipsative data are useful in studying commonalities in intraindividual differences, as in Q-technique factor analysis.

The Q-sort protocol is appealing because the protocol yields data for each participant that are exactly equally distributed (i.e., data for each participant that are symmetrical and have the same skewness and kurtosis). As Glass and Hopkins (1984) noted, "r can equal 1.0 only when the marginal distributions of X and Y have precisely the same shape" (p. 91). Thus, having data with exactly the same distributional shapes is appealing because when the participants are correlated, none of the person correlation coefficients is attenuated by differences in score distribution shapes, even if one is computing a matrix of Pearson r coefficients as the basis for the Q-technique factor analysis.

As an aside, it is interesting to note that researchers using Q-technique factor analysis are usually extremely careful to make sure that person scores have exactly the same distribution shapes, so that interperson relationship statistics are not attenuated at all by distributional differences. People using R-technique factor analysis, however, often do not pay much attention to these influences as regards intervariable relationships. These differences in conventional practices across the two techniques are striking, because the mathematics of the factor extraction process are identical in both cases. Apparently, either Q-technique researchers are being a bit too obsessive, or R-technique users are being a bit too glib.

Mediated Q Sort

The Q sort is appealing because the protocol allows participants to provide data regarding a lot of variables without being cognitively overwhelmed. For example, it is not reasonable to ask participants to rank order more than 15–20 variables with no ties. The task of rank ordering 90 items would irritate and confuse even the most patient and brilliant participant.

However, Thompson (1980a) proposed a *two-stage* measurement protocol that does yield data that are rank ordered with no ties. First, participants complete a conventional Q-sort protocol. Second, participants are asked to rank order the statements within each of the Q-sort categories. Without being cognitively overwhelming, this strategy yields more variance in responses and so theoretically should allow the isolation of more stable factors of participants.

Unnumbered Graphic Scale

Normative measurement (Cattell, 1944) allows participants to rate (as against rank) data; the response to one item does not in any way mechanically constrain participants' responses to other items. Likert scales are an example of normative measurement. The only constraints are self-imposed psychological (nonmechanical) constraints in the event that participants elect to respond consistently to items containing roughly the same content.

What drives the reliability of scores is having greater variance in the data (Reinhardt, 1996; Thompson & Vacha-Haase, 2000). Traditionally, there was considerable debate about whether it might be desirable in attitude measurement to use a 1–7 Likert scale, as against a 1–5 scale, whether a 1–9 scale might be more preferable still, and so forth. Certainly, more response alternatives allow participants to provide more variable responses, if they wish to do so. As Nunnally (1967) explained, "it is true that, as the number of scale points increases, the error variance increases, but at the same time, the true-score variance increases at an even more rapid rate" (p. 521). Thus, Guilford (1954) suggested that "it may pay in some favorable situations to use up to 25 scale divisions" (p. 291).

Yet as Thompson (1981) noted, "use of a large number of scale steps . . . becomes undesirable when participants become confused or irritated at being confronted with a cognitively overwhelming number of response alternatives" (p. 5). Confused or irritated participants may not pay as much attention to rating tasks and may therefore provide less reliable data.

Thompson (1981), however, also described a response format that may reduce cognitive pressure on participants while still yielding normative data that are highly variable. This format has been labeled an *unnumbered graphic scale* (cf. Freyd, 1923). Participants are presented with a straight line drawn between two antonyms (e.g., "disagree" and "agree") and are asked to draw a mark through the line at the position that best indicates the extent of their agreement with a given statement. These marks are subsequently scored by the researcher using an equal-interval measurement scale with a relatively large number of categories (e.g., 1–15). This protocol puts a limited cognitive burden on participants but can still yield more variable scores.

Of course, using normative data means that the bivariate correlation coefficients analyzed in a Q-technique factor analysis will inherently be attenuated by variations in the distribution shapes of scores for different individuals and that these differences affect the identification of the factors extracted from the correlations. The assumption that distributions of scores are the same across people is perfectly met with both Q-sort and *mediated Q-sort* measurement protocols, as noted previously, but is not perfectly met with normative data. It is conceivable, however, that tolerating some deviations in distribution shapes does not devastate the factor-analytic solution and may be worthwhile if eliminating the requirement of forced choices yields more accurate reflections of participants' feelings or beliefs.

An Illustrative Example

Table 6.2 presents a hypothetical heuristic data set that is used to illustrate the process of addressing the three questions typically posed in a Q-technique factor analysis. The example involves family-focused residential psychological interventions for adolescents. Presume that a psychologist has decided to explore parents' and adolescents' perceptions of these interventions, as regards the three questions that may be posed in a Q-technique factor analysis.

The data set involves three parents and two adolescent clients. The five participants hypothetically ranked 15 variables describing early intervention programs. Here I forego describing the important issues of exactly how the participants and the variables were selected, except to posit that the data were collected in a mediated Q sort. It should also be noted that the typical Q study would involve both more people and more variables (although there should always be at least twice as many variables as people, so that the person factors are reasonably stable, as emphasized previously).

In the data set, the five participants hypothetically ranked the 15 intervention features from 1 (*most important*) to 15 (*least important*). The scaling of responses subsequently becomes very important to the interpretation of the results; that is, the scaling direction is arbitrary, but

Table 6.2

Illustrative Data for a Q-Technique Intervention Study

Parents			Clients			
1	2	3	1	2	Intervention	Feature
1	2	1	13	13	1	Emphasis on long-term intervention goals
2	1	4	7	7	2	Consideration of all family members' needs re intervention design
3	9	6	5	3	3	Involvement of family members in providing assessment information
4	3	5	6	6	4	Participation of other family members in intervention
5	7	7	2	1	5	Consideration of social cohesion within intervention setting
6	5	3	12	12	6	Access of intervention re scheduling
7	6	8	1	2	7	Emphasis on social skills as intervention goal
8	4	2	10	10	8	Quantity of intervention personnel per child
9	8	9	15	15	9	Access of intervention re geographic location
10	11	10	4	14	10	Use of technology in intervention
11	10	11	14	4	11	Formal education of intervention personnel
12	12	12	8	8	12	Having a manageable number of intervention goals
13	13	15	11	11	13	Attractiveness of intervention setting
14	15	14	9	9	14	Costs of interventions
15	14	13	3	5	15	Practicality and logistics of interventions

whatever scaling direction is selected must be considered as part of the interpretation process.

Because the participants all ranked 15 items with no ties, the distribution characteristics are identical. For example, here the five person means are all 8.0, the five person standard deviations are all 4.47, and all five coefficients of skewness are 0. In a Q-technique study using either Q-sort or mediated-ranking data-collection methods, correct data entry can be confirmed partly by examining the person means (here all 8.0) and person standard deviations (here all 4.47) to ensure that they are exactly equal across persons.

Table 6.3 presents the correlation coefficients of the persons with each other. A quick perusal of this small 5 × 5 matrix would suggest that the data delineate two discrete person factors: one involving the three parents and one involving the two adolescent clients. Of course, with real data the correlation matrix would be larger and the coefficients would be less homogeneous within groups and more heterogeneous across groups, so the person factors would be less obvious from mere inspection of the bivariate correlation matrix. This is why in real practice one estimates the person factors empirically rather than through mere subjective examination of a large and ambiguous correlation matrix.

Table 6.4 presents the two person factors extracted from the Table 6.3 interperson correlation matrix. These results address the first research question, "How many types (factors) of people are there?" Here the answer is two.

Because the factors were rotated orthogonally and are therefore uncorrelated (see Gorsuch, 1983; or for a shorter but still thorough treatment, Hetzel, 1996), these coefficients are termed *pattern/structure*

Table 6.3

***N* × *N* Correlation Matrix for Table 6.2's Q-Technique Data**

	Person				
Person	PARENT1	PARENT2	PARENT3	CLIENT1	CLIENT2
PARENT1	1.000				
PARENT2	0.882	1.000			
PARENT3	0.871	0.921	1.000		
CLIENT1	0.050	−0.086	−0.093	1.000	
CLIENT2	0.107	−0.018	−0.075	0.625	1.000

Table 6.4

***N* × *N* Varimax-Rotated Pattern/Structure Coefficient Matrix**

	Factor	
Person	I	II
PARENT1	.954	.106
PARENT2	.970	−.050
PARENT3	.965	−.088
CLIENT1	−.038	.899
CLIENT2	.018	.902

coefficients (Thompson & Daniel, 1996). Each coefficient represents the correlation of a given person with a given person factor. For example, as reported in Table 6.5, PARENT2 is most highly correlated with (most prototypic of) person Factor I (r_S = .970).

Thus, the Table 6.5 results also address the second research question, "Which people are most associated with each person factor?" The three parents define the first person factor, while the two adolescent clients define the second person factor.

A very helpful graphic representation of the factors can also be developed by plotting the factor pattern/structure coefficients. This is illustrated for these data in Figure 6.2. Such visual representations of factor-analytic results can be very useful in synthesizing and communicating results in a nonnumeric fashion.

The third Q-technique research question is "Which variables were and were not useful in differentiating the various person types or factors?" This question is addressed by consulting the factor scores computed as part of the analysis. Each variable has a factor score on each person factor. These factor scores are in *z-score form* (i.e., have a mean of 0 and a standard deviation and variance of 1.0).

The factor scores can be conceptualized as a prototypic ranking of the variables with regard to the persons defining a given person factor. For example, Table 6.5 presents the sorted factor scores on person Factor I.

It often is useful to invoke cutscores to interpret Q-technique factor scores, and often $|1.0|$ is used for this purpose. Thus, for the Table 6.5 results, the five variables most useful for defining person Factor I were Variables 1 (−1.533), 2 (−1.318), 13 (1.323), 15 (1.385), and 14 (1.473).

Table 6.5

Sorted Factor Scores on Factor I From the $V \times F$ Factor Score Matrix

Item	Factor score	Program	Feature
1	−1.533	1	Emphasis on long-term intervention goals
2	−1.318	2	Consideration of all family members' needs re intervention
3	−0.934	4	Participation of other family members in intervention
4	−0.770	8	Quantity of intervention personnel per child
5	−0.763	6	Access of intervention re scheduling
6	−0.477	3	Involvement of family members in providing assessment information
7	−0.406	5	Consideration of social cohesion within intervention setting
8	−0.248	7	Emphasis on social skills as intervention goal
9	0.174	9	Access of intervention re geographic location
10	0.568	10	Use of technology in intervention
11	0.598	11	Formal education of intervention personnel
12	0.928	12	Having a manageable number of intervention goals
13	1.323	13	Attractiveness of intervention setting
14	1.385	15	Practicality and logistics of interventions
15	1.473	14	Costs of interventions

Because the *smallest number* (1) in the Table 6.2 ranking of the 15 program features reflected the most important feature, this means that the three parents most associated with person Factor I found most important the program features with the smallest factor scores:

- emphasis on long-term intervention goals (−1.533)
- consideration of all family members' needs as regards to intervention (−1.318).

The three parents most associated with person Factor I found least important the program features with the largest factor scores:

- attractiveness of the intervention setting (1.323)
- practicality-logistics of interventions (1.385)
- costs of interventions (1.473).

The same process would be used with the factor scores from person

Figure 6.2

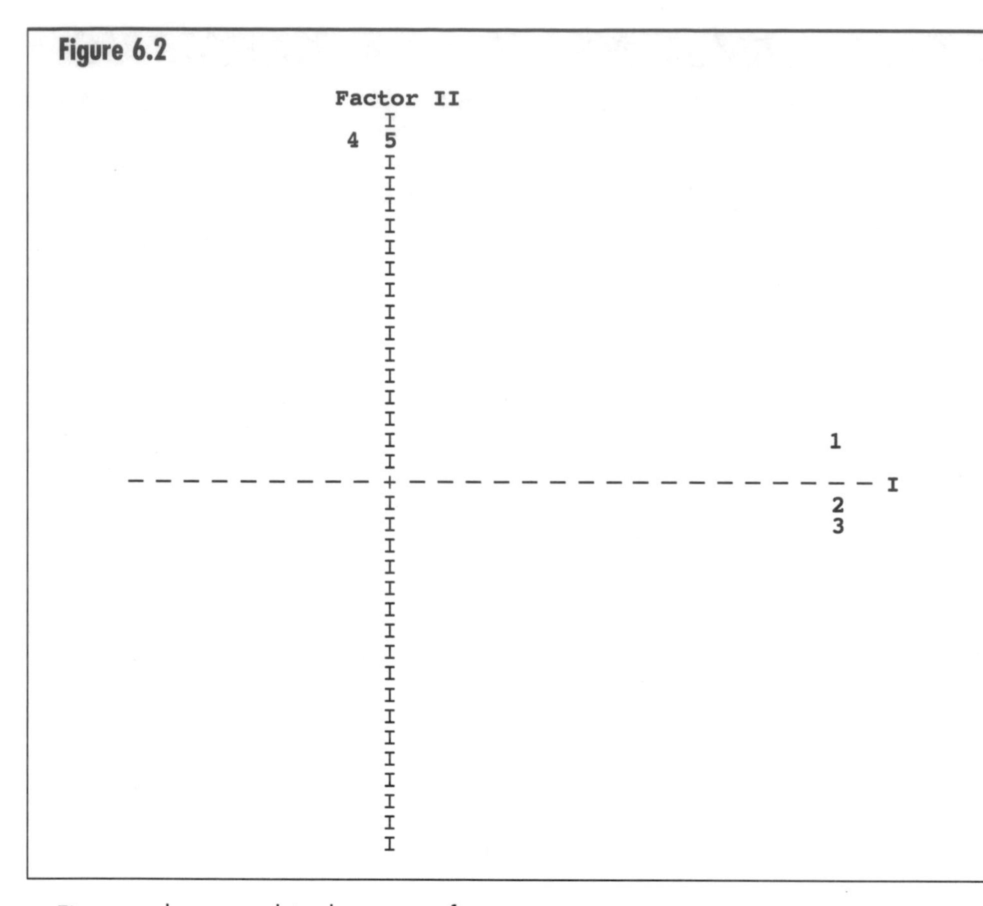

Five people arrayed in the person factor space. 1 = PARENT1, 2 = PARENT2, 3 = PARENT3, 4 = CLIENT1, 5 = CLIENT2.

Factor II to determine what variables were most salient to the persons delineating this factor. Another analysis can be conducted to isolate those variables that were most salient or relevant across the set of person factors and which items were least salient across the set of person factors. This can be done by computing the average absolute value of the factor scores of a given variable across all the factors. For example, the program feature that was most salient (highest average absolute value) across both person factors was the item called emphasis on long-term intervention goals ($[|-1.533| + |1.202|]/2 = 1.367$). The item most irrelevant to defining the two person factors (nearest to zero average absolute value) was the program feature called use of technology in intervention (.408). For further analysis of factor scores within Q-technique factor analysis, see Thompson (1980b).

Discussion

Q-technique factor analysis can be useful in psychological research because the method addresses questions about person types, and psychologists are often more interested in people than in variables. Too many researchers use R-technique factoring of variables in a vain attempt to address questions about types of people. The heuristic example illustrated the potential application of Q-technique methods to address three research questions.

In typical applications, Q technique uses a response format that is ipsative, as previously noted. This analytic model is most useful in studying aspects of reality that are themselves inherently forced choice. For example, there may be myriad personal values or personality traits that are desirable, but practical realities constrain our pursuit of all possible values or the full development of all possible personality traits, and so choices must be made. The Q-technique methods described here are well suited to studying phenomena in which there are numerous ideals present in a reality in which only a limited number of ends or means can be realistically pursued.

Suggestions for Further Reading

Regarding the association statistics (e.g., correlation coefficients) from which person factors are extracted, Dolenz-Walsh (1996) provided a comprehensive and readable explanation of the data features that do and do not influence various coefficients. Hetzel (1996) also provided a short but thorough treatment of factor analysis (also see Bryant & Yarnold, 1995). For excellent in-depth treatments of Q-technique factor analysis, see Stephenson's (1953) classic book and chapters within Kerlinger (1986, chap. 32) and Gorsuch (1983) as well as Carr's (1992) excellent short treatment. Or for a more comprehensive discussion, see Campbell's (1996) very readable treatment.

Glossary

FACTOR A set of weights (i.e., factor pattern coefficients) that can be applied to the measured variables to derive scores (i.e., factor scores) on latent or synthetic variables; when the factors are uncorrelated, a factor pattern/structure coefficient represents the correlation between the factored entity (e.g., variable, person) and the factor.

IPSATIVE Data collected in such a manner that responses on one item constrain responses on other items (e.g., ranking an item as first constrains any other items from being ranked first).

MEDIATED Q SORT A two-stage *Q-sort* strategy that yields ranked data with no ties in a manner that makes the ranking task cognitively manageable; participants first perform a conventional *Q sort* and then are asked to rank order the stimuli within each of the *Q-sort* categories.

NORMATIVE Data collected in such a manner that responses on one item do not physically constrain responses on other items (e.g., rating an item a 5 on a Likert scale does not constrain rating responses to any other items, even though participants may feel psychologically compelled to respond in a logically consistent manner on similar items).

Q SORT An *ipsative* data-collection procedure in which participants sort stimuli (e.g., attitude items, pictures) into various categories, under a restriction that a predetermined number of stimuli must be placed in each category, usually to create a quasinormal distribution; a Q sort is a data-collection procedure and should not be equated with *Q-technique factor analysis*, which is an analytic method.

Q-TECHNIQUE FACTOR ANALYSIS One of the 6 two-mode factor analysis techniques identified in Cattell's (1966) "data box"; this two-mode technique identifies types of people (i.e., person factors) on the basis of their varying perceptions of variables.

STRUCTURED Q SORT A *Q sort* that presents a given number of stimuli from each of several underlying dimensions (e.g., seven items from each of six hypothesized types of love; Lee, 1973/1976).

UNNUMBERED GRAPHIC SCALE A data-collection procedure in which participants are presented with an unnumbered straight line drawn between two antonyms (e.g., "disagree" and "agree") and are asked to draw a mark through the line at the position that best indicates their perceptions regarding a given statement; these marks are subsequently scored by the researcher using an equal-interval measurement scale with a relatively large number of categories (e.g., 1–15).

UNSTRUCTURED Q SORT A *Q sort* that presents stimuli without sampling a given number of such stimuli from each of several underlying dimensions.

References

Bryant, F. B., & Yarnold, P. R. (1995). Principal-components analysis and exploratory and confirmatory factor analysis. In L. G. Grimm & P. R. Yarnold (Eds.), *Reading and understanding multivariate statistics* (pp. 99–136). Washington, DC: American Psychological Association.

Bump, W. (1991, January). *The normal curve takes many forms: A review of skewness and kurtosis.* Paper presented at the annual meeting of the Southwest Educational Research Association, San Antonio, TX. (ERIC Document Reproduction Service No. ED 342 790)

Campbell, T. (1996). Investigating structures underlying relationships when variables are not the focus: Q-technique and other techniques. In B. Thompson (Ed.), *Advances in social science methodology* (Vol. 4, pp. 207–218). Greenwich, CT: JAI Press.

Carr, S. (1992). A primer on Q-technique factor analysis. *Measurement and Evaluation in Counseling and Development, 25,* 133–138.

Cattell, R. B. (1944). Psychological measurement: Normative, ipsative, interactive. *Psychological Review, 51,* 292–303.

Cattell, R. B. (1953). A quantitative analysis of the changes in the culture pattern of Great Britain, 1837–1937, by P-technique. *Acta Psychologia, 9,* 99–121.

Cattell, R. B. (1966). The data box: Its ordering of total resources in terms of possible relational systems. In R. B. Cattell (Ed.), *Handbook of multivariate experimental psychology* (pp. 67–128). Chicago, IL: Rand McNally.

Daniel, L. G. (1989, November). *Stability of Q-factors across two data collection methods.* Paper presented at the annual meeting of the Mid-South Educational Research Association, Little Rock, AR. (ERIC Document Reproduction Service No. ED 314 438)

Dolenz-Walsh, B. (1996). Factors that attenuate the correlation coefficient and its analogs. In B. Thompson (Ed.), *Advances in social science methodology* (Vol. 4, pp. 21–32). Greenwich, CT: JAI Press.

Frankiewicz, R. G., & Thompson, B. (1979, April). *A comparison of strategies for measuring teacher brinkmanship behavior.* Paper presented at the annual meeting of the American Educational Research Association, San Francisco, CA. (ERIC Document Reproduction Service No. ED 171 753)

Freyd, M. (1923). The Graphic Rating Scale. *Journal of Educational Psychology, 14,* 83–102.

Gillaspy, A., Campbell, T., & Thompson, B. (1996, January). *Types of lovers: A Q-technique factor analysis of counselor person-factors defined by counselors' love attitudes.* Paper presented at the annual meeting of the Southwest Educational Research Association, New Orleans, LA.

Glass, G. V, & Hopkins, K. D. (1984). *Statistical methods in education and psychology* (2nd ed.). Englewood Cliffs, NJ: Prentice-Hall.

Gorsuch, R. L. (1983). *Factor analysis* (2nd ed.). Hillsdale, NJ: Erlbaum.

Guilford, J. P. (1954). *Psychometric methods.* New York: McGraw-Hill.

Hendrick, C., & Hendrick, S. S. (1990). A relationship-specific version of the Love Attitudes Scale. *Journal of Social Behavior and Personality, 5,* 239–254.

Hetzel, R. D. (1996). A primer on factor analysis with comments on analysis and interpretation patterns. In B. Thompson (Ed.), *Advances in social science methodology* (Vol. 4, pp. 175–206). Greenwich, CT: JAI Press.

Jones, H. L., Thompson, B., & Miller, A. H. (1980). How teachers perceive similarities and differences among various teaching models. *Journal of Research in Science Teaching, 17,* 321–326.

Kerlinger, F. N. (1986). *Foundations of behavioral research* (3rd ed.). New York: Holt, Rinehart & Winston.

Lee, J. A. (1976). *The colors of love: An exploration of the ways of loving.* Don Mills, Ontario, Canada: New Press. (Original published 1973)

Nunnally, J. C. (1967). *Psychometric theory.* New York: McGraw-Hill.

Reinhardt, B. (1996). Factors affecting coefficient alpha: A mini Monte Carlo study. In B. Thompson (Ed.), *Advances in social science methodology* (Vol. 4, pp. 3–20). Greenwich, CT: JAI Press.

Spearman, C. (1904). "General intelligence," objectively determined and measured. *American Journal of Psychology, 15,* 201–293.

Stephenson, W. (1953). *The study of behavior: Q-technique and its methodology.* Chicago, IL: University of Chicago Press.

Thompson, B. (1980a). Comparison of two strategies for collecting Q-sort data. *Psychological Reports, 47,* 547–551.

Thompson, B. (1980b). Validity of an evaluator typology. *Educational Evaluation and Policy Analysis, 2,* 59–65.

Thompson, B. (1981, January). *Factor stability as a function of item variance.* Paper presented at the annual meeting of the Southwest Educational Research Association, Dallas, TX. (ERIC Document Reproduction Service No. ED 205 553)

Thompson, B., & Borrello, G. M. (1987a, January). *Comparisons of factors extracted from the correlation matrix versus the covariance matrix: An example using the Love Relationships Scale.* Paper presented at the annual meeting of the Southwest Educational Research Association, Dallas, TX. (ERIC Document Reproduction Service No. ED 280 862)

Thompson, B., & Borrello, G. (1987b). Concurrent validity of a love relationships scale. *Educational and Psychological Measurement, 47,* 985–995.

Thompson, B., & Borrello, G. (1992a). Different views of love: Deductive and inductive lines of inquiry. *Current Directions in Psychological Science, 1,* 154–156.

Thompson, B., & Borrello, G. M. (1992b). Measuring second-order factors using confirmatory methods: An illustration with the Hendrick-Hendrick Love Instrument. *Educational and Psychological Measurement, 52,* 69–77.

Thompson, B., & Daniel, L. G. (1996). Factor analytic evidence for the construct validity of scores: An historical overview and some guidelines. *Educational and Psychological Measurement, 56,* 213–224.

Thompson, B., Levitov, J. E., & Miederhoff, P. A. (1982). Validity of the Rokeach Value Survey. *Educational and Psychological Measurement, 42,* 899–905.

Thompson, B., & Miller, L. A. (1984). Administrators' and evaluators' perceptions of evaluation. *Educational and Psychological Research, 4,* 207–219.

Thompson, B., & Vacha-Haase, T. (2000). Psychometrics *is* datametrics: The test is not reliable. *Educational and Psychological Measurement, 60,* 174–195.

Tucker, L. R. (1967). Three-mode factor analysis of Parker-Fleishman complex tracking behavior data. *Multivariate Behavioral Research, 2,* 139–151.

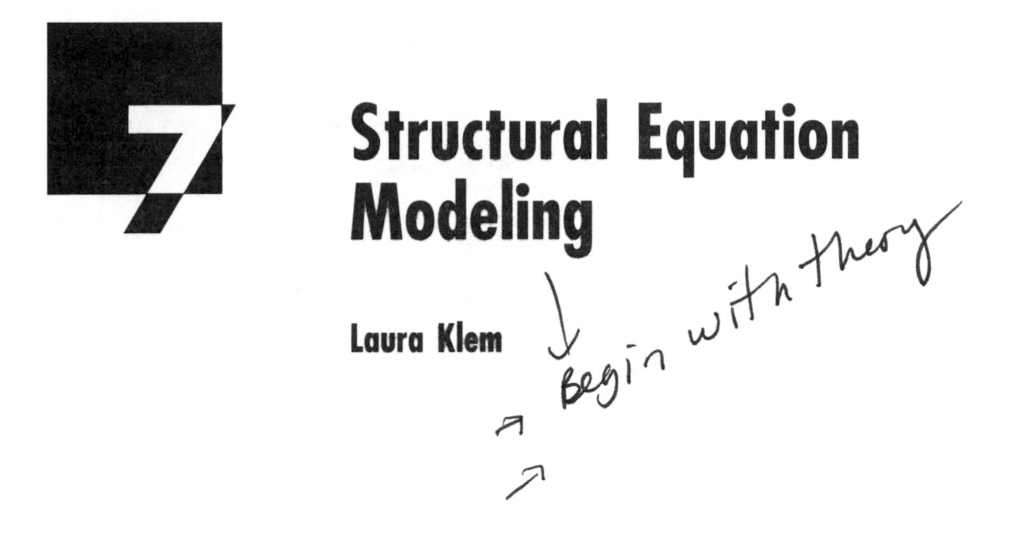

7 Structural Equation Modeling

Laura Klem

Suppose that you want to test the plausibility of a theory about the causal relationships among four concepts: depression, anxiety, creativity, and achievement. What technique should you use? Classic techniques are not suitable for this problem. *Factor analysis*, which deals with concepts, does not allow you to specify causal relationships among concepts. *Path analysis*, which allows you to specify causal relationships, deals only with observed variables. A solution to the problem, if certain assumptions about the data are met, is the relatively new structural equation approach.

Structural equation modeling (SEM), in the sense in which the term is meant today, began in 1970, when Swedish statistician Karl Jöreskog had the idea of combining features of econometrics and psychometrics into a single model. His work was based on that of many scholars and was a turning point in the history of SEM, partly because it was soon followed by the first popular SEM computer program, LISREL. The original idea is still the basis of SEM: In its most general form, SEM can be viewed as a combination of path analysis and factor analysis. In path analysis, the concern is with the predictive ordering of *measured variables*. For example, X → Y → Z is a path analysis model in which X, Y, and Z are measured variables and the arrows represent hypothesized causal effects. In a full structural equation model, the concern is with the predictive ordering of factors. A structural equation model in which the Fs are factors is F1 → F2 → F3, and just as in path analysis, the arrows represent hypothesized causal effects. *Factors* are unmeasured, or latent, variables that relate to measured variables. SEM joins measurement sophistication, achieved by using factors, with models that specify causal relationships.

I have two main goals for this chapter. The first is to introduce the reader to the many possible uses of structural equation models (maybe for his or her own data). The second is to provide enough explanation about the language and strategies of SEM so that the reader can confidently and critically read journal articles that use this procedure.

The chapter is divided into four sections. In the first section, Overview, I use a contrived example to introduce the elements of SEM and the language that describes them. These ideas are then illustrated using two research examples. The second section, Technical Issues, covers evaluating models, model modification, and assumptions. In the third section, I discuss five useful types of models. Finally, in Concluding Comments, I provide a brief guide to SEM software and offer suggestions for further reading.

Overview

A Structural Equation Model

The data used for the central example in this chapter come from Monitoring the Future, a continuing study of the lifestyles and values of youth (Johnston, Bachman, & O'Malley, 1993). The study, which was sponsored by the National Institute of Drug Abuse, was conducted at the University of Michigan by a team of investigators led by Lloyd Johnston, Jerald Bachman, and Patrick O'Malley. Each spring, beginning with the class of 1975, data has been collected from a national representative sample of high school seniors. Typically, data are collected for about 1,300 variables from about 16,000 seniors. The content of the questionnaire was divided into different forms, such that not all respondents are asked all questions.

A structural equation model using 11 variables from the 1993 Monitoring the Future data set (Johnston et al, 1993) is shown in Figure 7.1 (see the appendix for a list of the variables). The model expresses a set of hypotheses about the relations between variables. The 11 variables from the data set, which are measured variables, are represented by rectangles. The ovals represent unmeasured, or latent, variables, which are abstract concepts. Straight lines in the diagram indicate hypothesized direct effects. For example, the unmeasured variable Self-Esteem

Figure 7.1

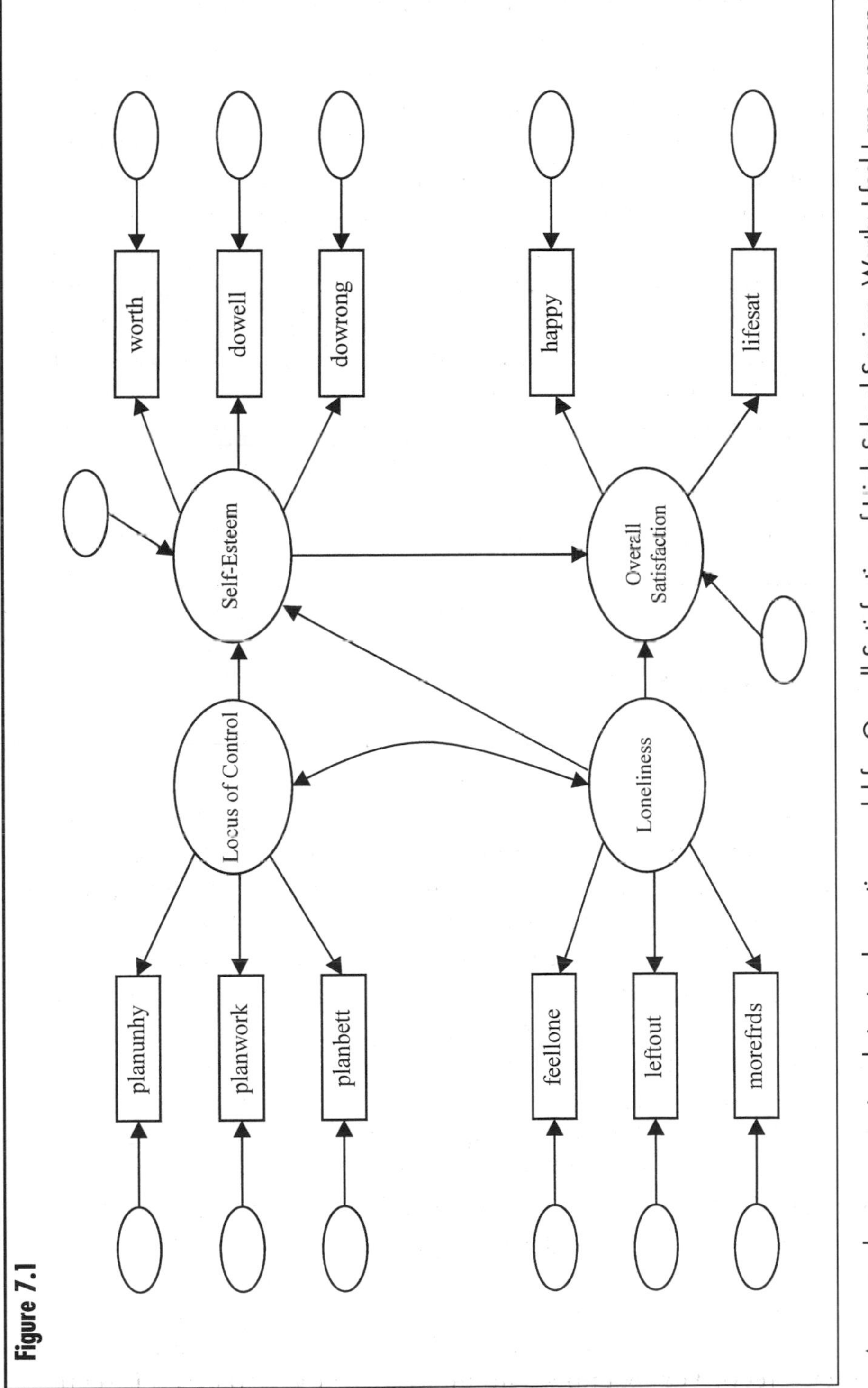

A proposed measurement and structural equation model for Overall Satisfaction of High School Seniors. Worth: I feel I am a person of worth. Dowell: I am able to do things as well as most other people. Dowrong: I feel I can't do anything right. Happy: Taking all things together, how would you say things are these days? Lifesat: How satisfied are you with your life as a whole? Planunhp: Planning only makes a person unhappy. Planwork: When I make plans, I am almost certain that I can make them work. Planbett: Planning ahead makes things turn out better. Feellone: A lot of times I feel lonely. Leftout: I often feel left out of things. Morefrds: I often wish I had more good friends.

is hypothesized to have a direct effect on three measured variables: worth, do well, and do wrong. (The expectation is that the effect on do wrong will be negative.) For reasons that will become clear in the course of this chapter, this simple model should be regarded only as an example.

Variables such as Self-Esteem, unmeasured variables that are hypothesized to cause the covariation among a set of measured variables, are often called *factors*. Factors are depicted in large ovals in Figure 7.1. The measured variables associated with factors are often called *indicators*. Thus, Loneliness, Locus of Control, Self-Esteem, and Overall Satisfaction are factors that are each associated with a set of indicators. The four straight lines in the middle of the figure represent hypothesized direct effects of one factor variable on another. For example, the factor Self-Esteem is hypothesized to have a direct effect on the factor Overall Satisfaction. The absence of a straight line in a model diagram such as Figure 7.1 represents a hypothesis that there is no effect. To illustrate, the missing links in Figure 7.1 demonstrate the following hypotheses: (a) The factor Loneliness does not directly affect a student's answer to the measured variable planwork, and (b) the factor Locus of Control does not directly affect the factor Overall Satisfaction.

As suggested earlier, a full structural equation model, such as the one in Figure 7.1, can be viewed as a combination of a factor analysis model and a path analysis model. (These techniques are discussed in Grimm & Yarnold, 1995.) The measurement part of the model corresponds to factor analysis and depicts the relationships of the latent variables to the observed variables. The structural part of the model corresponds to path analysis and depicts the direct and indirect effects of the latent variables on each other. In ordinary path analysis one models the relationships between observed variables, whereas in SEM one models relationships between factors. This SEM methodology is applicable to a range of models, especially in the social and behavioral sciences.

Elements of Models: Variables and Parameters

In the previous section, I distinguished between two major types of variables in structural equation models: measured and unmeasured variables. I now make some further distinctions are among types of variables, beginning with distinctions among unmeasured variables. As noted earlier, unmeasured variables that affect measured variables are factors. There are two types of factors: *exogenous* factors, which the model does not attempt to explain; and *endogenous* factors, which are affected by one or more other latent variables. In a model diagram, if a factor has a straight line with an arrow pointing toward it, it is endogenous; otherwise, it is exogenous. In Figure 7.1, Locus of Control and Loneliness are exogenous factors, whereas Self-Esteem and Overall Satisfaction are endogenous factors.

There is another distinction to be made among unmeasured variables. I have not yet mentioned the 13 exogenous unmeasured variables that are pictured as small ovals in Figure 7.1. They are errors of various kinds and are not interesting per se. (As seen, however, in the discussion of parameters below, their variances, and possibly their covariances, are important.) Consider the measured variable planunhy (planning only makes a person unhappy), in the upper left of the model in Figure 7.1. The model hypothesizes that part of the variance of planunhy is explained by the Locus of Control factor and the rest by error. Note that this is a broad view of error: In SEM, error variables include effects of variables omitted from the model as well as effects of measurement error.

Turning now to consideration of the observed variables in Figure 7.1, there is a single distinction: Six of the observed variables (at the left edge of the diagram) are attached to exogenous factors, whereas five of them (at the right edge of the diagram) are attached to endogenous factors. Altogether, there are seven types of variables in SEM. These are listed in Table 7.1; for convenience in reading literature, particularly books and articles written in the 1970s and 1980s, their LISREL designations are shown in the column on the right. Not all SEM software, nor all problems, requires making a distinction among all seven types of variables. Table 7.1 represents the maximum degree of complexity for variables.

It is also useful to distinguish among types of SEM parameters. Altogether, there are eight types. As seen in Figure 7.1, looking at the straight lines, four kinds of direct effects can be estimated in a full structural equation model: (a) the effect of an exogenous factor on a measured variable (e.g., the effect of Locus of Control on planunhy); (b) the effect of an endogenous factor on a measured variable (e.g., the effect of Self-Esteem on worth); (c) the effect of an exogenous factor on an endogenous factor (e.g., the effect of Locus of Control on Self-Esteem); and (d) the effect of an endogenous factor on another endogenous factor (e.g., the effect of Self-Esteem on Overall Satisfaction). These four types of direct effects are analogous to the coefficients

Table 7.1

Types of Variables in Structural Equation Modeling

Variable	LISREL notation
Dependent (endogenous) variable: Unmeasured (unobserved, latent)	η (eta)
Independent (exogenous) variable: Unmeasured (unobserved, latent)	ξ (xi or ksi)
Indicator of dependent variable: Measured (observed, manifest)	y
Indicator of independent variable: Measured (observed, manifest)	x
Error in observed dependent variable	ε (epsilon)
Error in observed independent variable	δ (delta)
Sources of variance in η not included among the ξ's ("disturbances," "error in equations")	ζ (zeta)

(βs or betas) from a multiple regression. One reason for using SEM is to estimate these effects.

The four remaining types of parameters of interest are all variances or covariances. In SEM one can obtain estimates of variances and covariances of unmeasured variables. For example, in Figure 7.1, the curved double-headed arrow linking the Locus of Control and Loneliness factors represents their covariance, and that covariance can be estimated by SEM software. SEM software can also estimate the variances and covariances of the unmeasured variables that represent errors. The variance of errors, in particular, is interesting because it is unexplained variance. For example, the variance of the error variable represented by the small oval in the upper left corner of Figure 7.1 is the variance of planunhy that is unexplained by Locus of Control. Similarly, the variance of the error linked to the factor Self-Esteem is the variance in Self-Esteem that is not explained by the model.

Table 7.2 lists eight types of parameters along with their LISREL designations. There are other parameters involved in SEM, but the eight listed are especially important because they are featured in LISREL software. The many distinctions among kinds of variables and parameters may seem burdensome and are sometimes unnecessary, but there is elegance to the exactness.

Table 7.2

Types of Parameters in SEM

Parameter	LISREL notation
Coefficients relating unmeasured dependent variables to measured dependent variables	Λ_y (lambda)
Coefficients relating unmeasured independent variables to measured independent variables	Λ_x (lambda)
Coefficients interrelating unmeasured dependent variables	B (beta)
Coefficients relating unmeasured independent variables to unmeasured dependent variables	Γ (gamma)
Variances and covariances among unmeasured independent variables	Φ (phi)
Variances and covariances among disturbances	Ψ (psi)
Variances and covariances among errors in measured dependent variables	Θ_ε (theta)
Variances and covariances among errors in measured independent variables	Θ_δ (theta)

Note. Individual LISREL coefficients are organized by matrix. Uppercase Greek letters name matrices that contain all coefficients of a given type. Thus, for example, the Λ_y matrix contains individual λ_y coefficients.

Data

Data for an SEM program consist of a covariance or correlation matrix. Two assumptions that underlie traditional SEM programs are (a) that the variables on which the matrix coefficients are based are intervally scaled and (b) that the variables have a multivariate normal distribution. These requirements of scale and normality are often difficult to meet with social sciences data. For two reasons, however, the requirements are not always prohibitive. First, maximum likelihood estimation, the estimation method most often used in SEM, has been found to be fairly robust to violation of normality (Chou & Bentler, 1995). Second, recent developments, now available as program options in SEM software, provide possible remedies for nonnormal variables. The remedies, some of which carry their own assumptions, differ depending on whether the cause of nonnormality is poorly distributed continuous variables or coarsely categorized continuous variables (for a thorough discussion of SEM with nonnormal variables, see West, Finch, & Curran, 1995).

Sample size is an important consideration when using SEM. The necessary sample size for reliable results depends on the complexity of the model, the magnitude of the coefficients, the number of measured variables associated with the factors, and the multivariate normality of the variable distributions. (More cases are required for complex models, models with weak relationships, models with few measured variables per factor, and nonnormal distributions.) Only a scattering of condition combinations have been studied. Loehlin (1992) and Schumacker and Lomax (1996) each gave brief summaries of sample-size research and included references to the original articles. The original studies indicate that the input matrix should be based on at least 100–150 cases. Assuming a minimum of 150 cases, one rough rule of thumb is to have between 5 and 10 cases per parameter estimated (Bentler & Chou, 1987). Bentler and Chou recommended the higher ratio of cases to parameters (i.e., 10) if the distribution of the variables is not multivariate normal.

Estimation of Parameters

The first step in modeling is the specification of a model, which should be based as much as possible on previous knowledge. Sometimes theory suggests competing models, in which case they should each be specified. The initial specification can take the form of a diagram, such as Figure 7.1, or a series of equations. Thus, one can indicate by a drawing that feellone (a lot of times I feel lonely) is affected by Loneliness plus error, or one can write $X1 = aF_{lo} + U_{x1}$, where X1 is feellone, a is a coefficient to be estimated, F_{lo} is Loneliness, and U_{x1} is the error associated with feellone. Similarly, one can write an equation that indicates that Overall Satisfaction (F_{OV}) is affected by Loneliness (F_{LO}) and Self-Esteem (F_{SE}) and by a disturbance (U_{OV}): $F_{OV} = aF_{LO} + bF_{SE} + U_{OV}$.

After a model is specified, the next step is to obtain parameter estimates, that is, estimates of the coefficients representing direct effects and of the coefficients representing variances and covariances of unmeasured variables. For this step in the modeling process, it is necessary to use special SEM software, such as LISREL (Jöreskog & Sörbom, 1993), EQS (Bentler, 1992), or AMOS (Arbuckle, 1997). (More about these software packages is included in a later section.) To use SEM software, the researcher must provide a program with a description of the model (the programs have ingenious setup rules that make this easy) and a matrix of correlations (or covariances) among the observed variables. To get estimates for the model in Figure 7.1, I entered into LISREL (Version VII) the correlation matrix shown in Table 7.3 along with a set of LISREL commands that described the model; 27 param-

Table 7.3

Bivariate Pearson Product–Moment Correlations for 11 Monitoring the Future Variables

	1	2	3	4	5	6	7	8	9	10	11
1. Worth	1.00										
2. Dowell	0.55	1.00									
3. Dowrong	−0.36	−0.36	1.00								
4. Happy	0.17	0.15	−0.20	1.00							
5. Lifesat	0.26	0.21	−0.23	0.39	1.00						
6. Planunhp	−0.16	−0.11	0.21	−0.15	−0.12	1.00					
7. Planwork	0.34	0.37	−0.28	0.18	0.23	−0.26	1.00				
8. Planbett	0.22	0.22	−0.16	0.11	0.14	−0.43	0.34	1.00			
9. Feellone	−0.20	−0.20	0.36	−0.32	−0.31	0.17	−0.18	−0.14	1.00		
10. Leftout	−0.25	−0.23	0.39	−0.23	−0.26	0.15	−0.22	−0.10	0.56	1.00	
11. Morefrds	−0.12	−0.15	0.23	−0.19	−0.14	0.07	−0.12	−0.01	0.32	0.42	1.00

Note. $N = 2503$. Worth: I feel I am a person of worth. Dowell: I am able to do things as well as most other people. Dowrong: I feel I can't do anything right. Happy: Taking all things together, how would you say things are these days? Lifesat: How satisfied are you with your life as a whole? Planunhp: Planning only makes a person unhappy. Planwork: When I make plans, I am almost certain that I can make them work. Planbett: Planning ahead makes things turn out better. Feellone: A lot of times I feel lonely. Leftout: I often feel left out of things. Morefrds: I often wish I had more good friends.

eters were designated to be estimated. The mathematics that SEM programs use to arrive at the parameter estimates are beyond the scope of this chapter. The basic idea, however, is simple: An SEM program determines the estimates that will most nearly reproduce the matrix of observed relationships. For example, in the Monitoring the Future data there is a correlation of −.12 (see Table 7.3) between the variables planunhy and lifesat (How satisfied are you with your life as a whole?). By looking at the diagram, one can see three paths that link those two measured variables. (One path is through Self-Esteem, one is though Loneliness, and one is through both Loneliness and Self-Esteem.) These paths involve seven coefficients. An SEM program finds estimates for those seven coefficients that will most nearly reproduce the correlation of −.12, while simultaneously taking into account the closest possible reproduction of the all the other correlations in the input matrix. Usually a program cannot find coefficients that exactly reproduce all the relationships in the observed matrix.

The parameter estimates for the model of Overall Satisfaction are shown in Figure 7.2. The numbers beside the lines represent the magnitude of the effects. The numbers at the tails of arrows represent the variances of errors. In this model, Self-Esteem has a stronger effect than Loneliness on Overall Satisfaction. Thirty-eight percent of the variance in Overall Satisfaction is explained by the model; sixty-two percent of the variance is unexplained.

Two Examples

Two examples from research illustrate the use of SEM. Fullagar, Gallagher, Gordon, and Clark (1995) investigated the effects of institutional and individual practices on union commitment and participation. Their proposed model, shown in Figure 7.3, is similar in form to the model for predicting Overall Satisfaction of High School Seniors; the conventions for displaying latent and observed variables in ovals and rectangles and the Greek (i.e., LISREL) notation used to distinguish parameters in Figure 7.3 should seem familiar. The form of the model differs from the previous example in one respect: Note that one of the latent variables has only a single indicator. Having only one measured variable for a concept poses no problem for SEM. Sometimes a single measured variable is scale-score constructed (before using SEM) from several variables, such as socioeconomic status or a score on a depression scale. Other times, a single measured variable is just a simple variable, such

Figure 7.2

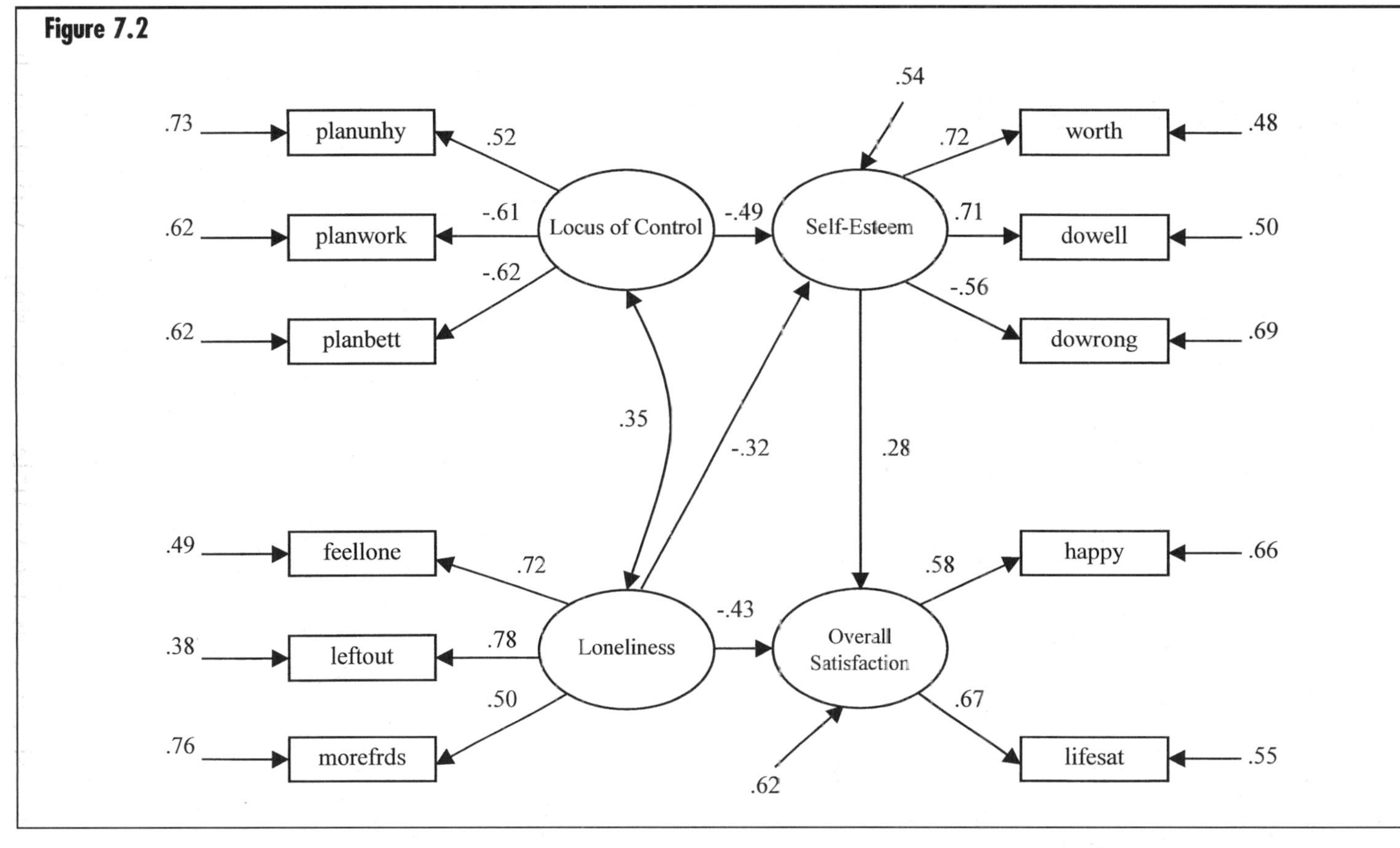

Model of Overall Satisfaction of High School Seniors with Standardized Parameter Estimates Obtained From LISREL VII. The numbers at the tails of arrows are error variances. Worth: I feel I am a person of worth. Dowell: I am able to do things as well as most other people. Dowrong: I feel I can't do anything right. Happy: Taking all things together, how would you say things are these days? Lifesat: How satisfied are you with your life as a whole? Planunhp: Planning only makes a person unhappy. Planwork: When I make plans, I am almost certain that I can make them work. Planbett: Planning ahead makes things turn out better. Feellone: A lot of times I feel lonely. Leftout: I often feel left out of things. Morefrds: I often wish I had more good friends.

Figure 7.3

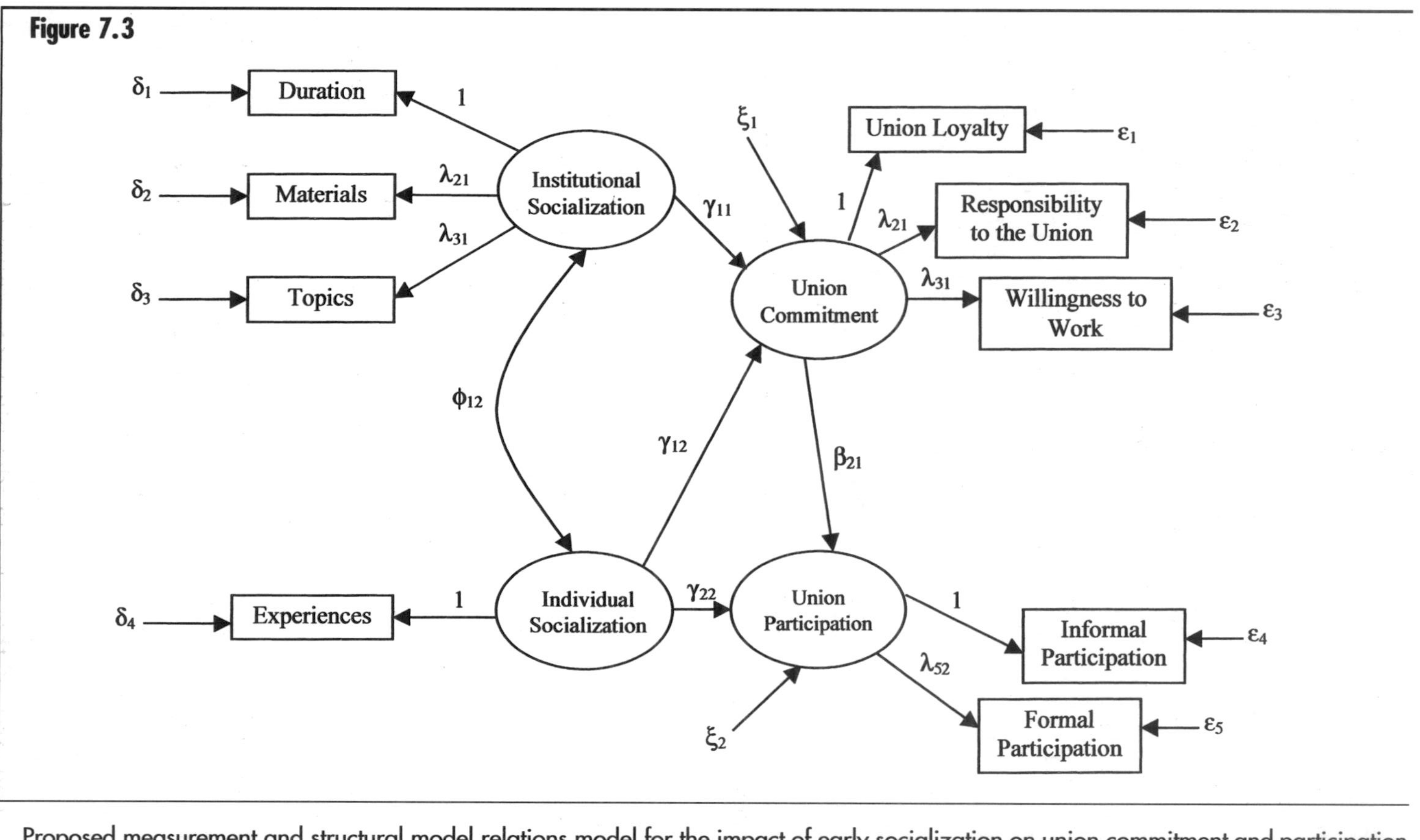

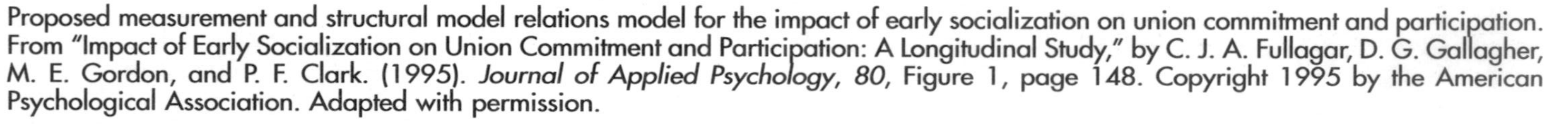
Proposed measurement and structural model relations model for the impact of early socialization on union commitment and participation. From "Impact of Early Socialization on Union Commitment and Participation: A Longitudinal Study," by C. J. A. Fullagar, D. G. Gallagher, M. E. Gordon, and P. F. Clark. (1995). *Journal of Applied Psychology, 80,* Figure 1, page 148. Copyright 1995 by the American Psychological Association. Adapted with permission.

as age or grade point average. In Figure 7.3, the variable experiences was the sum of 14 possible experiences. (In Figure 7.3, there are four *1*s where the reader might expect the lambda notation. The *1*s simply convey information about how the LISREL setup was done.) In a more complicated example of SEM, Vinokur, Price, and Caplan (1996) tested models in which financial strain was hypothesized to affect depression and the relationship satisfaction of unemployed job seekers and their spouses. The structural part of their model is shown in Figure 7.4. The measurement model, which involved 24 measured variables, is not shown in the figure but was discussed in detail in their article. Separate presentations of the measurement and structural models are a common way to handle what would otherwise be a messy diagram. The numbers on the paths and in the circles are parameter estimates that resulted from using SEM software. The figure caption contains information about the "fit" of the model. I turn to the matter of fit, and other slightly technical considerations, in the next section.

Technical Issues

In this section of the chapter, I discuss evaluating SEM results, the modification of models, and the assumptions of SEM. Careful modeling requires the evaluation of a model (both before and after estimation), judicious modification of a model (if any modifications at all is undertaken), and attention to the assumptions of the technique.

Evaluation of Results

Evaluating SEM results involves theoretical criteria, statistical criteria, and assessment of fit. Although the issue of fit is discussed in greater detail than the other issues, remember that fit is of no interest unless the results meet theoretical and statistical criteria.

A model submitted to an SEM program should be based, as much as possible, on theory. After the parameters of a model have been estimated by an SEM program, they should be assessed from a theoretical perspective; for example, the signs and magnitudes of the coefficients should be consistent with what is known from research. Results should be theoretically sensible.

Besides the theoretical criteria, there are also two major statistical criteria. The first of these is the identification status of the model. A

Figure 7.4

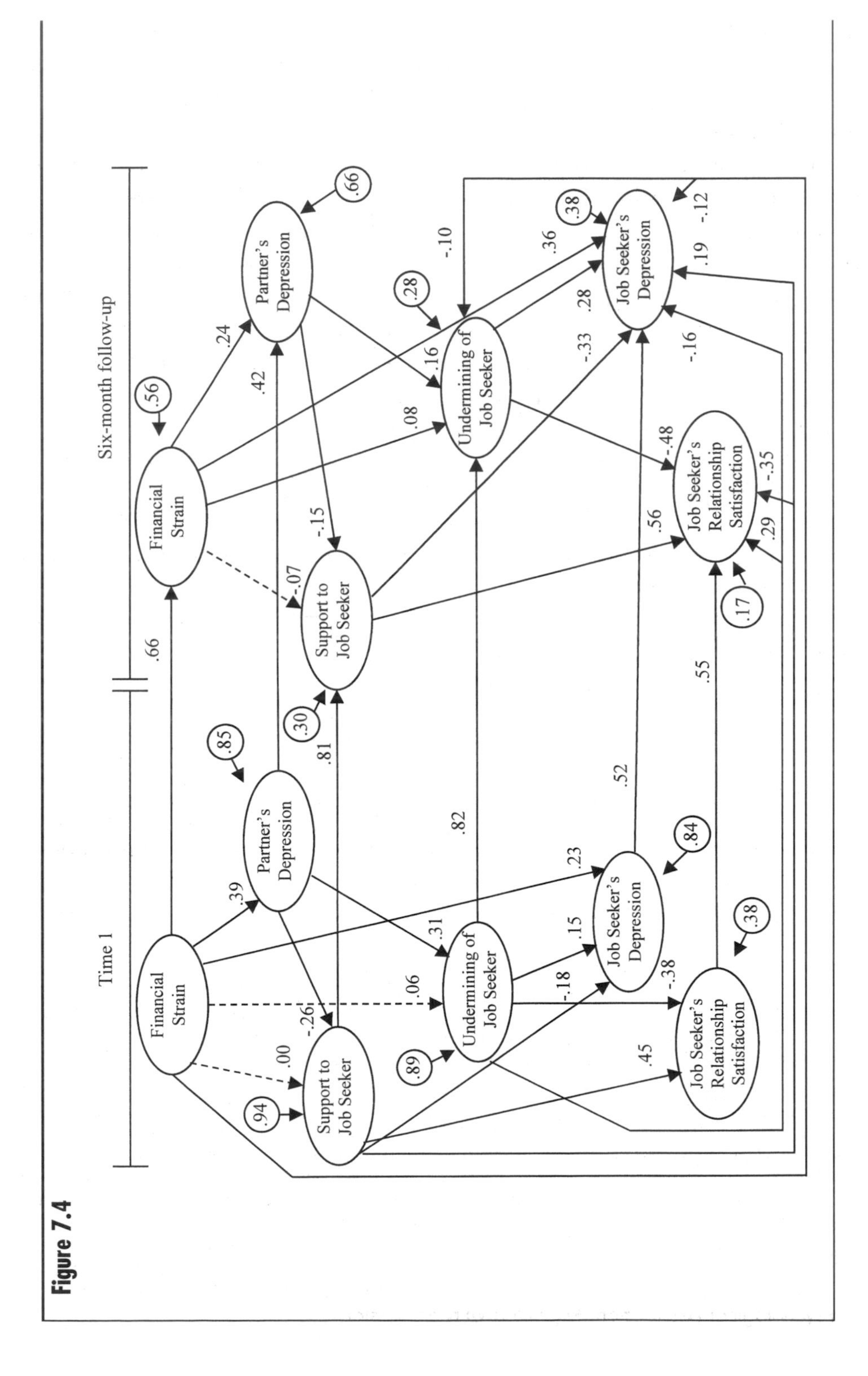
Time 1
Six-month follow-up
Financial Strain
Partner's Depression
Support to Job Seeker
Undermining of Job Seeker
Job Seeker's Depression
Job Seeker's Relationship Satisfaction
Financial Strain
Partner's Depression
Support to Job Seeker
Undermining of Job Seeker
Job Seeker's Depression
Job Seeker's Relationship Satisfaction
.39
.00
-.26
.06
.31
.23
.15
-.18
-.38
.45
.94
.89
.85
.84
.38
.66
.81
.82
.52
.55
.42
.24
.56
-.07
-.15
.30
.08
.16
.28
-.10
.36
.28
-.33
.38
.66
.56
-.48
-.16
.19
-.12
-.35
.29
.17

model is identified if there is a unique solution for each parameter in the model. The model's identification status should be considered both before and after the model is estimated. One way to determine the identification status of a model (seldom done in practice) is to prove identification algebraically. A more useful approach is to learn the patterns of relationships that result in identified models and the patterns that require special attention to be identified. It is often helpful to consider the identification status of the measurement and structural parts of the model separately. For guidelines on the identification status of measurement models, Bollen's (1989) text is particularly helpful. For guidelines on the identification of the structural part of the model, see Long (1983b) and Rigdon (1995).

Even if a model is thought to be identified, it may not be, either because a mistake was made in thinking it was or because of empirical underidentification. *Empirical underidentification* occurs when actual data values make a model that should be identified underidentified. For example, in certain circumstances a zero correlation between two variables that the researcher expected to have a nonzero correlation can cause underidentification. Fortunately for the researcher, modern SEM programs almost always detect and report identification problems. Computer printouts should be carefully examined for warning messages. Sometimes a model is nearly underidentified; if so, there are typically clues in the printout that signal a problem (e.g., high standard errors). Hayduk (1987, p. 149) suggested a clever procedure to use if it is suspected that a particular parameter is not identified. Jöreskog and Sörbom (1989, p. 18) explained a way to use SEM software to reliably check the identification status of a whole model.

The second major statistical criterion has to do with the statistical

(*Opposite page*) Structural equation model of the effects of financial strain, social support, and social undermining on depression and relationship satisfaction. $\chi^2(165, N = 693) = 432$, $p < .001$, NFI = .97, NNFI = .97, CFI = .98, and RMR = .03. Solid and broken lines represent, respectively, statistically significant ($p < .05$) and nonsignificant paths that were included in the model. The solid lines at the bottom of the figure represent paths from Time 1 (T1) to follow-up. Numbers in small circles are residual variances. Not shown are (a) three longitudinal paths, (b) the correlations between the disturbances of social support and undermining at T1 and at follow-up, (c) correlations between job seeker's depression and relationship satisfaction at T1, and (d) correlations between partners' depression at T1. From "Hard Times and Hurtful Partners: How Financial Strain Affects Depression and Relationship Satisfaction of Unemployed Persons and Their Spouses," by A. D. Vinokur, R. H. Price, and R. D. Caplan (1996). *Journal of Personality and Social Psychology, 71*, Figure 1, page 173. Copyright 1996 by the American Psychological Association. Reprinted with permission.

reasonableness of the parameters. A model that is misspecified can result in improper results, such as negative variances and correlations greater than one. The extensiveness of checking for inadmissible results varies across different computer programs, as do the rules for proceeding if such results are detected.

If the results of estimating a structural equation model look theoretically sensible, if the model is identified, and if there are no signs of statistically improper estimates, then the question arises "Does the data fit the model?" Recall that the estimates produced by an SEM program are the estimates that will most nearly reproduce the relationships in the original data matrix. The matrix of correlations (or covariances) that is implied by the model (i.e., the reproduced relationships) is called the $\hat{\Sigma}$ *(sigma-hat) matrix.* The original data matrix is called the *S matrix.* Fit has to do with how closely these two matrices match; that is, how well the relationships implied by the model match the observed relationships.

For the model of Overall Satisfaction, the matrix of model-implied correlations, copied from the LISREL printout, is shown in Table 7.4. Not shown here but available on SEM printouts is the matrix of residuals, the differences between the observed and implied relationships. Goodness-of-fit tests are based on these residuals (i.e., the difference between the S and $\hat{\Sigma}$ matrices).

The classic chi-square goodness-of-fit test assesses the size of the discrepancies between the S and $\hat{\Sigma}$ matrices. The researcher, assuming he or she hopes that the hypothesized model is consistent with the data, hopes that the discrepancies are small and that they are not significant. The researcher wants a high probability that the discrepancies could have occurred by chance, say, .30 or .40.

There are problems, however, with using the chi square associated with a model as a guide to the model's adequacy. One problem is that chi square is sensitive to sample size. For small samples big discrepancies are not significant, whereas for large samples even trivial discrepancies are significant. A second problem is that the chi square is very sensitive to an assumption of multivariate normality of the measured variables; departures from normality tend to increase chi square.

Although the chi-square test itself is not a satisfactory test of the fit of a model (except possibly for a simple model with 200–400 cases), it is the basis for *fit indices.* Over the past 20 years or so, at least 24 fit indices to supplement chi square have been proposed, studied, and compared. One important difference among indices is that some of

Table 7.4

Estimated Bivariate Pearson Product–Moment Correlations for Model of Overall Satisfaction of High School Seniors

	1	2	3	4	5	6	7	8	9	10	11
1. Worth	1.00										
2. Dowell	0.51	1.00									
3. Dowrong	−0.40	−0.39	1.00								
4. Happy	0.21	0.20	−0.16	1.00							
5. Lifesat	0.24	0.23	−0.18	0.39	1.00						
6. Planunhp	−0.23	−0.22	0.18	−0.10	−0.11	1.00					
7. Planwork	0.27	0.26	−0.21	0.12	0.13	−0.32	1.00				
8. Planbett	0.27	0.26	−0.21	−0.12	0.13	−0.32	0.38	1.00			
9. Feellone	−0.25	−0.25	0.20	−0.24	−0.27	0.13	−0.16	−0.16	1.00		
10. Leftout	−0.28	−0.27	0.21	−0.26	−0.30	0.14	−0.17	−0.17	0.56	1.00	
11. Morefrds	−0.18	−0.17	0.14	−0.16	−0.19	0.09	−0.11	−0.11	0.36	0.39	1.00

Note. $N = 2{,}503$. Worth: I feel I am a person of worth. Dowell: I am able to do things as well as most other people. Dowrong: I feel I can't do anything right. Happy: Taking all things together, how would you say things are these days? Lifesat: How satisfied are you with your life as a whole? Planunhp: Planning only makes a person unhappy. Planwork: When I make plans, I am almost certain that I can make them work. Planbett: Planning ahead makes things turn out better. Feellone: A lot of times I feel lonely. Leftout: I often feel left out of things. Morefrds: I often wish I had more good friends.

them take into account the number of parameters that were estimated to achieve the degree of fit. Another difference is that some indices compare the fit of the estimated model to some related model (e.g., the null model, which specifies no relationship among the measured variables). For a few fit indices, distribution properties are known; this allows for the construction of confidence intervals and the testing of hypotheses. LISREL 8 prints values for 18 fit indices, EQS 5 prints values for 8, and AMOS prints values for 24. There is no general agreement on which index or indices are best; most writers who offer advice have suggested reporting a few (see, e.g., Hoyle & Panter, 1995; and Raykov, Tomer, & Nesselroade, 1991). A model usually fits better according to some criteria than to others; the researcher should be guided by the preponderance of the evidence in reaching a conclusion about the adequacy of the model.

For the model of Overall Satisfaction of High School Seniors, the $\chi^2(39, N = 2503) = 656.65$, $p < .00$ was unacceptable. The χ^2/df, another measure of fit, is 16.8, which is also unacceptable. A χ^2/df of 2 or 3, or at least one less than 5, would be better. There are no established guidelines for this fit measure. These thresholds were mentioned by Bollen, 1989, p. 278). The goodness-of-fit and adjusted goodness-of-fit are .95 and .92, respectively, which are both acceptable (> .9 is the rough rule of thumb for these indices). The root mean square residual (RMR), .06, a measure of the average of the residuals, seems a bit high. (The RMR should be interpreted with respect to the size of the coefficients in the input matrix, which were Pearson product-moment correlations for this model.) Taken together, these are signs of a mediocre fit. For the model of Union Participation in Figure 7.3, the fit was "adequate" (Fullagar, Gallagher, Gordon, & Clark, 1995, p. 152). Although chi square was significant, other evidence, including three satisfactory fit indices, indicated a fairly good fit. It is interesting to note that the linkage from Institutional Socialization to Union Commitment was not significant. For the model of Job Seeker's Depression and Relationship Satisfaction in Figure 7.4, the fit, detailed in the figure caption, is good. The chi square is high, but the sample is large. All other evidence favors acceptance of the model: The four fit indices are each well above .90, and the RMR is modest.

A useful strategy for assessing fit is to test a *nested model*; Model A is nested within Model B if Model A can be derived from Model B by removing one or more linkages. Imagine the Overall Satisfaction model in Figure 7.1 with the linkage from Loneliness to Overall Satisfaction removed. This new model is nested within the original model. Often a hierarchical series of nested models is examined. Elegant tests can be performed on nested models that allow for a definitive statement that one model is significantly better than others.

Bagozzi and Yi (1988) and Bentler and Chou (1987) offered broad perspectives on evaluation that included such considerations as inadmissible results. On the narrower subject of fit indices, see Hu and Bentler (1995) for an additional review. The chapters in the edited book *Testing Structural Equation Models* (Bollen & Long, 1993) concentrate on testing model fit and respecification.

Modification of Models

The question of whether and how to modify a model naturally follows from the examination of results. SEM programs provide printed material that suggests changes that could be made to a model. Printouts identify both new linkages that could be added to improve the fit of a model and linkages that are in the model that turned out not to be significant. A modification to a model on the basis of the results of a previous model is a "data-driven" modification. There is the possibility of capitalizing on chance; that is, of fitting the model to peculiarities of the particular sample. The probability values associated with a model including data-drive modifications are not accurate (to an unknown extent). If results from a final model based on post hoc modifications are presented, they should be distinguished from the results of the original theory-based model. An excellent practice, if the sample size is sufficiently large, is to randomly divide the sample into two halves before the original theoretical model is estimated. Then the final model on Sample 1, which may include modifications, can be re-estimated using Sample 2 to assess the model's adequacy. Sample 1 and Sample 2 are sometimes referred to as "training samples" and "holdout samples," respectively. For a carefully documented application of this split sample procedure, see Bottorff, Johnson, Rainer, and Hayduk (1996).

Assumptions

As mentioned above in the discussion of appropriate data for SEM, the usual approaches are based on the assumption that the variables for which the input coefficients are computed are intervally scaled and have multivariate normal distributions. West et al. (1995), discussed both remedies if data do not meet assumptions and the effects of violations of assumptions. In general, nonnormal data leads to inflated chi-square values and the underestimation of standard errors. In addition to the scaling and normality assumptions, keep in mind that structural equation models are sensitive only to linear relationships. Unless special precautions are taken, they are not sensitive to interaction effects.

Note that one assumption is missing from the usual list. SEM does not assume the absence of measurement error in the measured variables. The accommodation of error in models is one of the great attractions of SEM. Models can incorporate error that is unique to a variable. (Recall the small ovals in Figure 7.1 associated with measured variables; they represent unmeasured random error and unique variance.) Models can also include factors that represent systematic error, such as a method effect, or bias factor. SEM is often used for multimethod–multitrait (MMMT) models, which include method factors.

The correct specification of the model is critical. Relevant variables should be included, and the direction of causal flow should be correct. The parameter estimates from SEM are conditional on the model being correct. I noted earlier that the model of Overall Satisfaction of High School Seniors should be regarded only as an example. This is partly because the data do not strictly meet the assumptions for SEM. The overwhelming reason for discounting the example, however, is that the model is certainly misspecified. The directions of the hypothesized effects may be wrong. Relationships may be reciprocal. Certainly relevant variables have been omitted.

Five Useful Models

Up to this point in the chapter, I have discussed full measurement–structural models that include all elements of a structural equation models (although not all complications). Often a research problem involves only part of a model. I begin this section by backtracking, discussing two relatively simple, widely used submodels: confirmatory factor analysis (CFA) and path analysis. Then, I introduce two extensions of the full model that involve relatively advanced application of SEM. These are models for longitudinal data and for analyzing, simultaneously, data from several samples. Finally, I discuss an alternative way to model the linkages between an unmeasured variable and its associated measured variables that is sometimes more appropriate than the conventional one.

Confirmatory Factor Analysis Models

CFA involves only a measurement model, that is, a model for the direct effects of the factors on the measured variables, the covariances among the factors, and the errors of measurement. Unlike a full structural equation model, a CFA does not include specification of a causal model that relates the factors to each other. The major difference between CFA and a garden-variety exploratory factor analysis (EFA) is that in CFA the researcher specifies that particular factors affect, or load on, particular measured variables, whereas in EFA all factors affect all measured variables. Figure 7.5 shows a CFA model for the 11 MTF variables used to model Overall Satisfaction of High School Seniors in Figure 7.1. There are additional ways, besides the particularity of factor loadings, in which a CFA model may be specified with more flexibility than an EFA model; for example, in a CFA some factors may be hypothesized to correlate, while others do not. CFAs are extensively used in the development of new scales and in the reexamination of established scales.[1] Another use for CFA is as the first step of a two-step plan for developing a full model. Jöreskog and Sörbom (1993, p. 128) took this approach in their outline for a strategy for generating a model.

A CFA model can be tested to see if it adequately fits the data. Estimation of the CFA model in Figure 7.5 resulted in only a mediocre fit with χ^2 (38,2503) = 638.10, $p < .00$. This chi square, which is almost as large as the one for the full model of Overall Satisfaction, indicates that much of the problem with the full model stems from the measurement model. (This is not surprising: There is only a single degree-of-freedom difference between the models, representing the linkage between Locus of Control and Overall Satisfaction.)

Path Analysis Models

In a full structural equation model, the structural part of the model involves the relationships between factors, or unmeasured variables. In a path analysis, such as the one in Figure 7.6, the model involves struc-

---

[1]For a modern exploration of an old problem, see Marsh (1996). There is an ongoing debate on whether the factors associated with positively and negatively worded self-esteem items represent a substantively meaningful distinction or an artifact of response style. Marsh used CFA to examine various models.

Figure 7.5

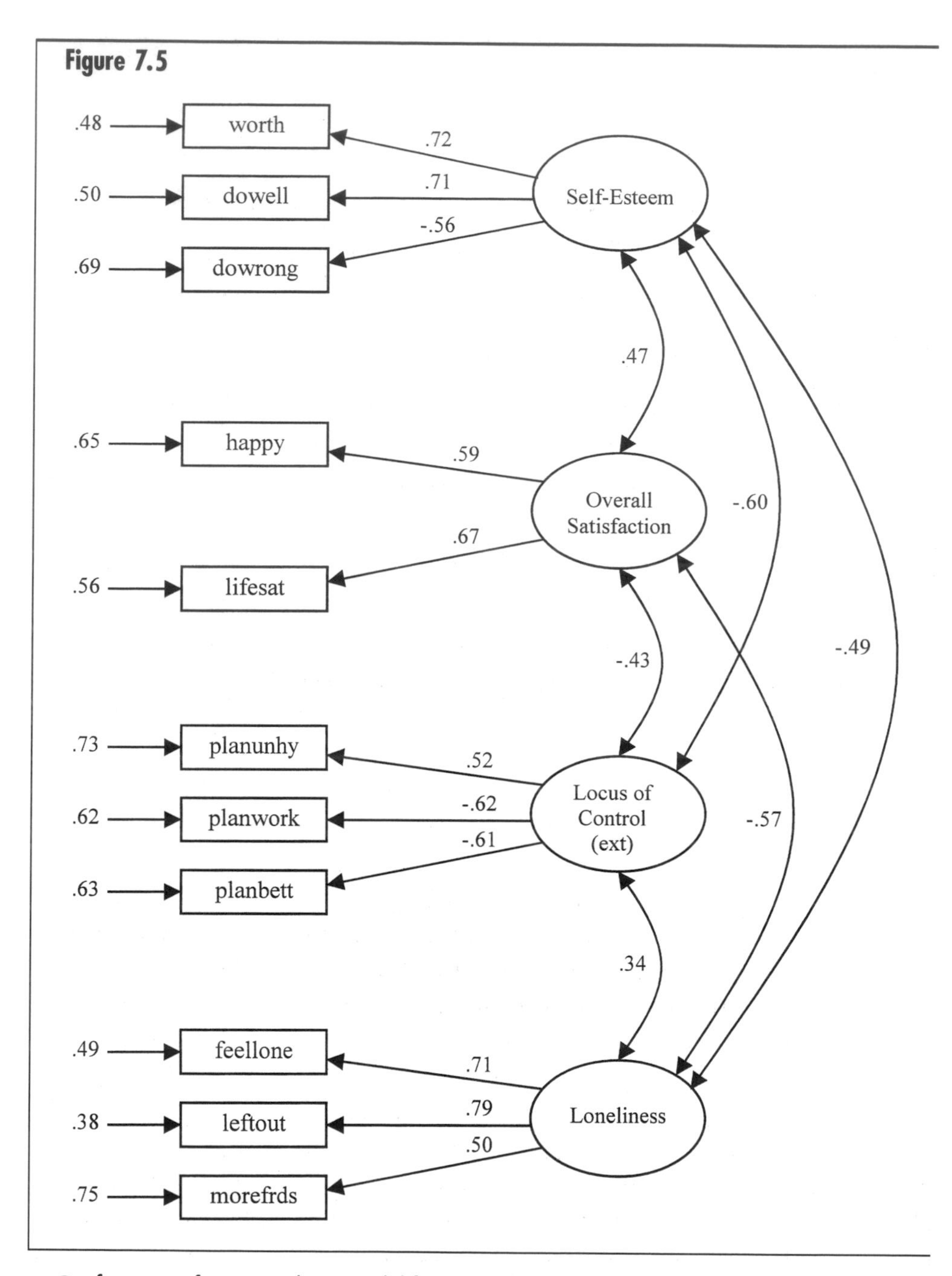

Confirmatory factor analysis model for 11 measured variables. Standardized parameter estimates were obtained from LISREL VII. The numbers at the tails of the arrows are error variances. $\chi^2(38,2503) = 638.10$, $p < .00$, GFI = .95, AGFI = .92, RMR = .06. Worth: I feel I am a person of worth. Dowell: I am able to do things as well as most other people. Dowrong: I feel I can't do anything right. Happy: Taking all things together, how would you say things are these days? Lifesat: How satisfied are you with your life as a whole? Planunhp: Planning only makes a person unhappy. Planwork: When I make

Figure 7.6

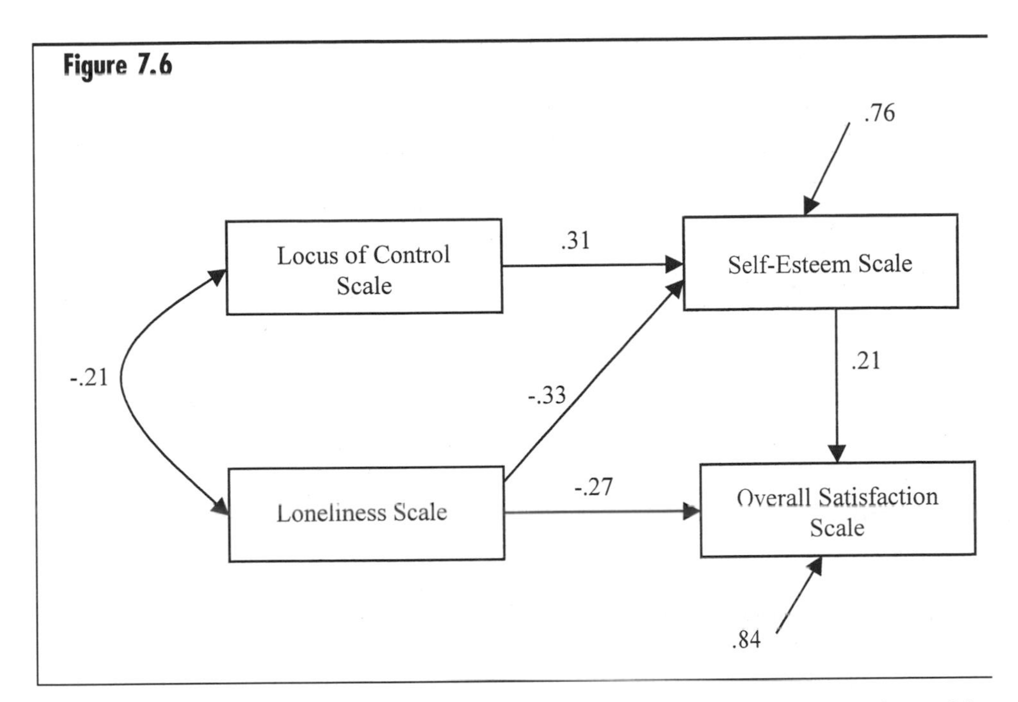

Path analysis of Overall Satisfaction of High School Seniors. The measured variables are scales (see text). $\chi^2(1,2503) = 37.02$, $p < .00$, GFI = .99, AGFI = .93, RMR = .03.

tural relationships between measured variables. Traditionally, parameters of a model such as the one in Figure 7.6 were estimated by a series of multiple regressions. SEM software, however, can be used to estimate the parameters of a path analysis model by equating the measured and unmeasured variables: In SEM path analysis, each unmeasured variable in the model is considered to be measured by a single observed variable. To replicate the results of a traditional path analysis, which assumes that the variables are perfectly measured, the coefficients linking pairs of measured and unmeasured variables must be fixed at 1.0. The parameter estimates from a conventional path analysis and a comparable SEM analysis are identical. There is a great advantage to using SEM, however, because the printout from an SEM analysis provides crucial information about the overall fit of the model. For some analyses there is also a second advantage: the opportunity in SEM to relax the assumption of perfect measurement. This is done by linking the unmeasured variables

plans, I am almost certain that I can make them work. Planbett: Planning ahead makes things turn out better. Feellone: A lot of times I feel lonely. Leftout: I often feel left out of things. Morefrds: I often wish I had more good friends.

to the measured variables by coefficients less than 1.0. The coefficients less than 1.0 are chosen to reflect the assumed or known reliability of the measured variables. It is extremely simple to run a path analysis using SEM, and it is the preferred method of doing such analyses.

The variables in Figure 7.6 were scales constructed (outside of SEM) by taking the means of the associated indicators. The substantive conclusions from the path analysis are similar to those for the full model (see Figure 7.2). The effects and explanatory power, however, are generally less in the path analysis. Probably this is partly because scales based on means were used for the path analysis, whereas factors with differential variable weights were used for the full model. Undoubtedly more important in accounting for the attenuation of effects is the assumption in the path model that the measured variables (which happen to be scales in this example) are measured without error. In the path analysis model in Figure 7.6, there is no counterpart to the 11 error variances in Figure 7.2 that are associated with measured variables.

Longitudinal Models

SEM is extremely useful in the area of longitudinal analysis, particularly for analysis of *panel data,* a kind of longitudinal data that is characterized by measurements on the same units (e.g., people) at two or more time points. Because a model of panel data includes the data at two or more different times, it is often possible to make stronger inferences about cause and effect than is possible with data from a single time point. If three or more time points are available, complicated models can be estimated. For example, it is possible, using SEM, to estimate models with both cross-lagged and synchronous effects. Suppose, for example, that a researcher were studying marital happiness. Husbands' happiness at earlier time points might affect wives' happiness at later time points and vice versa (thus creating cross-lagged effects). Also husbands' happiness and wives' happiness might have a reciprocal relationship at a given time point (thus, synchronous effects). As Finkel (1995) wrote, structural equation methods are now "routine tools for panel analysis" (p. 3). Finkel provided a readable practical guide to SEM for panel data (for a book-length discussion, see Kessler & Greenberg, 1981).

Multisample Models

A structural equation model can be fit to two or more groups simultaneously, allowing for any degree of difference between the groups. For example, one could test if the model in Figure 7.1 for Overall Satisfaction of High School Seniors was exactly the same for boys and girls, or one could test if it was the same for both groups, except for a few (or even a single) parameters. Typically, a hierarchy of nested hypotheses is tested and interactions (instances of differing relationships between groups) are pinpointed and tested precisely. The input for a multisample analysis is a covariance matrix for each group. For further information on multisample analysis, see Lomax (1983) and the Bollen (1989) and Hayduk (1987) texts. For an example of multisample analysis, see Scott-Lennox and Lennox (1995); they examined sex-race differences in social support and depression among disadvantaged rural elderly people.

Induced Variable Models

The above sections addressed factor variables or unmeasured variables that are hypothesized to cause the covariation of measured variables. In some models there is another kind of variable, an *induced variable*, which in some ways is very much like a factor. It is similar in that it is an unmeasured variable that affects other unmeasured variables in the model. The difference is that an induced variable is hypothesized as constituted of ingredient variables. For example, Abbey and Andrews (1985) modeled stress as composed of role ambiguity, negative life events, and social conflict. Another example appears in Liang (1986); he modeled chronic disease as composed of eight conditions (respiratory, circulatory, etc.). Figure 7.7 shows a diagram of an induced variable: Note that the arrows point from the ingredient variables (A, B, C) to the induced variable. Also note that there is no error variable associated with the induced variable: There is none because the model does not represent an attempt to explain variance in the induced variable. (Some formulations of an induced variable include an error term; for a discussion, see MacCallum & Browne, 1993.) The ingredients that cause an induced variable may or may not correlate; in contrast, the measured variables associated with a factor variable should correlate because they have a common cause, namely, the factor. Induced variables are sometimes termed sheaf variables, formative variables, or block variables. Bollen (1989) distinguished what are referred to here as factors and induced variables by whether their associated variables were "cause" variables or "effect" variables; his is an especially useful perspective if one is trying to decide the appropriate way to model partic-

Figure 7.7

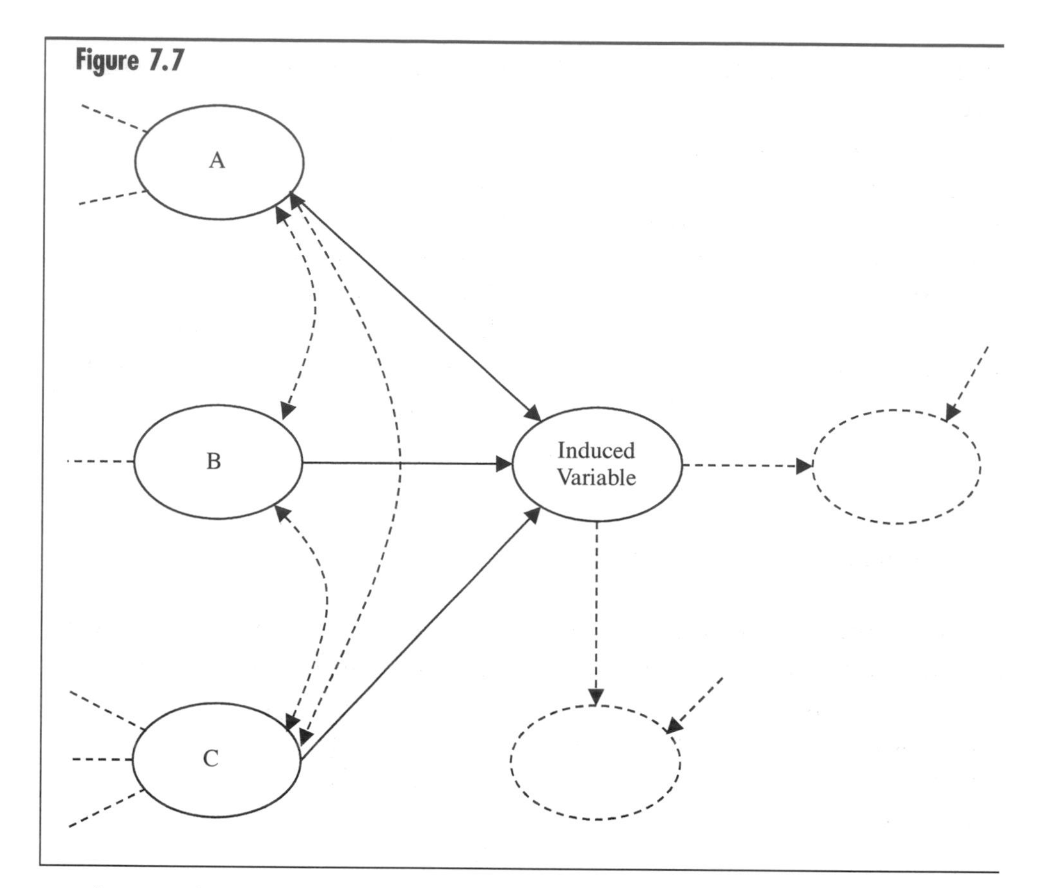

Schematic diagram of an induced variable. The dotted lines and ovals are intended to suggest that the induced variable is embedded in a larger model. A, B, and C represent the names of the variables that cause the induced variable.

ular data. If the associated variables are causes of the unmeasured variable, then the unmeasured variable should be modeled as an induced variable; if the associated variables are effects of the unmeasured variable, then the unmeasured variable should be modeled as a conventional factor.

Software Programs

LISREL was the first full-blown SEM program. Over the past 20 years or so, there have been constant improvements to LISREL, with the most recent version being LISREL 8 (Jöreskog & Sörbom, 1993). Meanwhile, other excellent SEM programs have arrived on the scene, including

EQS (Bentler, 1992), AMOS (Arbuckle, 1997), and CALIS (SAS Institute, 1991).

LISREL and EQS have been the most popular SEM software for the past decade. Both have a full range of SEM options, and from the perspective of the problems that can be analyzed, they are essentially equivalent. CALIS is noteworthy because it is part of the SAS System; however, it does not have an option for multisample analysis. AMOS, which has a full range of options and has recently been adopted by the SPSS computer package, is becoming very popular. Schumacker and Lomax (1996) provided telephone and fax numbers for these four software packages as well as for others.

Suggestions for Further Reading

There is now a continual flow of articles and books that makes SEM easy to understand and use appropriately. Four important texts are Bollen (1989), Hayduk (1987), Loehlin (1992), and Maruyama (1998). Bollen's book (which assumes that readers have experience with matrix algebra) provides the most advanced treatment, is comprehensive, and is widely cited. Hayduk's text is at an intermediate level and is less mathematically demanding than Bollen's. Loehlin's book, pleasantly informal in style, is intended as an introduction to SEM. Maruyama's book is intended to provide a gentle approach to the basis; it is an excellent choice, especially for someone without matrix algebra experience. For a monograph-length introduction to LISREL, see Long (1983b). Hoyle (1995) is an edited volume of chapters on concepts, issues, and applications in SEM; chapters are relatively free of technical jargon and equations and provide practical recommendations. Other particularly useful books include Bollen and Long (1993), Schumacker and Lomax (1996), and Hayduk (1996).

For introductions to CFA, see the chapter by Bryant and Yarnold (1995), in which they discussed both EFA and CFA, and a monograph by Long (1983a), which includes a brief introduction to the use of SEM for MMMT models. For a discussion of the historical background of SEM, again see Bollen (1989), which traces three lines of research that contributed to the development of SEM.

Many critiques have been written about the SEM technique itself, its practice, or both. Freedman (1987) discussed the limitations of path analysis, focusing on the failure of researchers to pay enough attention to the assumptions of the technique. Cliff (1983) and Breckler (1990) provided critiques of SEM that are useful because they indicated ways that practitioners can improve their research. (Cliff, 1983, wrote in his final paragraph: "Finally, let it be emphasized that the programs such as LISREL and its relatives provide completely unprecedented opportunities to do this sort of research well" [p. 125].) One characteristic of good modeling mentioned by both Cliff and Breckler is the consideration of equivalent models, that is, alternative models that fit the data equally well. MacCallum, Wegener, Uchino, and Fabrigar (1993) provided recommendations for alleviating and managing the problem of equivalent models.

Structural Equation Modeling: A Multidisciplinary Journal, which first appeared in 1994, is devoted to work from all disciplines interested in SEM and publishes articles on new developments and applications and reviews books and software. A recent issue published a 70-page annotated bibliography of current theoretical and technical contributions (Austin & Calderon, 1996). Among the numerous more general journals that publish articles on SEM are *Multivariate Behavioral Research, Psychological Methods, Educational and Psychological Measurement,* and *Psychometrika.*

I did not discuss in this chapter how to use SEM computer programs. The manuals for SEM software are generally excellent and, of course, need to be consulted to use the software. Even if the use of SEM is not in a research plan at this time, reading one of the manuals, all of which include many examples, is a good way to learn about SEM (see, e.g., Jöreskog & Sörbom, 1989; Bentler, 1992; and Arbuckle, 1997).

Glossary

ENDOGENOUS VARIABLE A variable that is determined in the model. In a model diagram, an endogenous variable will have one or more straight arrows pointed at it. An endogenous variable can be either measured or unmeasured. The designation of a variable as endogenous is model specific.

ERROR The error associated with variable X, which in SEM includes the effects of all sources of influence on X that are not included in the model (i.e., and measurement error; error includes effects of omitted variables). In structural equation models, there are errors associated with *measured variables* and with dependent (endogenous) *unmeasured variables.*

EXOGENOUS VARIABLE A variable that is determined outside of the model. Its causes are not specified in the model. An exogenous variable can be either measured or unmeasured. The designation of a variable as exogenous is model specific.

FACTOR ANALYSIS A technique that relates *measured variables* to *unmeasured variables* (factors). In exploratory factor analysis (EFA), the researcher makes few specifications about the model. In confirmatory factor analysis (CFA), which is a submodel of a *full structural equation model,* the researcher specifies a particular model. In a CFA model, each variable is usually hypothesized to be affected by only some of the unmeasured variables, rather than all of them as in EFA.

FIT Describes how consistent the observed data is with the model. In SEM, measuring the degree of fit involves comparing the correlations (or covariances) implied by the model with the actual coefficients in the data.

FIT INDICES Descriptive indices to evaluate the *fit* of a model. Dozens of indices have been proposed; SEM programs typically print several.

FULL STRUCTURAL EQUATION MODEL A model that combines a *measurement model* with a *structural model.* It is a model that includes both causal relationships among *unmeasured variables* and the relations of the unmeasured variables to *measured variables.*

INDUCED VARIABLE An *unmeasured variable* that is hypothesized to be caused by two or more *measured variables.* Like a conventional factor, it is a conceptual, composite variable; the difference between an induced variable and a conventional factor lies in the direction of the assumed causation between the unmeasured variable and its associated measured variables.

LONGITUDINAL DATA Data that are collected at two or more time points. The data may be on the same set of cases at each time point (i.e., panel data) or collected on different, but comparable, cases (i.e., a repeated cross-section of cases). Panel data can be analyzed using SEM.

MEASURED VARIABLE A variable for which there are values in the data set. Alternative names are *manifest variable, observed variable,* and *indicator.*

MEASUREMENT MODEL The part of a full model that relates *unmeasured variables* to *measured variables.*

MODEL IDENTIFICATION The status of a model based on the identification status of its parameters. A parameter is identified if it can be estimated uniquely, that is, if there is only one possible solution for its value. A model is identified if each of the parameters is identified. A model is not identified if one or more parameters is not identified. A subset of identified models is overidentified models; a model is overidentified if for at least one parameter there is more than one way to obtain its solution.

MODEL MODIFICATION The practice of changing an original model based on the results from the original specification. Sometimes a model is made simpler by deleting insignificant parameters. Sometimes the *fit* is improved by adding parameters. Because such changes may be at least partly determined by chance characteristics of the sample data, they may not generalize to another sample.

NESTED MODEL Often the parameters to be estimated in one model are a subset of the parameters to be estimated in a second model. In such a case, the first model is nested in the second model. More generally, a model that is a restricted version of another is nested within the less-restricted mode.

PANEL DATA Data collected at two or more time points for the same set of cases. The cases constitute a "panel."

PARAMETER A constant that measures the magnitude of the relationship between variables. SEM software estimates the parameters of a model.

PATH ANALYSIS Traditionally, a technique for fitting a causal model to *measured variables.* It is a submodel of a full structural model that does not involve any *unmeasured variables.* Path analyses can be done using SEM software.

STRUCTURAL EQUATION MODELING (SEM) Defining and estimating a general class of models that involves cause-and-effect relationships between variables, some of which may be unmeasured.

SPLIT SAMPLE A sample that is randomly divided into two groups. The groups are sometimes called the *training group* and the *holdout group.* A model can be developed on one group and tested on the other; this is a way of assessing the reliability of a model.

STRUCTURAL MODEL The part of a full model that summarizes the causal relationships between *unmeasured variables.*

UNMEASURED VARIABLE A conceptual or hypothesized variable. Alternative names are *latent variable, unobserved variable,* and *factor.*

References

Abbey, A., & Andrews, F. M. (1985). Modeling the psychological determinants of life quality. *Social Indicators Research, 16,* 1–34.

Arbuckle, J. L. (1997). *Amos user's guide, Version 3.6.* Chicago, IL: Smallwaters.

Austin, J. T., & Calderon, R. F. (1996). Theoretical and technical contributions to structural equation modeling: An updated annotated bibliography. *Structural Equation Modeling, 3,* 105–175.

Bagozzi, R. P., & Yi, Y. (1988). On the evaluation of structural equation models. *Journal of the Academy of Marketing Science, 16,* 74–95.

Bentler, P. M. (1992). *EQS structural equation manual.* Los Angeles, CA: BMDP Statistical Software.

Bentler, P. M., & Chou, C. P. (1987). Practical issues in structural modeling. *Sociological Methods and Research, 16,* 78–117.

Bollen, K. A. (1989). *Structural equations with latent variables.* New York: Wiley.

Bollen, K. A., & Long, J. S. (Eds.). (1993). *Testing structural equation models.* Newbury Park: Sage.

Bottorff, J. L., Johnson, J. L., Rainer, P. A., & Hayduk, L. A. (1996). The effects of cognitive-perceptual factors on health promotion behavior maintenance. *Nursing Research, 45,* 30–36.

Breckler, S. J. (1990). Applications of covariance structure modeling in psychology: Cause for concern? *Psychological Bulletin, 107,* 260–273.

Bryant, F. B., & Yarnold, P. R. (1995). Principal-components analysis and exploratory and confirmatory factor analysis. In L. G. Grimm & P. R. Yarnold (Eds.), *Reading and understanding multivariate statistics* (pp. 99–136). Washington, DC: American Psychological Association.

Chou, C., & Bentler, P. M. (1995). Estimates and tests in structural equation modeling. In R. H. Hoyle (Ed.), *Structural equation modeling: Concepts, issues, and applications* (pp. 37–55). Thousand Oaks, CA: Sage.

Cliff, N. (1983). Some cautions concerning the application of causal modeling methods. *Multivariate Behavioral Research, 18,* 115–126.

Finkel, S. E. (1995). *Causal analysis with panel data* (Sage University Paper series on Quantitative Applications in the Social Sciences, Monograph 07-105). Thousand Oaks, CA: Sage.

Freedman, D. A. (1987). As others see us: A case study in path analysis. *Journal of Educational Statistics, 12,* 101–128.

Fullagar, C. J. A., Gallagher, D. G., Gordon, M. E., & Clark, P. F. (1995). Impact of early socialization on union commitment and participation: A longitudinal study. *Journal of Applied Psychology, 80,* 147–157.

Grimm, L. G., & Yarnold, P. R. (Eds.). (1995). *Reading and understanding multivariate statistics.* Washington, DC: American Psychological Association.

Hayduk, L. A. (1987). *Structural equation modeling with LISREL.* Baltimore: Johns Hopkins University Press.

Hayduk, L. A. (1996). *LISREL: Issues, debates, and strategies.* Baltimore: Johns Hopkins University Press.

Hoyle, R. H. (Ed.). (1995). *Structural equation modeling: Concepts, issues, and applications.* Thousand Oaks, CA: Sage.

Hoyle, R. H., & Panter, A. T. (1995). Writing about structural equation models. In R. H. Hoyle (Ed.), *Structural equation modeling: Concepts, issues, and applications* (pp. 158–176). Thousand Oaks, CA: Sage.

Hu, L., & Bentler, P. M. (1995). Evaluating model fit. In R. H. Hoyle (Ed.), *Structural equation modeling: Concepts, issues, and applications* (pp. 76–99). Thousand Oaks, CA: Sage.

Johnston, L. D., Bachman, J. G., & O'Malley, P. M. (1993). *Monitoring the future: A continuing study of the lifestyles and values of youth* [Computer file]. Ann Arbor: University of Michigan, Survey Research Center, ICPSR [Ed.], Inter-University Consortium for Political and Social Research [Producer and distributor].

Jöreskog, K. G., & Sörbom, D. (1989). *LISREL 7: A guide to the program and applications* (2nd ed.). Chicago, IL: SPSS.

Jöreskog, K. G., & Sörbom, D. (1993). *LISREL 8: User's reference guide.* Chicago, IL: Scientific Software.

Kessler, R. C., & Greenberg, D. F. (1981). *Linear panel analysis.* New York: Academic Press.

Liang, J. (1986). Self-reported physical health among aged adults. *Journal of Gerontology, 41,* 248–260.

Loehlin, J. C. (1992). *Latent variable models: An introduction to factor, path, and structural analysis* (2nd ed.). Hillsdale, NJ: Erlbaum.

Lomax, R. G. (1983). A guide to multiple-sample structural equation modeling. *Behavior Research Methods and Instrumentation, 15,* 580–584.

Long, J. S. (1983a). *Confirmatory factor analysis: A preface to LISREL.* Beverly Hills, CA: Sage.

Long, J. S. (1983b). *Covariance structure models: An introduction to LISREL.* Beverly Hills, CA: Sage.

MacCallum, R. C., & Browne, M. W. (1993). The use of causal indicators in covariance structure models: Some practical issues. *Psychological Bulletin, 114,* 533–541.

MacCallum, R. C., Wegener, D. T., Uchino, B. N., & Fabrigar, L. R. (1993). The problem of equivalent models in applications of covariance structure analysis. *Psychological Bulletin, 114,* 185–199.

Marsh, H. W. (1996). Positive and negative global self-esteem: A substantive meaningful distinction or artifactors? *Journal of Personality and Social Psychology, 70,* 810–819.

Maruyama, G. M. (1998). *Basics of structural equation modeling.* Thousand Oaks, CA, Sage.

Raykov, T., Tomer, A., & Nesselroade, J. R. (1991). Reporting structural equation modeling results in *Psychology and Aging*: Some proposed guidelines. *Psychology and Aging, 6,* 499–503.

Rigdon, E. E. (1995). A necessary and sufficient identification rule for structural models estimated in practice. *Multivariate Behavioral Research, 30,* 359–383.

SAS Institute. (1991). *The CALIS procedure: Analysis of covariance structures.* Cary, NC: Author.

Schumacker, R. E., & Lomax, R. G. (1996). *A beginner's guide to structural equation modeling.* Mahwah, NJ: Erlbaum.

Scott-Lennox, J. A., & Lennox, R. D. (1995). Sex-race differences in social support and depression in older low-income adults. In R. H. Hoyle (Ed.), *Structural equation modeling: Concepts, issues, and applications* (pp. 199–216). Thousand Oaks, CA: Sage.

Vinokur, A. D., Price, R. H., & Caplan, R. D. (1996). Hard times and hurtful partners: How financial strain affects depression and relationship satisfaction of unemployed persons and their spouses. *Journal of Personality and Social Psychology, 71,* 166–179.

West, S. G., Finch, J. F., & Curran, P. J. (1995). Structural equation models with nonnormal variables: Problems and remedies. In R. H. Hoyle (Ed.), *Structural equation modeling: Concepts, issues, and applications* (pp. 56–75). Thousand Oaks, CA: Sage.

Appendix

Shown below are the items used in the model of Overall Satisfaction of High School Seniors. They were taken from Johnston et al. (1993), a data set distributed by the Inter-University Consortium for Political and Social Research. The short names on the left margin are mine.

Worth	I feel I am a person of worth, on an equal plane with others.
Dowell	I am able to do things as well as most other people.
Dowrong	I feel I can't do anything right.
Happy	Taking all things together, how would you say things are these days—would you say you're very happy, pretty happy, or not too happy these days?
Lifesat	How satisfied are you with your life as a whole these days?
Planunhp	Planning only makes a person unhappy since plans hardly ever work out anyway.
Planwork	When I make plans, I am almost certain that I can make them work.
Planbett	Planning ahead makes things turn out better.
Feellone	A lot of times I feel lonely.
Leftout	I often feel left out of things.
Morefrds	I often wish I had more good friends.

Happy was scored 1 to 3, not too happy to very happy. LIFESAT was scored 1 to 7, completely dissatisfied to completely satisfied. All other items were scored 1 to 5, disagree to agree.

Ten Commandments of Structural Equation Modeling

Bruce Thompson

Structural equation modeling (SEM; also variously termed "covariance structure analysis" and, somewhat speciously, "causal modeling") is increasingly being used within the social sciences. Indeed, it would be difficult to locate recent issues of social science journals in which some SEM applications were not reported. One new periodical, *Structural Equation Modeling: A Multidisciplinary Journal,* has been created that is exclusively devoted to SEM reports and issues. SEM has been called "the single most important contribution of statistics to the social and behavioral sciences during the past twenty years" (Lomax, 1989, p. 171). Similarly, Stevens (1996) argued that SEM techniques "have been touted as one of the most important advances in quantitative methodology in many years" (p. 415). Many would regard this as an understatement, although clearly SEM is sometimes used when much simpler methods would suffice (Wilkinson & APA Task Force on Statistical Inference, 1999).

Clearly also SEM is sometimes used incorrectly (Mueller, 1997). Of course, some misuses and errors are to be expected with a method that is relatively new, that is undergoing refinement at a seemingly exponential rate, and that many social scientists are still learning.

SEM has historical roots in two major classical traditions and is a hybrid of both. First, SEM always invokes a *measurement model,* specifying that the measured–observed variables reflect underlying latent–

A version of this chapter was presented at both the 1998 and the 1999 annual meetings of the U.S. Department of Education, Office of Special Education Programs (OSEP) Project Directors' Conference, Washington, DC.

synthetic variables, and is sometimes even used exclusively to investigate measurement issues (i.e., confirmatory factor analysis [CFA]). *Measured* (or observed) variables are scores that can be measured directly (e.g., how much people weigh, scores of clients on a self-concept measure). *Latent* (or synthetic or composite) variables are scores that are not directly measured (e.g., regression predicted *Y* scores, factor scores) but that are instead obtained by applying some system (e.g., a regression equation, a factor analysis factor) of weights (e.g., regression beta weights, factor pattern coefficients) to the measured variables. This measurement model aspect of structural modeling dates back to the factor analysis theory articulated by Spearman (1904).

Second, sometimes a regression structure among the latent-synthetic variables defined by the measurement models, called a "structural model," is also specified and tested. This second aspect of SEM can be traced back to classical path analysis methods (cf. Wright, 1921, 1934; see Klem, 1995).

The modern roots of SEM, however, can be traced especially to the theoretical developments formulated by Karl Jöreskog (cf. 1967, 1969, 1970, 1971, 1978) and to the computer program LISREL (analysis of *li*near *s*tructural *rel*ationships) developed by Jöreskog and his colleagues (e.g., Jöreskog & Sörbom, 1989). Today, modern SEM software is user friendly and allows microcomputer users to declare the models to test using software-aided drawings and point-and-click menus. Examples of such widely used software packages include EQS (Bentler, 1992a) and AMOS (Arbuckle, 1997).

In chapter 7, Laura Klem describes the uses and logic underlying SEM and outlines the steps involved in conducting SEM. The purpose of this chapter is to provide a primer regarding some of the basic issues that must be resolved to conduct SEM. First, seven key issues that must be considered in any SEM analysis are explained. Second, two common SEM misconceptions or incomplete understandings are confronted. Third, 10 commandments for proper SEM behavior are proffered.

Seven Key Decisions in SEM Analysis

1. Matrix of Associations to Analyze

It is widely known that all parametric statistical analyses are special cases within a single general linear model (GLM) family. In one of his innumerable seminal contributions, Jacob Cohen (1968) demonstrated that multiple regression subsumes all the univariate parametric methods (e.g., *t* test, analysis of variance [ANOVA], analysis of covariance) as special cases. Subsequently, Knapp (1978) presented mathematical theory showing that canonical correlation analysis subsumes all the parametric analyses, both univariate and multivariate, as special cases. Fan (1996a) and Thompson (1984, 1991) presented concrete demonstrations of these relationships.

SEM, however, represents an even bigger conceptual tent subsuming narrower special cases (Bagozzi, Fornell, & Larcker, 1981), including both canonical correlation analysis and multiple regression. Illustrations of these relationships were offered by Fan (1997) and by Thompson (1999a).

The GLM is a powerful heuristic device that can help researchers see three important commonalities that exist across various analytic methods. First, all of these methods use weights (e.g., regression beta weights, standardized canonical function coefficients) to optimize explained variance and minimize model error variance. Second, all the methods focus on the latent or synthetic variables (e.g., the regression $\hat{Y}$ variable, factor scores) created by applying the weights, such as beta weights, to scores on measured or observed variables, such as regression predictor variables; these latent or synthetic variables are taken as measures of our constructs. Third, all analytic methods are correlational (Knapp, 1978; Thompson, 1999a) and yield variance-accounted-for effect sizes analogous to r^2 (e.g., R^2, η^2, ω^2).

The commonality that all parametric methods apply weights to the measured or observed variables to compute latent or synthetic variables is obscured by the inherently confusing language of traditional statistics. As noted elsewhere, the weights in different analyses

> are all analogous, but are given different names in different analyses (e.g., beta weights in regression, pattern coefficients in factor analysis, discriminant function coefficients in discriminant analysis, and canonical function coefficients in canonical correlation analysis), mainly to obfuscate the commonalities of [all] parametric methods, and to confuse graduate students. (Thompson, 1992, pp. 906–907)

Indeed, both the weight systems (e.g., regression equation, factor, canonical function) and the synthetic or composite variables (e.g., the regression $\hat{Y}$ variable, factor scores, discriminant function scores) are also arbitrarily given different names across the analyses.

The first step of GLM analyses often involves the computation of a matrix of associations (e.g., Pearson product–moment correlation matrix, variance–covariance matrix) among the measured or observed variables. Indeed, because using only this matrix many GLM analyses can be replicated, editors frequently require that the association matrix be reported within SEM publications, so that readers may reproduce the analysis or explore alternative SEM analyses of the same data.

SEM analyses can be based on numerous matrices of association, such as product–moment correlation or polychoric correlation. Some researchers prefer to analyze Pearson correlation coefficients. These association coefficients are "scale free" because the standard deviations of a given pair of variables have been removed from the covariance of the two variables by division (i.e., $r_{XY} = COV_{XY}/[SD_X \times SD_Y]$). Thus, the weights derived from these correlations are themselves scale-free and can be more readily interpreted in relation to each other because all the measured variables have been effectively "standardized" by this process.

Most SEM theory, however, was developed for application with the matrix of associations among the measured or observed variables being a variance–covariance matrix (i.e., variances on the diagonal, covariances off the diagonal). It has also been established that while using the product–moment correlation matrix may be appropriate with some models, for other models some SEM statistics will be incorrect unless the variance–covariance matrix is used (Cudeck, 1989).

It is also important that the level of scale of the measured variables (e.g., categorical–nominal, ordinal–ranked, continuous–interval) is honored when one selects a given matrix of associations to be computed and analyzed. Of course, to some degree, judgments regarding measurement scale are subjective, and researchers may reasonably disagree as to some of these decisions. Data might be analyzed using a variety of plausible matrices of association, however, to confirm that results are not artifacts of methods choices.

2. Model Identification

When analyses are conducted, a model is being fit to the data, and the weights and other parameters (e.g., latent variable variances, latent variable covariances) associated with that model are being estimated. A critically important issue in this process involves determining whether the model is identified. A model is *identified* if, given the model and the data, a single set of weights and other model parameters can be computed (Byrne, 1998). If an infinite number of sets of weights and other parameters are equally plausible, then the parameters are mathematically indeterminate and the model is not identified (i.e., is *underidentified*). Mueller (1997, pp. 358–359) provided a fairly accessible summary of the conditions sufficient for model identification.

One key issue, with regard to identification, involves degrees of freedom. The notion of identification can be partially explored in the context of classical statistics, such as product–moment correlation or multiple regression. The degrees-of-freedom total in classical univariate analyses equals $n - 1$. If there are scores of only two people on only two variables, then the model degrees of freedom is one and the degrees-of-freedom error is zero; here, no matter what the scores are on the two variables, the r^2 value can only be one. This result can be computed, although the computation is a waste of time, because only one result is plausible when a model is *just-identified*. Similarly, scores of three people on one criterion variable and two predictor variables would yield a degrees of freedom error of 0 and an inescapable R^2 value of 1.0.

In SEM, the degrees of freedom total is a function of the number of nonredundant pieces of information present in the matrix of associations being analyzed (not the number of people). For example, with eight measured variables, there would be eight variances and 28 nonredundant (either below or above the diagonal) covariances ($[8 \times (8 - 1)]/2 = [8 \times 7]/2 = 56/2 = 28$). This would result in $8 + 28 = 36$ degrees of freedom being available for any structural equation model fit to these data.

In SEM, each parameter (e.g., weight, path coefficient, variance or covariance among latent or synthetic variables) that is estimated uses one degree of freedom. Thus, for the problem involving eight measured variables, if I specify a model involving the estimation of 36 model parameters, the model is just-identified. These parameters can be estimated; that is, these parameters are mathematically determined, with only one plausible set of estimates. The results from a just-identified structural equation model are just as trivial, however, as would be the results from an r^2 analysis involving scores of two people on two measured or observed variables, because such models always reproduce exactly the analyzed matrix of associations.

Researchers are most interested in models that spend fewer degrees of freedom (i.e., estimate fewer model parameters) and are thus more parsimonious. We prefer models that have more degrees of freedom (i.e., there are a lot more degrees of freedom total than the number of estimated parameters) but yield parameters that still do reasonably well at reproducing the matrix of associations (e.g., the correlation matrix) from which the parameters were all estimated are the preferred models. Such models receive the most interest because there are more ways in which these models are potentially falsifiable, and so such models represent more rigorous and persuasive tests of our conceptions of latent constructs (Mulaik, 1987, 1988; Mulaik et al., 1989). In other words, models that are considerably overidentified are preferred. Having more than zero degrees of freedom is a necessary, but not sufficient, condition for model identification. That is, one simply cannot estimate the parameters for any underidentified model.

SEM computer programs tend to run diagnostics that indicate when models have not been identified. When this occurs, some parameters for which estimates were initially requested (i.e., "freed" to be estimated) must be "fixed" as not estimated. For example, a weight or the latent–synthetic variable's variance that is "freed" might be fixed to equal 1.0, or the error variance of a measured or observed variable might be fixed to equal 0. Thus, if one's original structural equation model had more parameters to estimate (i.e., freed parameters) than were warranted given the number of measured or observed variables (i.e., the model was just- or underidentified), then it would be necessary to fix some of the freed parameters, so that they were no longer estimated, in order to achieve an overidentified model with sufficient degrees of freedom.

3. Parameter Estimation Theory

Classical univariate and multivariate parametric analyses (e.g., *t* tests, ANOVA, descriptive discriminant analysis) invoke a statistical theory of parameter estimation called "ordinary least squares." There are, in fact, numerous other statistical theories that can be invoked to estimate freed model parameters. Among these various alternatives are maximum likelihood (ML), generalized least squares (GLS), and asymptotically distribution-free (ADF; Browne, 1984) *estimation theories*; the various estimation theories differ with regard to both their assumptions and their theoretical properties.

For example, regarding assumptions, both ML and GLS estimations presume that the data have a multivariate normal distribution. Of course, this distributional assumption also invokes issues involving the measurement scale of the measured or observed variables because, for example, dichotomous variables cannot even be univariate normally distributed, even if the dichotomous variable scores are symmetrical. ADF estimation, however, does not require the assumption of multivariate normality. West, Finch, and Curran (1995) reviewed relevant issues and choices regarding distributional assumptions.

In most SEM computer programs, ML estimation is the default. Perhaps for this reason ML estimation is used with considerable frequency in SEM reports. ML estimates are used to seek parameters that best reproduce the estimated population variance–covariance matrix. Of course, this may be another reason for the frequent use of this estimation method because accurate estimates of population parameters in theory should yield results with the greatest replicability.

4. Multivariate Normality

A necessary but not sufficient condition for multivariate normality is bivariate normality of all pairwise combinations of the measured or observed variables. In turn, such a condition for bivariate normality is univariate normality of all the measured or observed variables.

Even simple univariate normality is a more elusive concept than most researchers realize, however (Bump, 1991). There is an infinite number of univariate normal data distributions, each differing in appearance. (Some researchers have been lulled into the misconception that all univariate normal distributions have a single classic "bell" shape because almost all textbooks only present graphs of the normal distributions of *z* scores. For data not in *z*-score form, however, there are numerous plausible symmetrical distributions that are all normal but that differ markedly in appearance.)

There is no distinctly superior method by which to establish that the multivariate normal distribution assumption has been met, so that certain estimation theories can then be used. Henson (1999) reviewed some of the available choices. One user-friendly method for evaluating multivariate normality invokes a graphical procedure and can readily be implemented using a short SPSS-for-Windows program. Thompson (1990; see also Fan, 1996b) described this method in more detail.

5. Model Misspecification and Specification Searches

All overidentified models portray the relationships among measured and latent variables. The goal is to specify these relationships for the population so that future samples from the same population will yield comparable findings. The model is overidentified partly so that the model is falsifiable but also because researchers seek simplifications of reality that remain useful and make the understanding of reality more manageable. As Mueller (1997) noted,

> a structural equation model is nothing more than an oversimplified approximation of reality, no matter how carefully conceptualized. A good model can be characterized as featuring an appropriate balance between efforts to represent a complex phenomenon in the simplest [most parsimonious] way and to retain enough complexity that [still] leads to the most meaningful [and true] interpretations possible. (p. 365)

A perfectly "specified" overidentified model exactly reproduces the associations among the measured or observed variables. Because the model is overidentified, however, the model never perfectly reproduces data in either the sample or the population. Thus, researchers must somehow evaluate whether the model is sufficiently adequate to remain both reasonably manageable and reasonably correct.

Model Misspecification

If the model is deemed not to be correct, then the model is called *misspecified.* Making this judgment is critical because the SEM parameter (e.g., weights, variances, covariances) estimation processes

> all fail to provide correct sample estimates, standard errors, and data-model fit chi-square statistics . . . if the model under consideration is misspecified and does not reflect at least a very close approximation to the true structure in the population. (Mueller, 1997, p. 359)

Of course, because a simplified model of reality is always at least partially misspecified, making the judgment as to when a model is misspecified can be challenging.

Through the years, myriad fit statistics have been developed to aid in making these judgments. Byrne (1998, pp. 109–119) reviewed some of the fit statistics provided by the SEM computer programs. Arbuckle (1997, pp. 551–572) summarized both some of the relevant formulae used to compute these statistics and a bit of the literature on the "rules of thumb" for interpreting these values.

The stark, harsh reality, however, is that there is still much to learn regarding both how these SEM-fit statistics operate under different conditions and what should be the criteria for declaring reasonable model fit. Indeed, until recently too much of the Monte Carlo simulation work on these issues did not use misspecified models, meaning that the stimulation results did not directly bear on the real-world situation in which the model is at least partially misspecified and in which the researcher does not know for certain which or how many features of the model specification are correct (cf. Fan, Thompson, & Wang, 1999; and Fan, Wang, & Thompson, 1997).

An important consideration in evaluating the fit of a given model involves the modeling context of this judgment. The most persuasive case that a model has been correctly specified is created when a researcher finds a differentially better fit of a given model as against the fit of numerous other defensible, thoughtfully formulated, rival plausible models; therefore, multiple models should be evaluated in any SEM project.

It is also critical to remember that even such findings do not conclusively establish that a single given model is definitively correct. Many, many models can fit a given data set. Thus, the fit of a single tested model may always be an artifact of not having tested all possible models.

In any case, also remember that an overidentified model is defined to simplify reality. Researchers seek a simplification that can be subjectively judged to be inherently somewhat inaccurate but still reasonably useful and manageable. What is not being sought is a single truth in the context of a simplification that inherently distorts some features of reality. Model fit statistics are used to assist in making these judgments, but the judgment that is made is inherently subjective. Researchers then must accept the responsibility for the construct definitions formulated (Mulaik, 1994), including both the measurement model definitions of how the constructs express themselves through the measured variables and the definitions of how the constructs relate to other constructs.

Model Fit Statistics

Given space limitations, only a few of the myriad model fit statistics can be reviewed here (see Bentler, 1994). A χ^2 *goodness-of-fit test statistic* can be computed to test the null hypothesis that the variance–covariance matrix reproduced by the freed model parameter estimates equals the variance–covariance matrix (i.e., the model exactly reproduces all relationships). This statistic is printed by all the SEM computer programs. Note that in contrast to most traditional statistical significance testing, the researcher hopes *not* to reject this null hypothesis so that the model can be taken as fitting the data.

Even though this application of statistical testing is a variant on usual practice, one of the numerous criticisms of classical statistical significance testing also applies: The result is partially an artifact of sample size (cf. Cohen, 1994; Thompson, 1996, 1998, 1999a, 1999c, 1999d). As Bentler and Bonett (1980) made clear,

> in very large samples virtually all models that one might consider would have to be rejected as statistically untenable. . . . This procedure cannot be justified, since the chi-square variate ν can be made small by simply reducing sample size. (p. 591)

The chi-square statistic, however, can be useful in comparing the fits of models for a given data set with a single sample size, particularly if the models are "nested" within each other (cf. Jöreskog & Sörbom, 1989, pp. 230–233).

The *Goodness-of-Fit Index* (GFI) and the *Adjusted Goodness-of-Fit Index* (AGFI) (Jöreskog & Sörbom, 1984) essentially compare the ability of a model to reproduce the variance–covariance matrix to the ability of no model at all to do so. The AGFI adjusts the GFI for the number of degrees of freedom expended in estimating the model parameters. Indices less than zero are treated as zero and range up to one, with one indicating perfect model fit. Most researchers expect values to be greater than .90 or .95 for correctly specified models.

The *root mean-square residual* (RMR) evaluates the average residual value for the variance–covariance matrix reproduced by the model parameters and the actual variance–covariance matrix. The RMR can range down to zero, which would indicate perfect model fit. A well-fitting model has values of, "say, .05 or less" (Byrne, 1998, p. 115).

Bentler and Bonett (1980) proposed a *Normed Fit Index* (NFI), which compares model fit to that of a model for the same data presuming independence of the measured or observed variables. NFI ranges between zero and one, with higher values indicating better fit. Usually values greater than .90 or .95 are considered reflective of adequate fit. The Bentler and Bonett article has been one of the most widely cited in psychological literature (see Bentler, 1992b). However, NFI has been shown to be an underestimate when small samples are used. Consequently, Bentler (1990) proposed an adjustment to the NFI, the *Comparative Fit Index* (CFI), which takes sample size into account. Some researchers have suggested that the CFI should be a fit statistic of choice in SEM research (Byrne, 1998).

Various *parsimony-weighted fit indices* have been proposed (see Mulaik et al., 1989, but also Marsh & Hu, 1998). These fit-statistic weights, which range up to one and down to zero for just-identified models, are multiplied by indices such as the NFI, to take model complexity into account and reward models that estimate fewer parameters.

Some fit indices focus on estimated population fit. Steiger and Lind (1980) proposed *root mean square error of approximation* (RMSEA) as one such index. As Byrne (1998) noted, RMSEA "has only recently been recognized as one of the most informative criteria in covariance structure modeling" (p. 112). Values approaching zero are desired, and a value of .08 or less would indicate a reasonable error of approximation (Browne & Cudeck, 1993).

The various fit indices provide a constellation of information about the competing models being considered in an SEM analysis. Because some of the fit indices evaluate different aspects of fit, it is important to evaluate fit based on multiple fit statistics so that judgments will not be an artifact of analytic choice. Furthermore, as Byrne (1998) correctly emphasized, "assessment of model adequacy must be based on multiple criteria that take into account theoretical, statistical, and practical considerations" (p. 119).

Specification Search

In addition to providing fit indices for a given model, SEM analyses also provide important information regarding exactly where potential model specification errors may have occurred. There are two possible types of errors, and different information is used to evaluate each of the possibilities.

First, model misspecification may involve having freed a parameter to be estimated when, in fact, the parameter is not useful in reproducing relationships and should instead have been fixed (e.g., two latent–synthetic variables should have been constrained to be uncorrelated in the model, or the measurement error variance of a measured or observed variable should have been constrained to be zero). In classical statistics, the ratio of a mean to the standard error of the mean can be computed and is the calculated test statistic, called *t*. For most sample sizes, a $t_{\text{calculated}}$ greater than two in absolute value is statistically significant at approximately the $\alpha = .05$ level.

In SEM, *t* (sometimes also called "Wold statistic" or the "critical ratio" [CR]) can be computed by dividing any given parameter estimate by its standard error. Any ratio less than $|2|$ suggests a possible model specification error in the form of freeing a parameter that instead might have been fixed.

Second, model misspecification may involve having fixed a parameter not to be estimated when, in fact, the parameter might be useful in reproducing relationships and should instead have been freed. SEM computer programs, on request, provide *modification indices* for each fixed model parameter, which indicate approximately how much smaller (i.e., better) the model chi-square statistic would become if a given fixed parameter were instead freed. Large values for these indices may indicate that freeing a given fixed parameter should be considered.

The process of modifying an a priori model based on such results is called a *specification search.* This practice is discouraged (see Mueller, 1997), unless the model is changed on the basis of statistical results for one sample and then the respecified model is evaluated in an independent (e.g., new or hold-out cross-validation) sample. Clearly, the more model features that are altered based on sample results, the greater is the likelihood that sampling error variance (i.e., the variability reflecting the idiosyncratic and nonreplicable features of a given sample) will be capitalized on, leading then to nonreplicable model fit. This is *not* an issue of statistical nit-picking, as MacCallum, Roznowski, and Necowitz (1992) and MacCallum, Wegener, Uchino, and Fabrigar (1993) powerfully demonstrated. Furthermore, model specification should never be based on blind, dust-bowl empiricism. Models should only be respecified in those cases in which the researcher can articulate a persuasive rationale as to why the modification is theoretically and practically defensible.

6. Sample Size

SEM is inherently a large-sample technique. At least four cases in which successively larger samples are needed can be noted. First, still larger samples are needed as more measured or observed variables are used. Second, even larger samples are required as more complex models with more parameters are evaluated. Third, even larger samples are needed when more elegant parameter estimation theories (e.g., asymptotically distribution-free estimation) are used. Fourth, even larger samples are needed if the researcher is going to perform any model specification search.

Some have suggested that sample size should be at least 200 (Baldwin, 1989). Similarly, Lomax (1989) suggested "a sample size of at least 100 (if not 200)" (p. 189). Furthermore, it has been suggested that the ratio of number of people to number of measured or observed variables (n:v) should be at least 10:1 (Mueller, 1997) if not 15:1 or 20:1. Thus, in even the most straightforward SEM applications, sample size should probably be the minimum of either (a) 100–200 people or (b) an n:v ratio of at least 10:1 or 15:1. MacCallum, Browne, and Sugawara (1996) provided some statistical methods for more precisely estimating the sample size necessary for a given SEM problem.

7. Measurement Model Adequacy

As noted previously, SEM structural models incorporate several measurement models in which measured or observed variables are taken as reflecting underlying latent constructs in the form of latent–synthetic variables. The regression path models of some of these latent–synthetic variables with each other are then estimated (e.g., Bryant & Yarnold, 1995, discussed CFA in this context). Researchers have increasingly recognized that the measurement models within SEM structural models have often been the weak links in SEM analyses.

Simply stated, if the specified measurement models do not fit the measured variables, then knowing the relationships among the latent–synthetic variables defined by these measurement models is essentially useless. Thus, some researchers (cf. Anderson & Gerbing, 1988) have recommended that SEM structural analyses should be approached as a two-step hierarchical process: First, confirm that the specified measurement models all fit their respective data; second, then and only then, explore the structural relationships among the latent–synthetic variables.

It has been generally agreed that it is useful to explore the measurement models embedded within structural models prior to evaluating the structural models. Some have argued, however, that measurement models may also be reasonably re-evaluated and perhaps respecified within the subsequent structural model analyses (see Hayduk, 1996).

A common error, which should be studiously avoided, is often made when interpreting measurement models with correlated factors. If factors are correlated within a measurement model or CFA, then measured variables may all be substantially correlated with all factors, even with those factors for which given measured variables were specified to have zero factor pattern coefficients (Thompson, 1997). Thus, when factors in the measurement model are correlated, the measured variables should not be interpreted as having zero correlations with correlated factors on which they have zero pattern coefficients.

Overall, clearly bad measurement models make the related structural models of questionable value. Some researchers have paid inadequate attention to the fit of the measurement models they have specified within their structural models.

Two Common Misconceptions

Two common misconceptions or, at least, incomplete misunderstandings, pervade much of SEM practice. The first involves the role of measurement error in SEM compared with that in classical statistical analyses. The second involves the view of SEM as causal modeling. Space precludes the presentation here of SEM analyses that would make this discussion concrete; however, Jöreskog and Sörbom (1989, pp. 151–156) presented analyses of data reported by Bagozzi (1980) that illustrate both sets of issues.

Role of Measurement Error Variance Within SEM

Too few researchers understand neither what reliability is (see Thompson & Vacha-Haase, 2000) nor, as the APA Task Force recently emphasized, how score reliability affects statistical results (Wilkinson & APA Task Force on Statistical Inference, 1999). For example, some researchers have persisted in erroneously referring to the "reliability of the test" (see Reinhardt, 1996; Thompson, 1994; and Vacha-Haase, 1998).

In classical statistical analyses (e.g., ANOVA, regression, canonical correlation analysis), measurement error affects parameter estimates and attenuates detected effect sizes (Thompson, 1994; Wilkinson & APA Task Force on Statistical Inference, 1999). In classical analyses, however, these measurement effects are not directly explored and evaluated. The primary distinguishing feature of SEM is that score reliability (i.e., [1 − measurement error variance]/total score variance) is directly considered as part of model fitting (Stevens, 1996, p. 415).

Jöreskog and Sörbom (1989, pp. 151–156) presented an instructive heuristic analysis of the Bagozzi (1980) data that illustrates (a) how measurement error variance estimates can affect parameter estimates throughout the model and (b) how these effects can be directly evaluated using SEM. One model posited that one measured variable, "verbal intelligence," was measured with perfect reliability (i.e., the error variance for this measured variable was fixed to equal 0). In a second model the error variance of verbal intelligence was posited to be $\delta_5 = 1.998$.

That is, based on previous reliability generalization research (Vacha-Haase, 1998) or theory, it was assumed that the reliability of the measured variable was .85 rather than 1.0; so in this second model, the error variance for verbal intelligence was instead fixed to equal 1 − .85 times the measured score variance of 13.323 (i.e., 1.998). (Of course, generally, the error variances of measured variables can also be freed so that these parameters are instead estimated.)

Jöreskog and Sörbom (1989, p. 155) then compared the estimates for the freed parameters across the two analyses to illustrate how just slightly changing the error variance for even one measured variable can change (at least slightly) the parameter estimates throughout the entire model. Thus, the consideration of error variances (i.e., whether these estimates are freed or fixed and, if fixed, fixed to what) is an integral part of SEM. The sensitivity of the results to one's choices here can be directly evaluated in SEM.

Classical statistical analyses (e.g., ANOVA, canonical correlation analyses) presume no measurement error variance for any of the measured variables, whereas SEM models are usually specified to estimate and to take into account measurement error variance for all or most of the measured or observed variables. Think how different the parameter estimates even for the same data may therefore be across SEM as compared with non-SEM analyses because usually all or none, respectively, of the score reliability coefficients are taken into account in making parameter estimates. Which analytic model best honors a reality where measured or observed variables are not measured with perfect reliability?

SEM as "Causal Modeling"

As noted at the outset of this chapter, historically SEM has been referred to by some as "causal modeling." Here, the mechanisms for this thinking are briefly described, but some strong cautions are also noted.

In the Jöreskog and Sörbom (1989, pp. 151–156) reanalyses of Bagozzi's (1980) data, one model specified that the latent–synthetic variable, "job performance," predicts the latent–synthetic variable, "job satisfaction," but not vice versa. A rival model was identical, except that job performance and satisfaction were presumed to reciprocally predict each other. Finally, yet another rival model specified that job satisfaction predicts job performance but not vice versa.

For the evaluation of the causal issues involved in these three rival models, the fit statistics for the three models can be compared. Additionally, the parameters for job performance predicting, reciprocally related to, and predicted by job satisfaction, respectively, can be examined in relation to their respective standard errors (i.e., *t* statistics). If for one model the fit statistics for the model were appreciably better and *t* for the path between the two latent–synthetic variables was considerably larger than for the other two models, some evidence regarding causality can be adduced from the analysis.

But does such an analysis prove a "causal model"? Certainly insight regarding causality might be inferred from the comparisons made here. Making such inferences, however, would be extremely controversial. The view taken here is that definitive causal evidence can only be extrapolated from thoughtfully designed true experiments. Given a nonexperimental design, such as the one that yielded these data, such correlational analyses yield inherently ambiguous causal results.

The argument can be framed by considering the "context-specificity" of all GLM weights (see Thompson, 1999a). If a single measured–observed variable was added or subtracted, all of the parameters might change dramatically. This is one aspect of model specification (i.e., are the exactly correct and only the exactly correct measured variables present?).

If it was certain that exactly (and only) the correct measured variables were present, then SEM might bear more powerfully on issues of causality. As Pedhazur (1982) noted, though, "the rub, however, is that the true model is seldom, if ever, known" (p. 229). As Duncan (1975) noted as well, "indeed it would require no elaborate sophistry to show that we will never have the 'right' model in any absolute sense" (p. 101).

Ten Commandments for Good SEM Behavior

Huberty and Morris (1988) observed that "as in all of statistical inference, subjective judgment cannot be avoided. Neither can reasonableness!" (p. 573). This is true throughout the panorama of statistical methods, but judgment and reasonableness are especially the sine qua non of SEM.

In the previous discussion, some basic precepts and principles were laid out to guide the novice modeler in exercising this judgment. Some of these principles can be summarized in the following 10 commandments for good SEM behavior, presented in reverse order of importance:

10. Do not use SEM with small samples.
9. Carefully consider the levels of scale and the distributions of measured or observed variables when selecting the matrix of associations to analyze.
8. All things being roughly equal for a given data set, prefer well-fitting, more parsimonious models because their fit is least an artifact of the model being nearly just-identified.
7. When using estimation theories requiring multivariate normality, use measured or observed variables that can be normally distributed (e.g., variables that are not dichotomous) and empirically evaluate whether the distributional assumption is met.
6. Use multiple fit statistics because several fit statistics consider different aspects or conceptions of fit, so that a judgment of correct specification is not a product of analytic choice and because there is still much to learn about the behavior of these statistics.
5. In evaluating model specification, in addition to considering statistical evidence, "assessment of model adequacy must be based on multiple criteria that [also] take into account theoretical . . . and practical considerations" (Byrne, 1998, p. 119; i.e., remember that we define the constructs that we use and that we are responsible for making and defending our decisions).
4. Individually evaluate the measurement models prior to evaluating a structural equation model (but still consider reformulating the measurement models if structural modeling then suggests this may be appropriate, notwithstanding the preliminary measurement model evaluations).
3. Test multiple plausible rival models, so that stronger evidence supporting the correct specification of a model can be adduced.
2. Regarding specification searches, require larger samples, test the respecified model with a hold-out or independent sample, and never change a specification unless a theoretical justification for the changes to the a priori model can be offered.
1. Never conclude that a model has been definitively proven because infinitely many models can fit any given data set (thus,

the fit of a single tested model may always be an artifact of having tested too few models).

Suggestions for Further Reading

Accessible short treatments of SEM have been provided by Baldwin (1989), Mueller (1997), and Lomax (1989). Extraordinarily good longer treatments, which include numerous examples and focus on EQS and LISREL, are the various works by Barbara Byrne (cf. 1994, 1998; also see Long, 1983a, 1983b). West, Finch, and Curran (1995) reviewed the relevant issues and choices regarding the distributional assumptions of various parameter estimation theories. Both Byrne (1998, pp. 109–119) and Arbuckle (1997, pp. 551–572) reviewed some of the fit statistics provided by the SEM computer programs. MacCallum, Browne, and Sugawara (1996) provided some statistical methods for more precisely estimating the sample size necessary for a given SEM problem.

Glossary

ESTIMATION THEORY The estimation of a model's parameters (e.g., weights, correlations of scores on the *latent–synthetic variables*) invokes one from among many statistical theories (e.g., ordinary least squares, maximum likelihood, asymptotically distribution free); the various estimation theories differ with regard to both their assumptions (e.g., normality) and their theoretical properties.

FIT STATISTIC Fit statistics are indices of the ability of the model to reproduce the relationships among the measured variables; these statistics vary regarding the aspects of fit being evaluated and as to whether better fit is indicated by larger (e.g., CFI) or smaller (e.g., RMSEA) values.

IDENTIFIED A model is identified if, given the data, only a single set of parameters can be estimated (vs. an infinite number of equally plausible estimates being available for an "unidentified" model).

LATENT–SYNTHETIC VARIABLES Variables involving scores that are not directly measured (e.g., regression-predicted *Y* scores, factor scores) but that are instead obtained by applying a system (e.g., a regression equation, a factor analysis factor) of weights (e.g., regression beta weights, factor pattern coefficients) to the measured variables.

LOADING Loading is an overly generic slang term sometimes used to refer to pattern or structure coefficients from a measurement model or to path coefficients from within the structural model. The interchangeable use of one term to refer to different results can confuse readers or confound interpretations (Gorsuch, 1983, p. 25; Thompson & Daniel, 1996).

MEASURED–OBSERVED VARIABLES Variables involving scores that are measured directly (e.g., how much people weigh, scores of clients on a self-concept measure), without applying weights (e.g., regression beta weights) to maximize or minimize a criterion (e.g., the sum of squares error).

MEASUREMENT MODEL The measurement model specifies how a given construct (defined as a *latent–synthetic variable*) expresses itself through the *measured–observed variables;* an SEM analysis consisting only of measurement models is usually referred to as a confirmatory factor analysis (CFA).

MODEL MISSPECIFICATION A model is misspecified if a parameter is estimated (i.e., "freed") when the parameter should not have been estimated or a parameter is not estimated (i.e., "fixed") when the parameter should have been estimated; misspecification may also include the incorrect specification of *latent–synthetic variables* that should or should not have been estimated or of paths or relationships among variables of either kind.

MODIFICATION INDICES These indices indicate approximately how much smaller (i.e., better) the model chi-square statistic would become if a given fixed parameter was instead freed.

PATTERN COEFFICIENTS The factor analysis or measurement model "weights" applied to *measured–observed variables* to estimate *latent–synthetic variable* scores.

SPECIFICATION SEARCH The controversial practice of modifying the model (e.g., parameters that are freed or fixed) on the basis of statistical results for one sample and then of evaluating the fit of the respecified model using the same sample.

STRUCTURAL MODEL The structural model specifies the relationships between the constructs in the form of the *latent–synthetic variables.*

STRUCTURE COEFFICIENT The bivariate correlation between a *measured–observed variable* and a *latent–synthetic variable* in SEM, just as in regression and other GLM techniques.

References

Anderson, J. C., & Gerbing, D. W. (1988). Structural equation modeling in practice: A review and recommended two-step approach. *Psychological Bulletin, 103,* 411–423.

Arbuckle, J. A. (1997). *Amos users' guide: Version 3.6.* Chicago, Smallwaters.

Bagozzi, R. P. (1980). Performance and satisfaction in an industrial sales force: An examination of their antecedents and simultaneity. *Journal of Marketing, 44,* 65–77.

Bagozzi, R. P., Fornell, C., & Larcker, D. F. (1981). Canonical correlation analysis as a special case of a structural relations model. *Multivariate Behavioral Research, 16,* 437–454.

Baldwin, B. (1989). A primer in the use and interpretation of structural equation models. *Measurement and Evaluation in Counseling and Development, 22,* 100–112.

Bentler, P. M. (1990). Comparative fit indices in structural models. *Psychological Bulletin, 107,* 238–246.

Bentler, P. M. (1992a). *EQS: Structural equations program manual.* Los Angeles, CA: BMDP Statistical Software.

Bentler, P. M. (1992b). On the fit of models to covariances and methodology to the *Bulletin. Psychological Bulletin, 112,* 400–404.

Bentler, P. M. (1994). On the quality of test statistics in covariance structure analysis: Caveat emptor. In C. R. Reynolds (Ed.), *Cognitive assessment: An multidisciplinary perspective* (pp. 237–260). New York: Plenum Press.

Bentler, P. M., & Bonett, D. G. (1980). Significance tests and goodness of fit in the analysis of covariance structures. *Psychological Bulletin, 88,* 588–606.

Browne, M. W. (1984). Asymptotically distribution-free methods for the analysis of covariance structures. *British Journal of Mathematical and Statistical Psychology, 37,* 62–83.

Browne, M. W., & Cudeck, R. (1993). Alternative ways of assessing model fit. In K. A. Bollen & J. S. Long (Eds.), *Testing structural equation models* (pp. 136–162). Thousand Oaks, CA: Sage.

Bryant, F. B., & Yarnold, P. R. (1995). Principal-components analysis and exploratory and confirmatory factor analysis. In L. G. Grimm & P. R. Yarnold (Eds.), *Reading and understanding multivariate statistics* (pp. 99–136). Washington, DC: American Psychological Association.

Bump, W. (1991, January). *The normal curve takes many forms: A review of skewness and kurtosis.* Paper presented at the annual meeting of the Southwest Educational Research Association, San Antonio, TX. (ERIC Document Reproduction Service No. ED 342 790)

Byrne, B. M. (1994). *Structural equation modeling with EQS and EQS/Windows: Basic concepts, applications, and programming.* Thousand Oaks, CA: Sage.

Byrne, B. M. (1998). *Structural equation modeling with LISREL, PRELIS, and SIMPLIS: Basic concepts, applications, and programming.* Mahwah, NJ: Erlbaum.

Cohen, J. (1968). Multiple regression as a general data-analytic system. *Psychological Bulletin, 70,* 426–443.

Cohen, J. (1994). The earth is round ($p < .05$). *American Psychologist, 49,* 997–1003.

Cudeck, R. (1989). The analysis of correlation matrices using covariance structure models. *Psychological Bulletin, 105,* 317–327.

Duncan, O. D. (1975). *Introduction to structural equation models.* New York: Academic Press.

Fan, X. (1996a). Canonical correlation analysis as a general analytic model. In B. Thompson (Ed.), *Advances in social science methodology* (Vol. 4, pp. 71–94). Greenwich, CT: JAI Press.

Fan, X. (1996b). A SAS program for assessing multivariate normality. *Educational and Psychological Measurement, 56,* 668–674.

Fan, X. (1997). Canonical correlation anaylsis and structural equation modeling: What do they have in common? *Structural Equation Modeling, 4,* 65–79.

Fan, X., Thompson, B., & Wang, L. (1999). The effects of sample size, estimation methods, and model specification on SEM fit indices. *Structural Equation Modeling, 6,* 56–83.

Fan, X., Wang, L., & Thompson, B. (1997, March). *Effects of data nonnormality on fit indices and parameter estimates for true and misspecified SEM models.* Paper presented at the annual meeting of the American Educational Research Association, Chicago, IL. (ERIC Document Reproduction Service No. ED 408 299)

Gorsuch, R. L. (1983). *Factor analysis* (2nd ed.). Hillsdale, NJ: Erlbaum.

Hayduk, L. A. (1996). *LISREL issues, debates, and strategies.* Baltimore: Johns Hopkins University Press.

Henson, R. K. (1999). Multivariate normality: What is it and how is it assessed? In B. Thompson (Ed.), *Advances in social science methodology* (Vol. 5, pp. 193–212). Stamford, CT: JAI Press.

Huberty, C. J., & Morris, J. D. (1988). A single contrast test procedure. *Educational and Psychological Measurement, 48,* 567–578.

Jöreskog, K. G. (1967). Some contributions to maximum likelihood factor analysis. *Psychometrika, 32,* 443–482.

Jöreskog, K. G. (1969). A general approach to confirmatory maximum likelihood factor analysis. *Psychometrika, 34,* 183–220.

Jöreskog, K. G. (1970). A general method for analysis of covariance structures. *Biometrika, 57,* 239–251.

Jöreskog, K. G. (1971). Simultaneous factor analysis in several populations. *Psychometrika, 36,* 409–426.

Jöreskog, K. G. (1978). Structural analysis of covariance and correlation matrices. *Psychometrika, 43,* 443–477.

Jöreskog, K. G., & Sörbom, D. (1984). *LISREL VI user's guide* (3rd ed.). Mooresville, IN: Scientific Software.

Jöreskog, K. G., & Sörbom, D. (1989). *LISREL 7: A guide to the program and applications* (2nd ed.). Chicago: SPSS.

Klem, L. (1995). Path analysis. In L. G. Grimm & P. R. Yarnold (Eds.), *Reading and understanding multivariate statistics* (pp. 65–97). Washington, DC: American Psychological Association.

Knapp, T. R. (1978). Canonical correlation analysis: A general parametric significance testing system. *Psychological Bulletin, 85,* 410–416.

Lomax, R. (1989). Covariance structure analysis: Extensions and developments. In B.

Thompson (Ed.), *Advances in social science methodology* (Vol. 1, pp. 171–204). Greenwich, CT: JAI Press.

Long, J. S. (1983a). *Confirmatory factor analysis: A preface to LISREL*. Thousand Oaks, CA: Sage.

Long, J. S. (1983b). *Covariance structure models: An introduction to LISREL*. Thousand Oaks, CA: Sage.

MacCallum, R. C., Browne, M. W., & Sugawara, H. M. (1996). Power analysis and determination of sample size for covariance structural modeling. *Psychological Methods, 1*, 130–149.

MacCallum, R. C., Roznowski, M., & Necowitz, L. B. (1992). Model misspecifications in covariance structure analysis: The problem of capitalization on chance. *Psychological Bulletin, 111*, 490–504.

MacCallum, R. C., Wegener, D. T., Uchino, B. N., & Fabrigar, L. R. (1993). The problem of equivalent models in applications of covariance structure analysis. *Psychological Bulletin, 114*, 185–199.

Marsh, H. W., & Hu, K. (1998). Is parsimony always desirable? *Journal of Experimental Education, 66*, 274–285.

Mueller, R. O. (1997). Structural equation modeling: Back to basics. *Structural Equation Modeling, 4*, 353–369.

Mulaik, S. A. (1987). A brief history of the philosophical foundations of exploratory factor analysis. *Multivariate Behavioral Research, 22*, 267–305.

Mulaik, S. A. (1988). Confirmatory factor analysis. In R. B. Cattell & J. R. Nesselroade (Eds.), *Handbook of multivariate experimental psychology* (2nd ed., pp. 259–288). New York: Plenum Press.

Mulaik, S. A. (1994). The critique of pure statistics: Artifact and objectivity in multivariate statistics. In B. Thompson (Ed.), *Advances in social science methodology* (Vol. 3, pp. 247–296). Greenwich, CT: JAI Press.

Mulaik, S. A., James, L. R., van Alstine, J., Bennett, N., Lind, S., & Stilwell, C. D. (1989). Evaluation of goodness-of-fit indices for structural equation models. *Psychological Bulletin, 105*, 430–445.

Pedhazur, E. J. (1982). *Mutiple regression in behavioral research: Explanation and prediction* (2nd ed.). New York: Holt, Rinehart & Winston.

Reinhardt, B. (1996). Factors affecting coefficient alpha: A mini Monte Carlo study. In B. Thompson (Ed.), *Advances in social science methodology* (Vol. 4, pp. 3–20). Greenwich, CT: JAI Press.

Spearman, C. (1904). The proof and measurement of association between two things. *Journal of Psychology, 15*, 72–101.

Steiger, J. H., & Lind, J. C. (1980, June). *Statistically based tests for the number of common factors.* Paper presented at the annual meeting of the Psychonomic Society, Iowa City, IA.

Stevens, J. (1996). *Applied multivariate statistics for the social sciences* (3rd ed.). Mahwah, NJ: Erlbaum.

Thompson, B. (1984). *Canonical correlation analysis: Uses and interpretation.* Thousand Oaks, CA: Sage.

Thompson, B. (1990). MULTINOR: A FORTRAN program that assists in evaluating multivariate normality. *Educational and Psychological Measurement, 50*, 845–848.

Thompson, B. (1991). A primer on the logic and use of canonical correlation analysis. *Measurement and Evaluation in Counseling and Development, 24*, 80–95.

Thompson, B. (1992). DISCSTRA: A computer program that computes bootstrap resampling estimates of descriptive discriminant analysis function and structure coefficients and group centroids. *Educational and Psychological Measurement, 52,* 905–911.

Thompson, B. (1994). Guidelines for authors. *Educational and Psychological Measurement, 54,* 837–847.

Thompson, B. (1996). AERA editorial policies regarding statistical significance testing: Three suggested reforms. *Educational Researcher, 25*(2), 26–30.

Thompson, B. (1997). The importance of structure coefficients in structural equation modeling confirmatory factor analysis. *Educational and Psychological Measurement, 57,* 5–19.

Thompson, B. (1998). Review of *What if there were no significance tests?*. *Educational and Psychological Measurement, 58,* 332–344.

Thompson, B. (1999a). Five methodology errors in educational research: A pantheon of statistical significance and other faux pas. In B. Thompson (Ed.), *Advances in social science methodology* (Vol. 5, pp. 23–86). Stamford, CT: JAI Press.

Thompson, B. (1999b). If statistical significance tests are broken/misused, what practices should supplement or replace them? *Theory & Psychology, 9*(2), 167–183.

Thompson, B. (1999c). Journal editorial policies regarding statistical significance tests: Heat is to fire as *p* is to importance. *Educational Psychology Review, 11,* 157–169.

Thompson, B. (1999d). Statistical significance tests, effect size reporting, and the vain pursuit of pseudo-objectivity. *Theory & Psychology, 9*(2), 191–196.

Thompson, B., & Daniel, L. G. (1996). Factor analytic evidence for the construct validity of scores: An historical overview and some guidelines. *Educational and Psychological Measurement, 56,* 213–224.

Thompson, B., & Vacha-Haase, T. (2000). Psychometrics *is* datametrics: The test is not reliable. *Educational and Psychological Measurement, 60,* 174–195.

Vacha-Haase, T. (1998). Reliability generalization: Exploring variance in measurement error affecting score reliability across studies. *Educational and Psychological Measurement, 58,* 6–20.

West, S. G., Finch, J. F., & Curran, P. J. (1995). Structural equation models with non-normal data. In R. H. Hoyle (Ed.), *Structural equation modeling* (pp. 56–75). Thousand Oaks, CA: Sage.

Wilkinson, L., & APA Task Force on Statistical Inference. (1999). Statistical methods in psychology journals: Guidelines and explanations. *American Psychologist, 54,* 594–604.

Wright, S. (1921). Correlation and causality. *Journal of Agricultural Research, 20,* 557–585.

Wright, S. (1934). The method of path coefficients. *Annals of Mathematical Statistics, 5,* 161–215.

9 Canonical Correlation Analysis

Bruce Thompson

Canonical correlation analysis (CCA) is an analytic method that can be used to investigate relationships among two or more variable sets. Each variable set usually consists of at least two variables (otherwise the canonical analysis is typically called something else, such as a *t* test or a regression analysis). Although in theory the canonical logic can be generalized to more than two variable sets (Horst, 1961), in practice most researchers use CCA in situations involving only two variable sets.

Canonical analysis was originally conceptualized by Hotelling (1935). Notwithstanding its long history, as Krus, Reynolds, and Krus (1976) noted, "dormant for nearly half a century, Hotelling's (1935) canonical variate analysis has come of age. The principal reason behind its resurrection was its computerization and inclusion in major statistical packages" (p. 725). Of course, having sophisticated statistical packages available does not in and of itself justify the use of CCA or any other analysis.

For two reasons, however, multivariate methods are being used with increasing frequency. First, multivariate methods control the inflation of experimentwise (Type I) error rates ($\alpha_{experimentwise}$) that can occur when several univariate tests are conducted with a single sample's data, even when the testwise error rate ($\alpha_{testwise}$) is very small. Thompson (1994d) provided further explanation of what experimentwise error is and how this error rate can be estimated.

Second, multivariate methods, such as CCA, best honor the nature of the reality that most researchers want to study because many of us believe that we live in a reality where most effects have multiple causes and most causes have multiple effects. As Tatsuoka (1973) emphasized,

> the often-heard argument, "I'm more interested in seeing how each variable, in its own right, affects the outcome" overlooks the fact that any variable taken in isolation may affect the criterion differently from the way it will act in the company of other variables. It also overlooks the fact that multivariate analysis—precisely by considering all the variables simultaneously—can throw light on how each one contributes to the relation. (p. 273)

That is, univariate and multivariate analyses of the same data can yield results that differ like night and day with regard to both statistical significance and effect sizes (R^2, eta^2, etc.), and the multivariate picture in such cases is the accurate portrayal. Fish (1988) provided an empirical example of how univariate and multivariate analyses of the same data can yield contradictory results.

Thompson's (1999a) example was even more dramatic. For his data, two univariate analyses of variance yielded statistically nonsignificant results (both p values were .774) with eta^2 variance-accounted-for effect sizes of both 0.5%. For the same data analyzed by multivariate analysis of variance (MANOVA), $p_{\text{calculated}}$ was .000239, and the multivariate eta^2 was 62.5%.

This second reason for the more frequent use of multivariate methods is the more noteworthy of the two. Some researchers avoid inflated experimentwise error rates by making so-called Bonferroni corrections (i.e., downward adjustments in α_{testwise} so as to moderate increases in $\alpha_{\text{experimentwise}}$), but this second reason still applies even when such adjustments are invoked. Furthermore, this second reason for using multivariate analyses is more noteworthy because the first reason involves statistical significance testing, and social scientists have been placing less emphasis on statistical significance testing (cf. Cohen, 1994; Thompson, 1996; Thompson & Snyder, 1997, 1998). Indeed, the APA Task Force on Statistical Inference has issued a report that greatly emphasizes the importance of focusing interpretations on effect sizes, particularly in relation to the previous effects found in related prior studies (Wilkinson & APA Task Force on Statistical Inference, 1999).

It is also important to emphasize that although some researchers incorrectly believe that they can appropriately first conduct multivariate tests and then conduct so-called "protected" univariate tests (Maxwell, 1992, pp. 138–140), again the second rationale for conducting multivariate analyses still exists (Thompson, 1994d). That is, univariate tests cannot reasonably be used to investigate and understand the patterns first isolated in multivariate analyses; only a multivariate analysis can explore a multivariate effect.

Thus, for these two reasons, CCA has been used in a variety of published research. Wood and Erskine (1976) and Thompson (1989) provided extensive bibliographies of applications of CCA. Example applications include those reported by Chastain and Joe (1987); Dunst and Trivette (1988); Estabrook (1984); Fowler and Macciocchi (1986); Fuqua, Seaworth, and Newman (1987); Pitts and Thompson (1984); and Zakaahi and Duran (1982). One particularly interesting application involves studies of multivariate test–retest score reliability or of multivariate criterion-related score validity (cf. Sexton, McLean, Boyd, Thompson, & McCormick, 1988).

The purpose of this chapter is to provide a primer on CCA. A longer and more technical treatment is provided by Thompson (1984). The chapter (a) explains the basic logic of CCA using a heuristic data set, (b) provides a brief explanation of how CCA is related to other commonly used univariate and multivariate parametric analyses, (c) illustrates the steps in interpreting canonical results, and (d) details some common errors to avoid in interpreting canonical analyses.

Basic Logic of Canonical Calculations

Problems With the Nonmultivariate Alternative

Imagine that the director of personnel for a national chain of department stores wishes to determine the relationship between characteristics of sales staff and indices of job performance, using a random selection of salespeople from various stores. The first set of variables is obtained from personnel files and is comprised of scores on three questionnaires: leadership potential, need for achievement, and empathy; and of scores on a fourth variable, years of previous sales experience. Call this variable set "employee attributes." A conceptually discrete second set of variables, called "job performance," might be comprised of scores on three measures: past year's sales, absenteeism, and numbers of suggestions previously submitted for improving store operations. How can the director of personnel make sense of these data?

The investigator could first separately examine the *intradomain* matrix of bivariate correlations between the variables in each set. For example, an examination of only the employee attributes variables could provide insight regarding the nature of their interrelatedness. If these measures were highly correlated, this would suggest that these scores measure one underlying dimension or construct; such an outcome might be unlikely for these variables. However, if only some of the measures correlated highly with each other, this would suggest that the use of only one dimension to represent all the employee attributes variables is inappropriate.

The investigator could also examine the bivariate relationships between scores on the measures across the two variable sets (i.e., the *interdomain* correlation coefficients). The director of personnel may wonder whether all of the measures of employee attributes correlate strongly with all of the measures of job performance. Of course, for these variables, this outcome is also unlikely.

The problem with examining only the 21 ($[7 \times (7 - 1)]/2 = (7 \times 6)/2 = 42/2$) unique bivariate correlation coefficients defined by the scores on the seven variables is that even for this relatively simple problem, the dynamics represented within the data are beyond the perceptual ability of nearly anyone unless the patterns within the data are ridiculously obvious (e.g., all 21 correlation coefficients are nearly +1). Furthermore, several different patterns of relationships may simultaneously underlie the system of relationships. These can be discovered only by using a multivariate analysis. Of course, even if the patterns within the data are blatantly obvious, the multivariate analysis will still detect them.

The same conclusion about why the bivariate approach does not work can be seen from the perspective of individual scores. In the simplest of all worlds, all of the employee attributes variables measure one underlying construct, and the same is true for the job performance measures. In this case, one single score (rather than four different scores) could be used to describe the employee attributes data, and one single score could be used to describe the job performance data. Assuming that the variables are measured using a common metric (e.g., *z* scores), a simple approach would then involve adding together the four employee attributes scores to obtain one overall "employee" score for each salesperson and adding together the three job performance scores to obtain one overall "performance" score for each person. The single bivariate Pearson correlation between these two scores could then reasonably be interpreted as the sole basis for understanding the relationship between the two variable sets.

If the variable sets cannot each be appropriately considered purely unidimensional, however, then computing this simple sum of scores is akin to adding apples and oranges. In addition, regardless of the dimensionality of each variable set, a simple sum of scores does not reflect the possibility that some variables are more important and thus need to be weighted more strongly than others when exploring relationships between the two variable sets. Also such a simple summation weighting system would not intentionally maximize the relationship between the variable sets. CCA is the method of choice in such situations.

In this example, CCA would determine the exact weighting scheme for computing one or more employee scores and for computing one or more performance scores. The specific combination of weights is called a *canonical function* (conceptually, a *function* consists of weights [multiplicative constants] similar to the beta [β] weights that constitute a regression equation), and the score obtained by applying the canonical function coefficients to a set of actual measured scores for a given person is known as a *synthetic score*. CCA determines weighting schemes that create synthetic–latent scores that are maximally correlated. That is, no other possible combination of weighting schemes can be devised that would ultimately lead to a higher correlation between the resulting two *synthetic–latent variables* on a given CCA function.

Heuristic Data Set

An actual data set is used to make this discussion concrete with regard to the basic logic of CCA. The illustration uses scores on five measured–observed variables (i.e., two in one set and three in the other set) from 301 cases from the Holzinger and Swineford (1939, pp. 81–91) data. These scores on ability batteries have classically been used as examples in both popular textbooks (Gorsuch, 1983, *passim*) and computer program manuals (Jöreskog & Sörbom, 1989, pp. 97–104) and thus are familiar to many readers.

The illustrative data involve five variables, each of which is intervally scaled. Other levels of scale can be used in canonical analyses, however, if data are still somewhat normally distributed (e.g., Cooley & Lohnes, 1976, p. 209). As Maxwell (1961) noted, "the theory of canonical variate analysis, widely used with continuous variables, can be employed when the variables are dichotomous" (p. 271; for more technical detail, see Thompson, 1984, pp. 16–18).

One set of measured variables involves scores on three measures of "perceptual abilities": visual perception (VP), lozenges flipped shapes (LFS), and counting groups of dots (CGD). The second, conceptually discrete set of variables measured "academic achievement": word meaning (WM) and mixed math fundamentals (MMF). The syntax commands to conduct the canonical analysis of these variables within the popular SPSS programs, including the Windows versions, are a bit tricky because within SPSS, CCA is run within the MANOVA procedure. The commands for this example are

```
MANOVA VP LFS CGD WITH WM MMF/
PRINT = SIGNIF(MULTIV EIGEN DIMENR)/
DISCRIM = STAN CORR ALPHA (.999)/
```

Table 9.1 presents the matrix of Pearson product–moment correlation coefficients for these data. For example, the interdomain bivariate correlation between VP (a member of the first variable set) and MMF (a member of the second variable set) is .2826. The correlation matrix is "symmetric" about the diagonal. The diagonal contains all *1*s, reflecting the correlation of each measured–observed variable with itself. The correlation between VP and MMF in the first row and last column is reproduced exactly as the correlation between MMF and VP in the first column and last row.

Of course, one could subject the Table 9.1 correlation matrix to a

Table 9.1

Bivariate Correlation Matrix

	Variable				
Variable	**VP**	**LFS**	**CGD**	**WM**	**MMF**
VP	1.0000	0.4407	0.2239	0.3568	0.2826
LFS	0.4407	1.0000	0.1860	0.1977	0.1668
		R11		**R12**	
CGD	0.2239	0.1860	1.000	0.1496	0.3111
WM	0.3568	0.1977	0.1496	1.000	0.4401
		R21		**R22**	
MMF	0.2826	0.1668	0.3111	0.4401	1.0000

Note. VP = visual perception; LFS = lozenges flipped shapes; CGD = counting groups of dots; WM = word meaning; MMF = mixed math fundamentals. The quadrant names are designated in bold; **R11**, for example, is an intradomain quadrant in that both subscripts are the same.

factor analysis to evaluate relationships among the variables. If the measured variables consist of theoretically discrete variable sets, however, or if the variables were measured at chronologically discrete times, such a factor analysis would not honor the view that the measured variables exist within meaningful sets.

Indeed, it is only when the researcher believes the measured–observed variables exist within meaningful variable sets that CCA would be an appropriate analysis. In the present example, the conceptualization of the two variable sets as being discrete seems reasonable. Of course, such classifications are matters of researcher judgment, and even reasonable researchers differ regarding such judgments (just as researchers may reasonably disagree about most aspects of the research endeavor).

CCA computer programs first partition the correlation matrix into quadrants associated with the variable sets, as illustrated in Table 9.1. Note that each quadrant is identified by a boldface **R** with two subscripts. **R11** includes the intradomain correlations between variables that are measures of perceptual abilities (variable set 1), and **R22** includes correlations between variables that are measures of academic achievement (variable set 2). Both **R12** and **R21** include the interdomain correlations between variables from across the two variable sets.

After the correlation matrix is computed, a quadruple-product matrix is then computed from the four quadrants, using the following matrix algebra formula:

$$\mathbf{R22}^{-1}_{2\times2}\ \mathbf{R21}_{2\times3}\ \mathbf{R11}^{-1}_{3\times3}\ \mathbf{R12}_{3\times2} = \mathbf{A}_{2\times2}.$$

This matrix, $\mathbf{A}_{2x2}$, is then subjected to a principal components analysis, and the results are expressed as standardized weights (Thompson, 1984, pp. 11–14 provides more detail) called standardized canonical function coefficients.

Although further discussion of the mathematical underpinnings of CCA is beyond the scope of this chapter, note that these function coefficients are directly akin to beta (β) weights in regression or the pattern coefficients from exploratory factor analysis. As noted shortly, these function coefficients are one important element within the process of result interpretation.

It may occur to the reader that many statistical analyses invoke weights but that different names are used for the same concepts across techniques (e.g., beta weights vs. pattern coefficients vs. function coefficients and equation vs. factor vs. function). Sometimes it appears that the sole purpose of such misnomers is to confuse graduate students into thinking that these analyses are unrelated to each other rather than all being part of one general linear model (GLM; Fan, 1996, 1997; Knapp, 1978; Thompson, 1991b).

Table 9.2 presents the standardized function coefficients for the current example. The number of functions (sets of weights) in a CCA is always equal to the number of variables in the smaller of the two variable sets (i.e., in this example, two). In Table 9.2, the first canonical function is labeled Function I and the second is labeled Function II. As is always the case, these functions are perfectly uncorrelated with each other and so are the scores on the latent or synthetic variables computed by applying the weights to the observed or measured variables (i.e., here five measured variables). Thus, synthetic–latent variables are never directly measured. Synthetic variables are obtained by applying weights to the measured variables. The synthetic variables are estimates of the latent constructs of interest and are the actual focus of all statistical analyses.

Computing synthetic variable scores is actually a simple matter. For example, for the variable set consisting of three variables, the scores of the first person in the data set were VP, 20; LFS, 3; and CGD, 115. For the second variable set, this person's scores were WM, 9, and MMF, 24. To compute the synthetic variables, raw scores must first be transformed to *z* scores (i.e., scores having a mean of 0 and a standard deviation of 1). The *z*-score equivalents of the first person's five scores were −1.373, −1.658, +.220, −.821, and −.056, respectively.

For the first participant, based on the first canonical function, two synthetic scores are computed. The first synthetic score, PRED1, is for the perceptual abilities variable set. The second synthetic score, CRIT1, is for the cognitive academic achievement variable set. In principal components analysis, this would be analogous to computing factor scores on Factor I; however, because in CCA there are two variable sets, one distinguishes synthetic variables by using two names, namely, PRED1 and CRIT1. If two canonical functions were present, then synthetic variables on the second factor would be labeled PRED2 and CRIT2.

For the first participant, the synthetic PRED1 score can be obtained by applying the standardized function coefficients to the *z* scores for the measured variables. For example, the synthetic score for first participant equals the *z* score for VP (−1.373) multiplied by the standardized function coefficient (0.733), plus the *z* score for LFS (−1.658) multiplied by the standardized function coefficient for LFS (0.091), plus

Table 9.2

Canonical Coefficients in the Format Recommended for CCA Reports

	Function I			Function II			
Variable–statistic	Function	r_s	r_s^2	Function	r_s	r_s^2	h^2
VP	0.733	0.879	77.26%	−0.622	−0.462	21.34%	98.61%
LFS	0.091	0.502	25.20%	−0.101	−0.206	4.24%	29.44%
CGD	0.473	0.654	42.77%	0.915	0.756	57.15%	99.93%
Adequacy			48.41%			27.58%	
Rd			8.71%			0.97%	
Rc^2			18.00%			3.50%	
Rd			12.94%			0.98%	
Adequacy			71.88%			28.06%	
WM	0.548	0.825	68.05%	−0.969	−0.565	31.92%	99.98%
MMF	0.629	0.870	75.69%	0.919	0.492	24.21%	99.90%

Note. CCA = canonical correlation analysis; VP = visual perception; LFS = lozenges flipped shapes; CGD = counting groups of dots; Rd = redundancy coefficient for a given variable set; Rc^2 = squared canonical correlation coefficient; WM = word meaning; MMF = mixed math fundamentals.

the *z* score for CGD (0.220) multiplied by the standardized function coefficient for CGD (0.473). The result of the computation for the first participant yields PRED1 = −1.053.

Similarly, the synthetic–latent variable score CRIT1 for the first participant would be computed as the *z* score on WM (−0.821) multiplied by the standardized function coefficient for WM (0.548), plus the *z* score for MMF (−0.056) multiplied by the standardized function coefficient for MMF (0.621). The result of this computation yields CRIT1 = −0.485. Note that all these synthetic scores are themselves in *z*-score form. That is, PRED1, PRED2, CRIT1, and CRIT2 are all *z* scores, with a mean of 0 and a standard deviation of 1.

Figure 9.1 presents a scatterplot of the 301 scores on PRED1 and CRIT1. For variables in *z*-score form, the best-fitting regression line has a slope, ß, that is equal to the correlation of the two synthetic variables (+.4246). This bivariate product–moment correlation coefficient is nothing more or less than the multivariate canonical correlation between the weighted variables in the two variable sets, Rc. Remember, no other possible weighting scheme can produce two synthetic variables that have a higher correlation between them: CCA maximizes Rc.

In Table 9.3, note that PRED and CRIT synthetic scores are uncorrelated with each other except when they are from the same function. As shown, PRED1 and CRIT1 are correlated with each other (the first Rc) and so are PRED2 and CRIT2 (the second Rc), but the remaining correlations among these different synthetic scores are all zero. This establishes that all the synthetic variable scores, except PRED1 with CRIT1 and PRED2 with CRIT2, are "bi-orthogonal." Note that the last row of Table 9.3 presents the bivariate product–moment correlation coefficients involving the four synthetic variables and, for illustrative purposes, one arbitrarily selected observed or measured variable, WM. This facilitates the understanding of a second coefficient (in addition to the standardized function coefficient), called a *structure coefficient,* which is important in all multivariate analyses and in univariate analysis (Thompson, 1997; Thompson & Borrello, 1985).

A structure coefficient (indicated as r_s in Table 9.2) is the bivariate product–moment correlation between scores on an observed or measured variable and scores on a synthetic or latent variable for that measured variable's variable set. Thus, because structure coefficients are correlation coefficients, they range from −1 to +1, inclusively; standardized function coefficients, however, are not usually correlation coefficients and have no definitive boundaries (see Thompson, 1984,

Figure 9.1

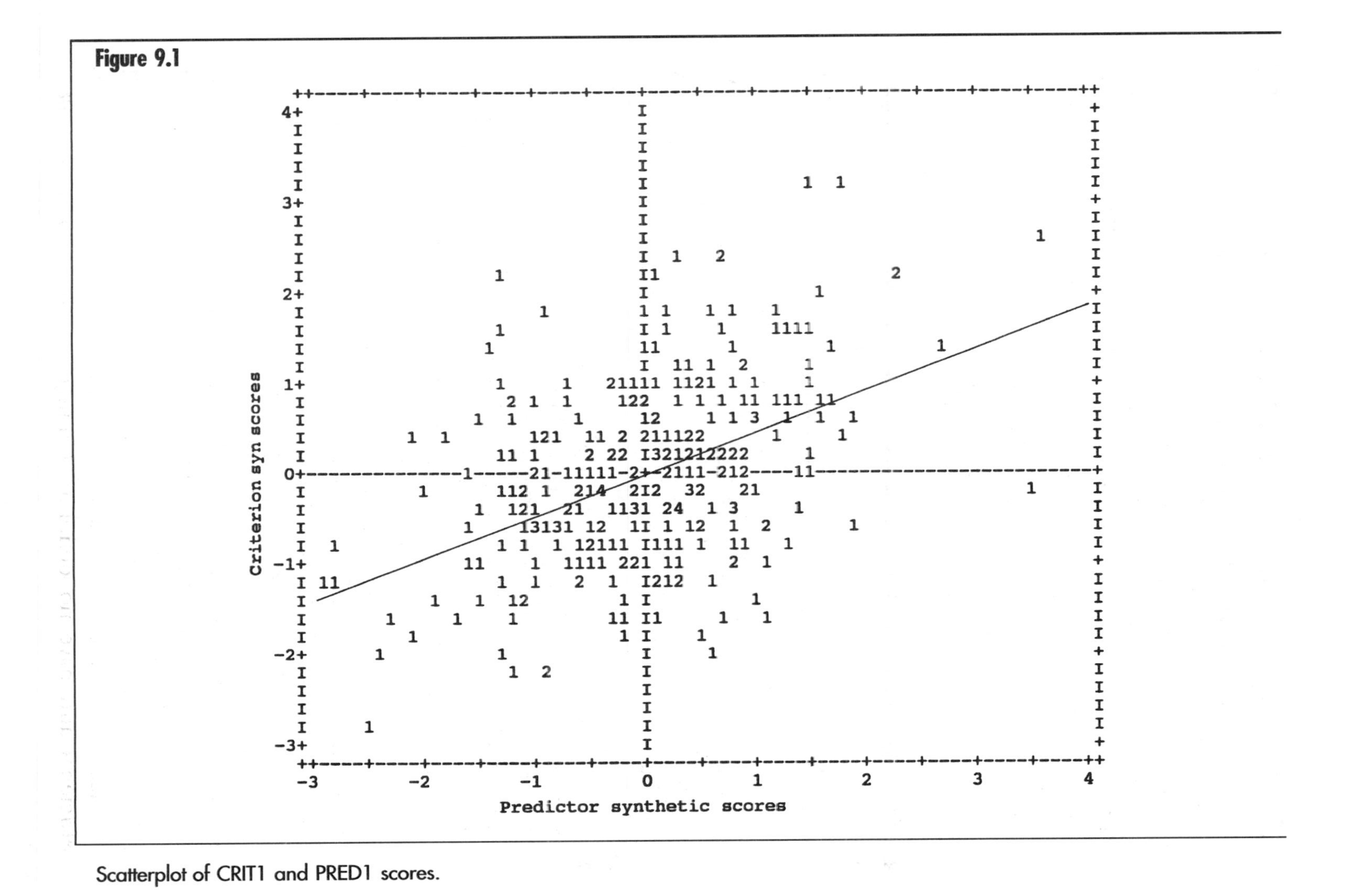

Scatterplot of CRIT1 and PRED1 scores.

Table 9.3

Product–Moment Correlation Coefficients Among the Four Synthetic Variables Scores and One Measured Variable

	Variable				
Variable	CRIT1	CRIT2	PRED1	PRED2	WM
CRIT1	1.0000				
CRIT2	0.0000	1.0000			
PRED1	0.4246[a]	0.0000	1.0000		
PRED2	0.0000	0.1863[b]	0.0000	1.0000	
WM	0.8253[c]	−0.5647[d]	0.3504[e]	−0.1052[e]	1.0000

Note. WM = word meaning. [a]Rc_I, as reported in Table 9.2. [b]Rc_{II}, as reported in Table 9.2. [c]The structure coefficient for measured variable WM on Function I, as reported in Table 9.2. [d]The structure coefficient for measured variable WM on Function II, as reported in Table 9.2. [e]The index coefficients for measured variable WM (see Thompson, 1984, pp. 30–31).

pp. 21–24). The use of function and structure coefficients are further explained when rubrics for result interpretation are presented.

Table 9.2 also presents other indices that (unlike function and structure coefficients, which are almost always essential to interpret) are at least in some instances important in evaluating and interpreting CCA results. The special circumstances when the following two coefficients are important are detailed in the subsequent section on interpreting results. A common error by less-experienced CCA researchers involves interpreting the following two coefficients when they are in fact irrelevant and do not to be interpreted.

A *canonical adequacy coefficient* indicates how adequately a given function, on average, reproduces the variance of a given set of measured variables. For example, consider the first canonical function and the academic achievement variable set. For WM, the structure coefficient is $r_s = .825$. Here, the squared structure coefficient is $r_s^2 = .6806$. This means that 68.06% of the observed WM variable is useful within this canonical function. Also here, the mean of the squared structure coefficients ([.6806 + .7569]/2 = .7188) is the canonical adequacy coefficient for the first function. Thus, on average, this canonical function reproduces approximately 72% of the variance of the WM and MMF variables. The largest this number can be is 100%, indicating that all the variance of the measured variables in the given set has been reproduced within the function and the associated synthetic variable.

Table 9.2 also presents a *redundancy coefficient* (Rd) for each canonical function and variable set (i.e., four Rds, two for each function). Rd is defined as the product of the canonical adequacy coefficient for a variable set multiplied by the squared canonical correlation, Rc^2 (for further discussion, see Thompson, 1991b). For example, for the academic achievement variable set, the redundancy coefficient for the first canonical function is Rd = .7188 × .1800, or Rd = .1294.

Finally, Table 9.2 also presents *canonical communality coefficients* (h^2), which unlike the previous two coefficients are often important to evaluate when interpreting results. Canonical commonality coefficients are equal to the sum of the squared structure coefficients for a given variable across the canonical functions. For example, for VP, h^2 = .7726 (i.e., r_s^2 for Function I) + .2134 (i.e., r_s^2 for Function II) = 0.9861. This indicates that the two synthetic variables, considered together, can reproduce 98.61% of the variance of the VP variable, or conversely, that 98.61% of this measured variable was useful in defining the function.

Note that h^2 was greater than 98% for all measured variables in the example, except for LFS, for which h^2 was less than 30%. This suggests that compared with the other measured variables, the synthetic variables obtained in this CCA had relatively less to do with scores on the LFS variable. Sometimes measured variables with anomalously low communality coefficients may be deleted from the analysis to obtain a more parsimonious solution (see Thompson, 1984, pp. 47–51).

Canonical Correlation Analysis in the General Linear Model

CCA is the most general case of the parametric general linear model (Baggaley, 1981; Fan, 1996; Fornell, 1978; Thompson, 1991b), unless one wishes to delineate an even broader general linear model that also directly takes into account measurement error (Bagozzi, Fornell, & Larcker, 1981; Fan, 1997; Thompson, 1999b). Knapp (1978) demonstrated these views in some mathematical detail and concluded that "virtually all of the commonly encountered tests of significance can be treated as special cases of canonical correlation analysis" (p. 410). Thus, Knapp's work was cited in a compilation of the seminal methodology publications produced during the past several decades (Thompson & Daniel, 1996).

Saying that CCA constitutes the parametric general linear model, subsuming all other parametric univariate and multivariate analyses, means that the other analyses are special cases of canonical analysis, and that the other analyses (e.g., *t* tests, analysis of variance [ANOVA], regression, MANOVA, descriptive discriminant analysis) can actually be conducted using the logic of CCA (Campbell & Taylor, 1996). This realization is of immense heuristic value for people trying to understand how various analytic methods are related to each other.

Three Realizations

Three important realizations can be extracted from the view of CCA as the general linear model. First, the GLM view forces researchers to understand that all analyses are correlational. Some designs are experimental, but all analyses are correlational, and it is the design (not the analysis) that enables the making of causal inferences.

Too many researchers use OVA methods (ANOVA, ANCOVA [analysis of covariance], MANOVA, and MANCOVA [multivariate analysis of covariance]) because they have come to associate making causal inferences with OVA methods; these erroneous associations are all the more pernicious because the associations tend to be made unconsciously. As Humphreys (1978) emphasized,

> the basic fact is that a measure of individual differences is not an independent variable, and it does not become one by categorizing the scores and treating the categories as if they defined a variable under experimental control in a factorially designed analysis of variance. (p. 873)

Similarly, Humphreys and Fleishman (1974) noted that categorizing variables in a nonexperimental design using an ANOVA or other OVA analysis "not infrequently produces in both the investigator and his audience the illusion that he has experimental control over the independent variable. Nothing could be more wrong" (p. 468).

So the use of OVA methods to analyze nonexperimental data can be unnecessary and is usually harmful if researchers discard variance on intervally scaled predictor variables, so that they can use OVA methods (Thompson, 1994c). As Cliff (1987) explained,

> such divisions are not infallible; think of the persons near the borders. Some who should be highs are actually classified as lows, and vice versa. In addition, the "barely highs" are classified the same as the "very highs," even though they are different. Therefore, reducing a reliable variable to a dichotomy makes the variable more unreliable, not less. (p. 130)

Second, the GLM view of canonical analysis correctly indicates that all parametric analyses either explicitly or implicitly invoke systems of weights applied to measured variables to produce synthetic variables. Again, these weight systems are often arbitrarily (and confusingly) given different names across different analyses (e.g., beta weights vs. pattern coefficients vs. function coefficients and equation vs. factor vs. function).

The fact that these weights are used to define synthetic–latent variables correctly suggests that it is the synthetic variables that are actually the focus in all analyses. That the synthetic variables are the analytic focus can be seen in the realization that the Rc equals the product–moment r between the synthetic variables for a given canonical function.

Third, the fact that all analyses are correlational implies that effect sizes ought to be reported and interpreted in all research studies. No knowledgeable researcher reporting bivariate or multiple correlation coefficients fails to comment on the magnitude of the squared correlation coefficients. Because all analyses are correlational, all researchers reporting t test, ANOVA, descriptive discriminant analysis, or any other analysis should always interpret uncorrected (e.g., eta^2) or corrected (e.g., omega^2) effect sizes or some other measures of effect size (Snyder & Lawson, 1993). The necessity of reporting such effect sizes is reinforced by the 1994 *Publication Manual of the American Psychological Association,* which notes that "neither of the two types of probability values reflects the importance or magnitude of an effect because both depend on sample size. . . . You are *encouraged* to provide effect-size information" (p. 18, emphasis added).

Unhappily, 11 empirical studies, each of either one or two volumes of 23 different journals, demonstrate that this "encouragement" has been ineffective (Vacha-Haase, Nilsson, Reetz, Lance, & Thompson, 2000). Thompson (1999c) observed that only encouraging effect-size reporting "presents a self-canceling mixed message. To present an 'encouragement' in the context of strict absolute standards regarding the esoterics of author note placement, pagination, and margins is to send the message, 'these myriad requirements count, this encouragement doesn't' " (p. 162). Given (a) a rationale as to why an encouragement is doomed to continued impotence and (b) empirical evidence that the encouragement indeed has been ineffective, editors of several journals have now come to require effect-size reporting (cf. Heldref Foundation, 1997; Murphy, 1997; Thompson, 1994a).

In fact, the recent report of the APA Task Force on Statistical Inference emphasized, "*always* provide some effect-size estimate when reporting a *p* value" (Wilkinson & APA Task Force on Statistical Inference, 1999, p. 599, emphasis added). Later the task force wrote that researchers should

> *always* present effect sizes for primary outcomes. . . . It helps to add brief comments that place these effect sizes in a practical and theoretical context. . . . We must stress again that reporting and interpreting effect sizes in the context of previously reported effects is *essential* to good research. (p. 599, emphasis added)

A Demonstration of the General Linear Model Concept

An illustration is provided to clarify how CCA subsumes other parametric analyses as special cases. The reader can consult other resources (e.g., Fan, 1996; Thompson, 1991b) for more comprehensive demonstrations of these various linkages, including SPSS or SAS syntax programs that conduct the related analytic proofs. Space precludes a complete treatment here, except for the following illustration. Specifically, the fact that CCA subsumes regression as a special case is demonstrated. This relationship is relatively straightforward because both analyses are explicitly correlational.

Figure 9.2 presents the SPSS for Windows regression printout for an analysis in which WM was treated as the dependent variable, whereas scores on VP, LFS, and CGD were used as predictor variables. Figure 9.3 presents the printout from the same analysis conducted using the logic of CCA.

The correlation coefficients from the analysis are the same (multiple R = +.36586 and Rc = +.366), except that the computer programmers arbitrarily elected to report results to different numbers of decimal places. However, the standardized weights (i.e., β and functional coefficients) from the two analyses seem to be different.

Actually, the weights are merely scaled differently, and such differences are not meaningful. Table 9.4 demonstrates how β weights can readily be converted into standardized canonical function coefficients and vice versa. Other resources show how CCA subsumes and can therefore be used to perform *t* tests, ANOVA, ANCOVA, MANOVA, and descriptive discriminant analyses (cf. Thompson, 1991b). In short, all analyses (a) are correlational, (b) invoke weights being applied to measured variables to estimate synthetic variables, and (c) yield variance-accounted-for effect sizes analogous to r^2.

Figure 9.2

```
Equation Number 1     Dependent Variable..    T9    WORD MEANING TEST

Block Number  1.  Method:  Enter       T1         T4         T12

Variable(s) Entered on Step Number
   1..     T12        SPEEDED COUNTING OF DOTS IN SHAPE
   2..     T4         LOZENGES FROM THORNDIKE--SHAPES FLIPPED
   3..     T1         VISUAL PERCEPTION TEST FROM SPEARMAN VPT

Multiple R            .36586
R Square              .13385
Adjusted R Square     .12511
Standard Error       7.17347

Analysis of Variance
                     DF       Sum of Squares       Mean Square
Regression            3          2361.87366         787.29122
Residual            297         15283.21604          51.45864

F =       15.29950       Signif F =  .0000

------------------ Variables in the Equation ------------------

Variable                   B          SE B         Beta          T  Sig T

T1                   .353005      .066737      .322413      5.289  .0000
T4                   .036160      .051249      .042660       .706  .4810
T12                  .026311      .021088      .069479      1.248  .2131
(Constant)          1.285475     2.632565                    .488  .6257
```

Abridged SPSS printout showing regression results.

Interpreting Canonical Results

Because all classical parametric analyses are special cases of CCA, the same interpretation strategy can be used for any parametric analyses. The interpretation is approached as a hierarchical, two-stage contingency model. Not to put the matter too technically, two questions are addressed:

1. Do I have anything?
2. Where does what I have originate?

One reaches the second question only if the answer to the first question is *yes*.

In addressing these two questions, the interpretation should be framed within the context of sample size. CCA is a large-sample method, potentially requiring 15–20 participants per measured variable (cf. Barcikowski & Stevens, 1975). When one is interpreting the results, more

Figure 9.3

```
- - - - - - - - - - - - - - - - - - - - - - - - - - - - - - - - - - - - -
Eigenvalues and Canonical Correlations

Root No.    Eigenvalue        Pct.    Cum. Pct.  Canon Cor.      Sq. Cor

      1           .155     100.000      100.000        .366         .134

- - - - - - - - - - - - - - - - - - - - - - - - - - - - - - - - - - - - -
Standardized canonical coefficients for DEPENDENT variables
          Function No.

Variable              1

T1                 .881
T4                 .117
T12                .190
```

Abridged SPSS printout showing regression performed as a canonical correlation analysis. Pct. = percent; Cum. Pct. = cumulative percent; Canon Cor. = canonical correlation; Sq. Cor = squared canonical correlation.

Table 9.4

Regression and Canonical Results

Variable	β	/	Rc	=	Function	×	R
	Coefficients						
VP	.322413	/	.366	=	.881	×	.36586
LFS	.042660	/	.366	=	.117	×	.36586
CGD	.069479	/	.366	=	.190	×	.36586

Note. VP = Visual perception; LFS = lozenges flipped shapes; CGD = counting groups of dots. The beta (β) and *R* coefficients came from the Figure 9.2 regression as regression results. The Rc and function coefficients came from the Figure 9.3 regression as canonical results.

confidence can be vested in result stability as sample size is larger. Furthermore, interpretations must be framed in the knowledge that paradoxically one can typically be more confident in the stability of the overall effect size than in conclusions regarding the specific origins of the effect (cf. Thompson, 1990, 1991a).

1. Do I Have Anything?

A researcher can select any combination of three sorts of evidence to address the question, "Do I have anything?" First, the researcher can interpret the statistical significance of the canonical correlation coefficients. Of course, because the calculated probability (p) for a given set of results is highly dependent on sample size, consideration of statistical significance provides limited information. Furthermore, notwithstanding common misconceptions to the contrary, statistical significance tests do *not* evaluate whether the sample results occur in the population or are likely to be replicated in future samples (cf. Cohen, 1994; Thompson, 1996; Snyder & Thompson, 1998; Thompson & Snyder, 1997, 1998).

Second, the researcher can interpret some measure of effect size (see Kirk, 1996). There are many choices, as partially enumerated by Snyder and Lawson (1993). In the canonical case, the squared canonical correlation coefficients can be interpreted, or so-called "adjusted" values of these coefficients (similar to the adjusted R^2 in multiple regression) can be interpreted (Thompson, 1990). The interpretation of effect sizes invokes the researcher's subjective judgment of the noteworthiness of results. This intimidates some researchers, who become afraid that they will say or write judgments with which others will disagree.

Some such researchers seek an atavistic escape from these fears by trying to use statistical significance tests as an objective standard for whether results are noteworthy. As Thompson (1993) explained, however, "if the computer package did not ask you your values prior to its analysis, it could not have considered your value system in calculating *p*'s, and so *p*'s cannot be blithely used to infer the value of research results" (p. 365). Clearly, research is in part an inherently subjective business, and researchers must inescapably make the necessary judgments.

There are no clear boundaries regarding what effect sizes are noteworthy. The judgment is made at the nexus of the researcher's value system and the substantive focus of the research. For example, the variance-accounted-for effect size associated with the effects of smoking on longevity is reportedly about 2% (cf. Gage, 1978, p. 21). This is a small number, but most people take any definitive decrease in their potential days on earth seriously and thus deem the effect noteworthy. Others, of course, may feel that the pleasures of smoking outweigh the possible costs, especially because most effects occur "on the average" (i.e., the relation between smoking and longevity does not apply equally to each individual who smokes).

Third, researchers can evaluate whether their effects are replicable. If science is about the business of identifying relationships that recur under specified conditions, so that knowledge is cumulated, then the replicability issue is a critical one. Because statistical significance tests do *not* evaluate result replicability (cf. Thompson, 1996; Snyder & Thompson, 1998; Thompson & Snyder, 1997, 1998), two classes of methods can be invoked to evaluate this important issue.

Ultimately, the best way to evaluate result replicability is to replicate a given study. This represents a so-called "external" replicability analysis because a completely new sample is drawn. For pragmatic reasons, however, most researchers feel unable to replicate all their studies (e.g., tenure or promotion or merit-raise decisions are approaching, one's spouse has threatened to abandon the doctoral student if the dissertation is not completed by date certain). In such cases, researchers can instead invoke so-called "internal" replicability analyses to try to evaluate replicability. These internal analyses are not as accurate as true replication efforts but are certainly better than no attempt to evaluate result replicability.

There are three basic logics for such empirical internal replicability analyses: (a) cross-validation, (b) the jackknife, and (c) the bootstrap. As Thompson (1996) explained, "basically, the methods combine the subjects in hand in different ways to determine whether results are stable across sample variations, i.e., across the idiosyncrasies of individuals which make generalization in social science so challenging" (p. 29). An expanded discussion of these three methods is beyond the scope of this treatment. Thompson (1994b) provided an overview. Crossman (1996) described canonical cross-validation. Thompson (1995) described a computer program, CANSTRAP, that implements a bootstrap CCA.

2. Where Does What I Have Originate?

Given that a decision has been made that the canonical results are noteworthy, the question then arises, "Where does what I have originate?" Many canonical coefficients can be interpreted to address this question (Thompson, 1984), but the two coefficients of primary importance are the standardized canonical function coefficients and the structure coefficients. Both must be interpreted (for a contrary view, see Harris, 1989).

Variables with function coefficients of zero on a given function clearly have no effect on defining the synthetic variables associated with the function. That is, multiplying the scores on an observed variable by the multiplicative constant of zero statistically kills that observed variable.

But it is critical to remember that a given observed variable can get a function coefficient of zero for either of two reasons: (a) The measured variable has nothing to contribute with regard to the relationship between the variable sets; or (b) whatever the measured variable has to contribute with regard to the relationship between the variable sets (which may even be quite a lot), one or more other variables also contain this variance or information and the given measured variable is arbitrarily denied any credit for providing this information. Obviously, the same thing can occur in regression, but certainly no reasonable researcher believes that in regression measured variables with ß weights of zero are inherently useless (Thompson & Borrello, 1985).

So measured variables with near-zero function coefficients may or may not be useless in creating the detected effects, and this ambiguity can only be resolved by consulting structure coefficients. As Meredith (1964) suggested, "if the variables within each set are moderately intercorrelated the possibility of interpreting the canonical variates by inspection of the appropriate regression weights [function coefficients] is practically nil" (p. 55). Similarly, Kerlinger and Pedhazur (1973) argued that, "a canonical correlation analysis also yields weights, which, theoretically at least, are interpreted as regression β weights. These weights [function coefficients] appear to be the weak link in the canonical correlation analysis chain" (p. 344). Levine (1977) was even more emphatic:

> I specifically say that one *has* to do this [interpret structure coefficients] since I firmly believe as long as one wants information about the nature of the canonical correlation relationship, not merely the computation of the [synthetic function] scores, one must have the structure matrix. (p. 20, emphasis in original)

One can determine that a measured variable is arbitrarily being denied credit for providing predictive information on a given function when the variable has a near-zero function coefficient but has a structure coefficient that is large in absolute value (i.e., approaching −1 or +1). For example, Sexton et al. (1988) presented a canonical analysis in which one variable had a function coefficient of +.02 on Function I but a structure coefficient of +.89.

Measured variables can also have near-zero structure coefficients but function coefficients that are large in absolute magnitude. Such results indicate the presence of so-called "suppression" effects explained in most regression textbooks. Horst (1966) provided the classic example. An accessible explanation is presented by Lancaster (1999).

Only measured–observed variables that have function and structure coefficients that are both near zero contribute nothing to defining a given function and its associated effects. Only measured variables with all function coefficients near zero and canonical communality coefficients near zero contribute nothing to the canonical solution as a whole.

A hybrid case, however, in which only one set of coefficients could be correctly interpreted should be noted (Thompson, 1984, pp. 22–23). The function and structure coefficient matrices for a given variable set are identical when the measured variables in the set are perfectly uncorrelated, as would be the case, for example, if the variables in a set consisted of factor scores on orthogonally rotated principal components.

A Brief Illustrative Interpretation

The Table 9.2 results can be interpreted to provide a model of the recommended interpretive procedure. First, the noteworthiness of detected effects is evaluated. Here, the squared canonical correlation coefficients (Rc^2) are 18.00% and 3.50%. One can presume that the researcher deemed only the first canonical function noteworthy, given personal values and the substantive context of the study.

Although it is important to judge which, if any, of the Rc^2 values (and the associated canonical functions) are deemed noteworthy to be retained for interpretation, a thorough researcher also assesses whether a similar outcome would result were the analysis repeated with an independent random sample. There are various ways to accomplish this (see Crossman, 1996; and Thompson, 1990, 1995). In reading applications of CCA in journals, look for whether the researcher has attempted to estimate replicability.

In the present context, for example, presume that the researcher first evaluated the replicability of the detected effect initially by applying the commonly used Ezekiel (1930) regression R^2 correction formula to compute an adjusted Rc^2 (Thompson, 1990) for Function I and considered the "shrunken" Rc^2 still noteworthy. Next, presume the researcher conducted either a canonical cross-validation (Crossman, 1996) or a canonical bootstrap analysis (Thompson, 1995) and empirically determined from this "internal" replicability analysis that the detected effect was reasonably stable across variations in the sample.

At this juncture, the researcher addresses the issue of where the effects originate. All five measured variables have function and structure coefficients on Function I that are all positive, as reported in Table 9.2. Thus, all the observed variables are positively related with the underlying synthetic variables on this function.

On the one hand, the LFS variable has a function coefficient of +.091. This variable has a structure coefficient of +.502 on the function, however, and this indicates that the other observed variables on this function are arbitrarily getting credit for the predictive power that LFS brings to the table.

On the other hand, LFS has a disproportionately low squared structure coefficient (25.20%) as against the remaining coefficients on this function, which range from 42.77% to 77.26%. Thus, the overarching pattern is that higher scores on the two perceptual tasks, VP and CGD, are predictive of higher scores on both achievement measures, WM and MMF scores.

These results can also be viewed from another intriguing perspective. Note that both the WM and MMF variables have function and structure coefficients that are both positive on Function I. The WM and MMF variables have function and structure coefficients with opposite signs on Function II. Function I can therefore be viewed as explaining aspects of positive covariation between the two measured achievement variables, which is most noteworthy (Rc^2 = 18.00%). In contrast, Function II evaluates aspects of differences between the two achievement variables, although here such differences are not very noteworthy (Rc^2 = 3.50%). This contrast illustrates the richness of canonical analysis, which, in this example, can be used to explore prediction of both the commonalities among and the differences between given measured variables.

Two Analytic Pitfalls to Avoid

For researchers interested in conducting CCA and for readers who must interpret the work of others who use this method, two potential analytic pitfalls can lead to erroneous conclusions and should be avoided. First, there is the problem of incorrect interpretation of the statistical tests performed in CCA. Most statistical packages that perform CCA produce multiple functions and multiple test statistics, but only the last test statistic is a test of the effect size associated with a single function.

For example, imagine a hypothetical problem in which there are three canonical correlations. In this instance, most programs provide three sets of test statistics, and this may lead one to believe that each test statistic can be used to evaluate each correlation coefficient independent of the others; however, this is not true. In this case, the first test statistic is used to evaluate all three canonical correlations (and their squared values as well); the second test statistic is appropriate for evaluating the second and third coefficients as a set; and only the third test statistic is a test of a single correlation coefficient, that is, the third and final canonical correlation. Most computer programs do not test each single canonical correlation, except the last one, and conducting tests of individual canonical correlations, other than the last one, is not a straightforward matter (see Stevens, 1992, pp. 411–412).

In the context of the illustrative problem, for example, that involved the two Rc^2 values reported in Table 9.2, most computer programs would report two test statistics (not shown in Table 9.2). For this example, the computer output reported an F statistic of 12.25 (degrees of freedom [df] = 6,592, $p < .001$) for "roots 1 to 2" and an F of 5.34 (df = 2,297, $p < .005$) for "roots 2 to 2." Only the last test is an appropriate test of a single canonical correlation (Rc = .1863, Rc^2 = 3.50%).

Second, researchers should generally avoid interpreting canonical redundancy coefficients (Rd; see Table 9.2). Stewart and Love (1968) conceptualized these statistics, and Miller (1975) developed a partial test distribution to test the statistical significance of redundancy coefficients.

Although Cooley and Lohnes (1971, p. 170) suggested that redundancy coefficients have great value, more recent thinking suggests that the interpretation of redundancy coefficients does not make much sense in a conventional canonical analysis. As Cramer and Nicewander (1979) clearly established, redundancy coefficients are not truly multivariate (see also Thompson, 1988). This is important to note because univariate results are generally not useful in interpreting multivariate effects.

Furthermore, canonical analyses optimize Rc (not Rd); it seems contradictory to emphasize statistics not optimized in a given analysis. If one were interested in redundancy coefficients, then a redundancy analysis should be performed rather than a CCA (see Thompson, 1984).

Indeed, a redundancy coefficient can only equal one when (a) the synthetic variables for the function represent all the variance of every variable in the set (i.e., all squared structure coefficients are one), and (b) the squared Rc also equals one. Such an outcome would be rare. In short, redundancy coefficients are useful only to test outcomes that rarely occur and that may even be unexpected (Thompson, 1984). Nevertheless, there can be exceptions to this rule, such as perhaps in multivariate test–retest reliability or multivariate concurrent validity studies, where one might expect redundancy coefficients to approach one. Sexton et al. (1988) reported just such an exception.

Conclusion

CCA can be a useful analytic tool, as noted previously, but interpreting canonical results may challenge even seasoned analysts. As Thompson (1980) noted,

> [one] reason why the technique is [too] rarely used involves the difficulties which can be encountered in trying to interpret canon-

> ical results. . . . The neophyte student of [CCA] may be overwhelmed by the myriad coefficients which the procedure produces. . . . [But CCA] produces results which can be theoretically rich, and if properly implemented the procedure can adequately capture some of the complex dynamics involved in educational reality. (pp. 1, 16–17)

Such difficulties can be mitigated by following the admonitions suggested in this chapter. As with most analytic methods, real understanding is best facilitated by practice in the context of actual analytic problems of intrinsic interest to a given researcher.

Suggestions for Further Reading

A good starting point for further reading would be Thompson (1991b), followed by Crossman (1996). Stevens (1992; or other editions of his book) provided more comprehensive treatments of canonical analyses. Next, the comprehensive and more technical treatment in Thompson (1984) would be useful. With regard to statistical testing in this and in other contexts, Cohen (1994), Kirk (1996), and Thompson (1996) are all strongly recommended.

Glossary

CANONICAL CORRELATION COEFFICIENT (Rc) The Pearson product–moment correlation between the two sets of *synthetic variable* scores computed for a given canonical function.

COMMUANLITY COEFFICIENT (h^2) The proportion or percentage of variance in a measured variable that is useful in defining the canonical solution; conversely, the proportion or percentage of variance in a *measured variable* that the CCA solution can reproduce.

EFFECT SIZE The measures of magnitudes of effect or relationship that can and should be calculated in all studies, which generally fall into two major classes: (a) variance-accounted-for effect sizes analogous to r^2 (e.g., Rc^2) and (b) standardized mean differences (see Kirk, 1996; Snyder & Lawson, 1993).

EXPERIMENTWISE ERROR RATE ($\alpha_{experimentwise}$) The probability of making one or more Type I errors in a set of hypothesis tests conducted in a single study, ranging from a minimum of ($\alpha_{testwise}$) to a maximum of $1 - (1 - \alpha_{testwise})^k$, where k is the number of hypotheses tested (see Thompson, 1994d).

EXTERNAL REPLICABILITY ANALYSIS An analysis evaluating result replicability in which new data are collected to determine the degree to which (a) the same effect sizes occur and (b) the effects originate with the same *measured variables*.

FUNCTION The set (in some analyses called "equation" or "factor") of weights (e.g., regression β weights, factor pattern coefficients, canonical function coefficients) applied to the *measured variables* to yield scores on synthetic variables (e.g., regression predicted Y scores, factor scores, canonical or discriminant function scores).

FUNCTION COEFFICIENT The multiplicative constant or *weight* applied to a given *measured variable* as part of the calculation of scores on *synthetic variables*; the weights are standardized if the measured variables to which they are applied are in *z*-score form.

GENERAL LINEAR MODEL (GLM) The concept that CCA subsumes all classical parametric methods (from *t* tests through MANOVA and descriptive discriminant analysis) as special cases and that therefore all analyses (a) are correlational, (b) invoke *weights* being applied to *measured variables* to estimate *synthetic variables,* and (c) yield variance-accounted-for effect sizes analogous to r^2 (see Knapp, 1978; Thompson, 1991b).

INTERDOMAIN CORRELATION The bivariate correlation between scores on two variables, both of which are members of two different conceptually discrete variable sets.

INTERNAL REPLICABILITY ANALYSIS An analysis (e.g., cross-validation, jackknife, or bootstrap) attempting to evaluate result replicability using the data in hand, without a true replication, thus resulting in a somewhat positively biased estimate of replicability (see Thompson, 1993, 1994b, 1995).

INTRADOMAIN CORRELATION The bivariate correlation between scores on two variables, both of which are members of a single conceptually discrete variable set.

MEASURED–OBSERVED VARIABLE A variable for which scores are derived by direct measurement by the researcher, as opposed to by applying *weights* to other variables.

REDUDANCY COEFFICIENT (Rd) A canonical coefficient in a squared metric that is not multivariate and that is useful in CCA only in unusual cases in which a "g" (general) function with a perfect effect size (Rc^2 = 100%) is expected.

STRUCTURE COEFFICIENT (r_S) The Pearson product–moment correlation, which should be reported and interpreted in all CCA analyses, between the scores on a given *measured variable* and the *synthetic variable* scores on a given function for the variable set to which the measured variable belongs.

SYNTHETIC–LATENT VARIABLE Estimates of latent constructs, and the actual focus of all statistical analyses, computed by applying *weights* to the *measured variables* (e.g., regression predicted Y scores, factor scores, discriminant or canonical function scores).

TESTWISE ERROR RATE ($\alpha_{testwise}$) The probability of making a Type I error in the test of a single hypothesis.

WEIGHT The multiplicative constants (e.g., regression β weights, factor pattern coefficients, canonical *function coefficients*) applied to the *measured variables* to yield scores on *synthetic variables* (e.g., regression predicted Y scores, factor scores, discriminant or canonical function scores).

References

American Psychological Association. (1994). *Publication manual of the American Psychological Association* (4th ed.). Washington, DC: Author.

Baggaley, A. R. (1981). Multivariate analysis: An introduction for consumers of behavioral research. *Evaluation Review, 5,* 123–131.

Bagozzi, R. P., Fornell, C., & Larcker, D. F. (1981). Canonical correlation analysis as a special case of a structural relations model. *Multivariate Behavioral Research, 16,* 437–454.

Barcikowski, R. S., & Stevens, J. P. (1975). A Monte Carlo study of the stability of canonical correlations, canonical weights and canonical variate–variable correlations. *Multivariate Behavioral Research, 10,* 353–364.

Campbell, K. T., & Taylor, D. L. (1996). Canonical correlation analysis as a general linear model: A heuristic lesson for teachers and students. *Journal of Experimental Education, 64,* 157–171.

Chastain, R. L., & Joe, G. W. (1987). Multidimensional relationships between intellectual abilities and demographic variables. *Journal of Educational Psychology, 79,* 323–325.

Cliff, N. (1987). *Analyzing multivariate data.* San Diego, CA: Harcourt Brace Jovanovich.

Cohen, J. (1994). The earth is round ($p < .05$). *American Psychologist, 49,* 997–1003.

Cooley, W. W., & Lohnes, P. R. (1971). *Multivariate data analysis.* New York: Wiley.

Cooley, W. W., & Lohnes, P. R. (1976). *Evaluation research in education.* New York: Irvington.
Cramer, E. M., & Nicewander, W. A. (1979). Some symmetric, invariant measures of multivariate association. *Psychometrika, 44,* 43–54.
Crossman, L. L. (1996). Cross-validation analysis for the canonical case. In B. Thompson (Ed.), *Advances in social science methodology* (Vol. 4, pp. 95–106). Greenwich, CT: JAI Press.
Dunst, C. J., & Trivette, C. M. (1988). A family systems model of early intervention with handicapped and developmentally at-risk children. In D. R. Powell (Ed.), *Parent education as early childhood intervention: Emerging directions in theory, research, and practice* (pp. 131–179). Norwood, NJ: Ablex.
Estabrook, G. E. (1984). A canonical correlation analysis of the Wechsler Intelligence Scale for Children–Revised and the Woodcock–Johnson Tests of Cognitive Ability in a sample referred for suspected learning disabilities. *Journal of Educational Psychology, 76,* 1170–1177.
Ezekiel, M. (1930). *Methods of correlational analysis.* New York: Wiley.
Fan, X. (1996). Canonical correlation analysis as a general analytic model. In B. Thompson (Ed.), *Advances in social science methodology* (Vol. 4, pp. 71–94). Greenwich, CT: JAI Press.
Fan, X. (1997). Canonical correlation analysis and structural equation modeling: What do they have in common? *Structural Equation Modeling, 4,* 65–79.
Fish, L. J. (1988). Why multivariate methods are usually vital. *Measurement and Evaluation in Counseling and Development, 21,* 130–137.
Fornell, C. (1978). Three approaches to canonical analysis. *Journal of the Market Research Society, 20,* 166–181.
Fowler, P. C., & Macciocchi, S. N. (1986). WAIS-R factors and performance on the Lauria-Nebraska's Intelligence, Memory, and Motor Scales: A canonical model of relationships. *Journal of Clinical Psychology, 42,* 626–635.
Fuqua, D. R., Seaworth, T. B., & Newman, J. L. (1987). The relationship of career indecision and anxiety: A multivariate examination. *Journal of Vocational Behavior, 30,* 175–186.
Gage, N. L. (1978). *The scientific basis of the art of teaching.* New York: Teachers College Press.
Gorsuch, R. L. (1983). *Factor analysis* (2nd ed.). Hillsdale, NJ: Erlbaum.
Harris, R. J. (1989). A canonical cautionary. *Multivariate Behavioral Research, 24,* 17–39.
Heldref Foundation. (1997). Guidelines for contributors. *Journal of Experimental Education, 65,* 95–96.
Holzinger, K. J., & Swineford, F. (1939). *A study in factor analysis: The stability of a bifactor solution* (No. 48). Chicago: University of Chicago.
Horst, P. (1961). Generalized canonical correlations and their applications to experimental data. *Journal of Clinical Psychology, 26,* 331–347.
Hotelling, H. (1935). The most predictable criterion. *Journal of Experimental Psychology, 26,* 139–142.
Humphreys, L. G. (1978). Doing research the hard way: Substituting analysis of variance for a problem in correlational analysis. *Journal of Educational Psychology, 70,* 873–876.
Humphreys, L. G., & Fleishman, A. (1974). Pseudo-orthogonal and other analysis of variance designs involving individual-differences variables. *Journal of Educational Psychology, 66,* 464–472.
Jöreskog, K. G., & Sörbom, D. (1989). *LISREL 7: A guide to the program and applications* (2nd ed.). Chicago: SPSS.

Kerlinger, F. N., & Pedhazur, E. J. (1973). *Multiple regression in behavioral research.* New York: Holt, Rinehart, & Winston.

Kirk, R. (1996). Practical significance: A concept whose time has come. *Educational and Psychological Measurement, 56,* 746–759.

Knapp, T. R. (1978). Canonical correlation analysis: A general parametric significance testing system. *Psychological Bulletin, 85,* 410–416.

Krus, D. J., Reynolds, T. S., & Krus, P. H. (1976). Rotation in canonical variate analysis. *Educational and Psychological Measurement, 36,* 725–730.

Lancaster, B. P. (1999). Defining and interpreting suppressor effects: Advantages and limitations. In B. Thompson (Ed.), *Advances in social science methodology* (Vol. 5, pp. 139–148). Stamford, CT: JAI Press.

Levine, M. S. (1977). *Canonical analysis and factor comparison.* Beverly Hills, CA: Sage.

Maxwell, A. E. (1961). Canonical variate analysis when the variables are dichotomous. *Educational and Psychological Measurement, 21,* 259–271.

Maxwell, S. (1992). Recent developments in MANOVA applications. In B. Thompson (Ed.), *Advances in social science methodology* (Vol. 2, pp. 137–168). Greenwich, CT: JAI Press.

Meredith, W. (1964). Canonical correlations with fallible data. *Psychometrika, 29,* 55–65.

Miller, J. K. (1975). The sampling distribution and a test for the significance of the bimultivariate redundancy statistic: A Monte Carlo study. *Multivariate Behavior Research, 10,* 233–244.

Murphy, K. R. (1997). Editorial. *Journal of Applied Psychology, 82,* 3–5.

Pitts, M. C., & Thompson, B. (1984). Cognitive styles as mediating variables in inferential comprehension. *Reading Research Quarterly, 19,* 426–435.

Sexton, J. D., McLean, M., Boyd, R. D., Thompson, B., & McCormick, K. (1988). Criterion-related validity of a new standardized developmental measure for use with infants who are handicapped. *Measurement and Evaluation in Counseling and Development, 21,* 16–24.

Snyder, P., & Lawson, S. (1993). Evaluating results using corrected and uncorrected effect size estimates. *Journal of Experimental Education, 61,* 334–349.

Snyder, P. A., & Thompson, B. (1998). Use of tests of statistical significance and other analytic choices in a school psychology journal: Review of practices and suggested alternatives. *School Psychology Quarterly, 13,* 335–348.

Stevens, J. (1992). *Applied multivariate statistics for the social sciences* (2nd ed.). Hillsdale, NJ: Erlbaum.

Stewart, D. K., & Love, W. A. (1968). A general canonical correlation index. *Psychological Bulletin, 70,* 160–163.

Tatsuoka, M. M. (1973). Multivariate analysis in educational research. In F. N. Kerlinger (Ed.), *Review of research in education* (pp. 273–319). Itasca, IL: Peacock.

Thompson, B. (1980, April). *Canonical correlation: Recent extensions for modelling educational processes.* Paper presented at the annual meeting of the American Educational Research Association, Boston, MA. (ERIC Document Reproduction Service No. ED 199 269)

Thompson, B. (1984). *Canonical correlation analysis: Uses and interpretation.* Newbury Park, CA: Sage.

Thompson, B. (1988, April). *Canonical correlation analysis: An explanation with comments on correct practice.* Paper presented at the annual meeting of the American Educational Research Association, New Orleans, LA. (ERIC Document Reproduction Service No. ED 295 957)

Thompson, B. (1989, August). *Applications of multivariate statistics: A bibliography of ca-*

nonical correlation analysis studies. Paper presented that the annual meeting of the American Psychological Association, New Orleans, LA. (ERIC Reproduction Service No. ED 311 070)

Thompson, B. (1990). Finding a correction for the sampling error in multivariate measures of relationship: A Monte Carlo study. *Educational and Psychological Measurement, 50,* 15–31.

Thompson, B. (1991a). Invariance of multivariate results. *Journal of Experimental Education, 59,* 367–382.

Thompson, B. (1991b). A primer on the logic and use of canonical correlation analysis. *Measurement and Evaluation in Counseling and Development, 24*(2), 80–95.

Thompson, B. (1993). The use of statistical significance tests in research: Bootstrap and other alternatives. *Journal of Experimental Education, 61,* 361–377.

Thompson, B. (1994a). Guidelines for authors. *Educational and Psychological Measurement, 54,* 837–847.

Thompson, B. (1994b). The pivotal role of replication in psychological research: Empirically evaluating the replicability of sample results. *Journal of Personality, 62*(2), 157–176.

Thompson, B. (1994c). Planned versus unplanned and orthogonal versus nonorthogonal contrasts: The neo-classical perspective. In B. Thompson (Ed.), *Advances in social science methodology* (Vol. 3, pp. 3–27). Greenwich, CT: JAI Press.

Thompson, B. (1994d, February). *Why multivariate methods are usually vital in research: Some basic concepts.* Paper presented as a featured speaker at the biennial meeting of the Southwestern Society for Research in Human Development (SWSRHD), Austin, TX. (ERIC Document Reproduction Service No. ED 367 687)

Thompson, B. (1995). Exploring the replicability of a study's results: Bootstrap statistics for the multivariate case. *Educational and Psychological Measurement, 55,* 84–94.

Thompson, B. (1996). AERA editorial policies regarding statistical significance testing: Three suggested reforms. *Educational Researcher, 25*(2), 26–30.

Thompson, B. (1997). The importance of structure coefficients in structural equation modeling confirmatory factor analysis. *Educational and Psychological Measurement, 57,* 5–19.

Thompson, B. (1999a, April). *Common methodology mistakes in educational research, revisited, along with a primer on both effect sizes and the bootstrap.* Invited address presented at the annual meeting of the American Educational Research Association, Montreal, Canada. (ERIC Document Reproduction Service No. ED 429 110)

Thompson, B. (1999b). Five methodology errors in educational research: A pantheon of statistical significance and other faux pas. In B. Thompson (Ed.), *Advances in social science methodology* (Vol. 5, pp. 23–86). Stamford, CT: JAI Press.

Thompson, B. (1999c). Journal editorial policies regarding statistical significance tests: Heat is to fire as *p* is to importance. *Educational Psychology Review, 11,* 157–169.

Thompson, B., & Borrello, G. M. (1985). The importance of structure coefficients in regression research. *Educational and Psychological Measurement, 45,* 203–209.

Thompson, B., & Daniel, L. G. (1996). Seminal readings on reliability and validity: A "hit parade" bibliography. *Educational and Psychological Measurement, 56,* 741–745.

Thompson, B., & Snyder, P. A. (1997). Statistical significance testing practices in the *Journal of Experimental Education. Journal of Experimental Education, 66,* 75–83.

Thompson, B., & Snyder, P. A. (1998). Statistical significance and reliability analyses in recent *JCD* research articles. *Journal of Counseling and Development, 76,* 436–441.

Vacha-Haase, T., Nilsson, J. E., Reetz, D. R., Lance, T. S., & Thompson, B. (2000). Reporting practices and APA editorial policies regarding statistical significance and effect size. *Theory & Psychology, 10,* 413–425.

Wilkinson, L., & APA Task Force on Statistical Inference. (1999). Statistical methods in psychology journals: Guidelines and explanations. *American Psychologist, 54,* 594–604. (Available online http://www.apa.org/journals/amp/amp548594.html)

Wood, D. A., & Erskine, J. A. (1976). Strategies in canonical correlation with application to behavioral data. *Educational and Psychological Measurement, 36,* 861–878.

Zakaahi, W. R., & Duran, R. L. (1982). All the lonely people: The relationships among loneliness, communication competence, and communication anxiety. *Communication Quarterly, 30,* 203–209.

10 Repeated Measures Analyses: ANOVA, MANOVA, and HLM

Kevin P. Weinfurt

Consider the following designs:

- *Example 1. Cognitive Psychology.* A psychologist was interested in how people's abilities to recognize stimuli are impaired when their attention is diverted. Each research participant indicated as soon as he or she identified a word flashed quickly on a computer screen; the experimenter measured the time between the presentation of the word and the participant's reaction. For each participant, this was conducted under three conditions. In the "full attention" condition, the participant performed the word identification task in silence. In the "distraction" condition, the participant performed the task while an audio recording of a man reading a story was played. In the "divided attention" condition, the participants performed the task while listening to another story. This time, however, the participants were required to count the number of times the reader used the word *but.* For each participant, the experimenter calculated the mean reaction time for each of the three conditions.
- *Example 2. Clinical Psychology.* A researcher wanted to examine the efficacy of cognitive-behavioral therapy (CBT) compared with treatment as usual for patients with major depression who come to primary care clinics. Patients screening positive for depression were randomized to either the CBT or control group. A baseline assessment of depression was made using the Beck Depression

I am grateful to Paul Yarnold and Jean Lennon Weinfurt for their helpful comments on an earlier draft of this chapter.

Inventory (BDI; Beck & Steer, 1987). The BDI was also administered to patients at 3, 6, and 12 months after treatment began. The question of interest was whether patients in the CBT group demonstrated a greater reduction in depressive symptoms over time than did patients in the control group.

- *Example 3. Social Psychology.* A psychologist studying racial prejudice believes that people perceive greater racial prejudice against their own race overall compared with prejudice directed toward themselves. That is, the amount of perceived prejudice will vary according to the level of group. In a survey of racial minorities at a liberal arts college, respondents were asked to rate the amount of prejudice using a 1 (*no prejudice shown*) to 10 (*extreme prejudice shown*) scale. Separate ratings were made for (a) "my ethnic group," (b) "members of my ethnic group on campus," and (c) "myself."

All three of these examples describe research studies that generate *repeated measures data,* which means that measurements are taken on each participant under each of several conditions. The levels of the conditions describe the *within-subjects (W-S) variable.* In Example 1, reaction time measurements were taken under three different conditions of attention for each participant. In Example 2, BDI scores were obtained from the patients at each of the four time points. In Example 3, ratings of racial prejudice were given for each of three levels of ethnic group.

Although all three examples describe repeated measures designs, they differ among themselves. The first two examples are true experiments, whereas the third is more typical of nonexperimental survey research. Also the depression study (Example 2) is *longitudinal* because the measurements are taken for each level of the variable "time." This is not the case with Examples 1 and 3. Finally, one might also notice that whereas Examples 1 and 3 contain only a W-S variable ("level of attention" and "level of group," respectively), Example 2 contains a W-S (time) and a *between-subjects (B-S) variable* (treatment group).

Thus, repeated measures data can arise from a variety of situations. The presence of repeated measures makes the statistical analysis of such data a little more complicated than for B-S designs. For this reason, consumers of research often have difficulty understanding the results of repeated measures analyses. In this chapter, I attempt to reduce the befuddlement by explaining the three major types of statistical models currently used to analyze repeated measures data: repeated measures analysis of variance (ANOVA), repeated measures multivariate analysis of variance (MANOVA), and hierarchical linear models (HLM). I also present a brief treatment of the various ways to analyze one repeated measures situation that can confuse researchers, namely, the pretest–posttest design.[1] By the time the reader has completed this chapter, he or she should be able to understand the results of repeated measures analyses in a research article and, as a critical consumer of research, should be able to evaluate the appropriateness of repeated measures models in particular cases. The chapter is designed for readers with a basic understanding of factorial ANOVA, MANOVA, and multiple regression. For helpful introductory treatments of these topics, see Cohen and Cohen (1983), Maxwell and Delaney (1990), Stevens (1996), and Weinfurt (1995). To help the reader understand each of the three types of analyses, I use an example drawn from the literature on memory and aging.

Method of Loci

Imagine that a researcher was interested in studying the ability of older adults to improve their verbal memory using the method of loci mnemonic technique (Bosman & Charness, 1996). The method involves associating each item to remember with a specific location from a highly familiar sequence of locations, such as the rooms of a person's house (e.g., for the word *hat*, the person might visualize a hat hanging on the hook in her front hallway). Because the method of loci relies heavily on the ability to manipulate visual images, it is expected that older adults who are or were involved in the visual arts of drawing, painting, sculpting, graphic design, and so forth will be better at using this method (see Lindenberger, Kliegl, & Baltes, 1992). To examine this possibility, the researcher first administers a test of verbal memory to 50 older-adult visual artists and 50 nonartist older adults, where the score of interest is the percentage correct. After the baseline assessment, both groups are trained in the method of loci and are retested four more times at 1-week intervals. Group means for each of the five measurement occasions are plotted in Figure 10.1.

Figure 10.1 illustrates one W-S variable (time) and one B-S variable

[1]The analyses described in this chapter are appropriate for repeated measurements on an interval or ratio scale. To learn about the analysis of categorical (nominal or ordinal) data over time, see Agresti (1990) and Bijleveld and van der Kamp (1998).

Figure 10.1

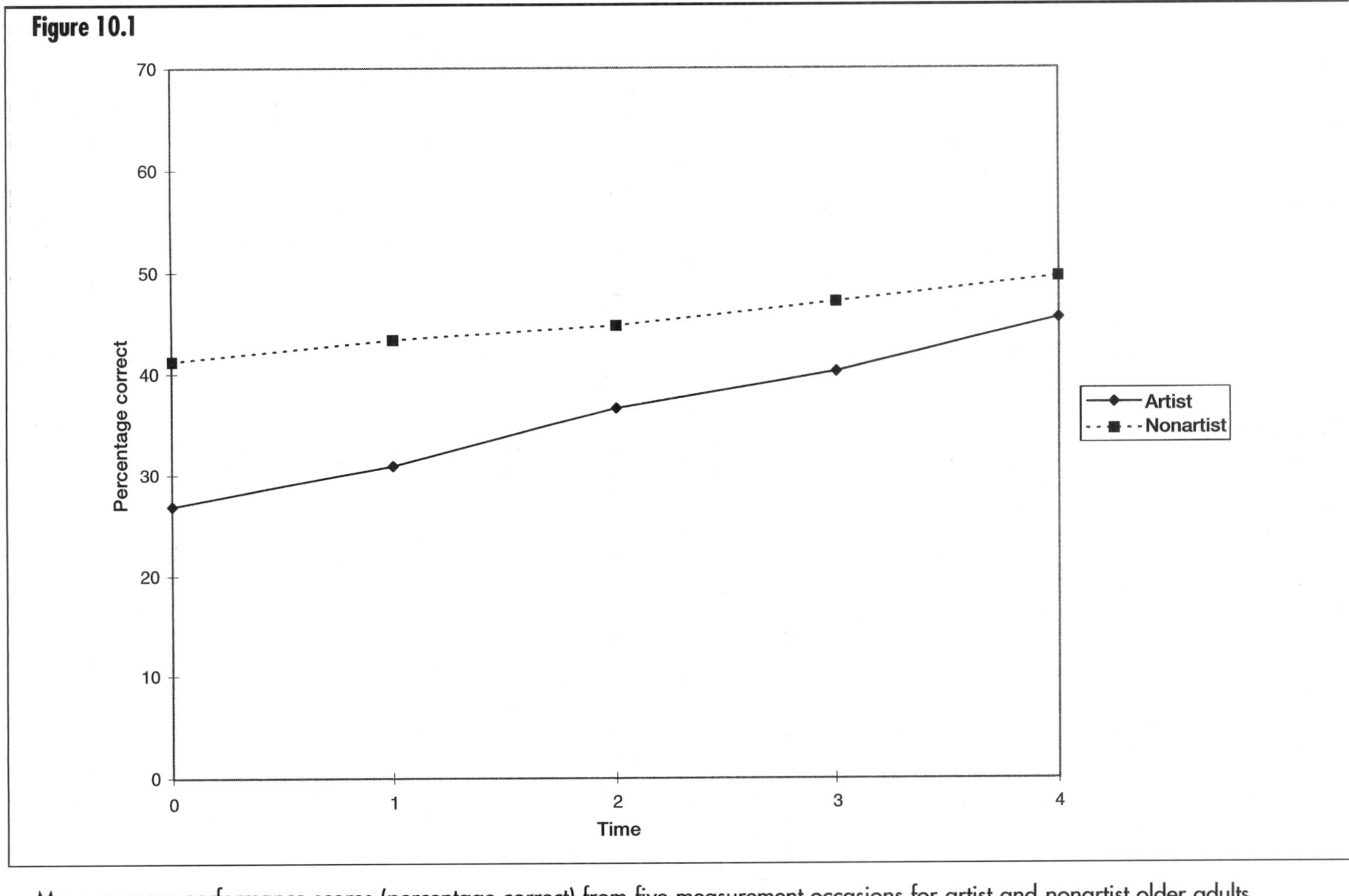

Mean memory performance scores (percentage correct) from five measurement occasions for artist and nonartist older adults.

(group). There are two a priori research hypotheses: (a) The method of loci should improve older people's memories over time, and (b) the rate of improvement should be greater for visual artist versus nonvisual artist older participants.

Repeated Measures Analysis of Variance

Overview of the Model

Each of the three statistical models presented in this chapter stresses different aspects of a repeated measures data set. The emphasis in repeated measures ANOVA is on the analysis of the mean values for each level of the W-S variable. To illustrate the basic principles of a repeated measures ANOVA, I begin with the simplest analysis using only a single W-S variable. Imagine that the method of loci study was conducted on a single sample of older adults (i.e., ignore the B-S variable of group). Table 10.1 displays a portion of the data and the overall means for each time point. The first hypothesis concerns whether the method of loci improves older people's memories over time. By conducting a repeated

Table 10.1

Subset of the Data for the Method of Loci Study

	Week					
Person	0 (Baseline)	1	2	3	4	Person's mean
1	42	47	55	59	65	53.6
2	38	39	37	28	30	34.4
3	27	41	45	58	70	48.2
4	27	24	31	36	33	30.2
5	34	36	32	31	39	34.4
6	39	44	37	44	43	41.4
(etc.)						
.						
.						
99	8	18	23	35	36	24.0
100	25	34	43	49	65	43.2
Mean	34.06	36.97	40.48	43.48	47.30	40.46
SD	14.57	13.74	14.94	17.45	22.04	14.39

measures ANOVA on these data, the researcher implicitly or explicitly reformulates the question as follows: Do the mean accuracy scores differ (or increase) across the five assessments? An examination of the means at the bottom of Table 10.1 suggests that they are different from one another, but it is not yet known whether these are statistically significant differences.

Why not conduct a simple one-way ANOVA to test whether the means for each time point differ? There are two answers to this question. First, a one-way ANOVA assumes that the observations are independent. Clearly this assumption would be violated because each participant in the study contributes more than one accuracy score to the data set (see Maxwell & Delaney, 1990). Violating this assumption can severely distort the results of the analysis. A second reason for not using a one-way ANOVA, which is discussed in greater detail later, is that the one-way ANOVA ignores useful information in the data set, namely, the systematic differences between individuals, as shown in the right-most column of Table 10.1. These differences between individuals in their average scores are counted as "errors" in the one-way ANOVA. One can reasonably assume, however, that these differences in scores between people are not simply random error; rather, they arise as a result of some people having better memories than others.

One way of accommodating the violation of independence and the presence of individual differences in memory scores is to use person as an independent variable along with time. To do this, "person" is considered a random effect and time a fixed effect. *Random* effects are those in which the levels of the variable are selected at random from a population of levels. In the case of the method of loci study, the 100 participants can be thought of as randomly selected from a population of older people. In this way, the 100 participants represent each of 100 levels of the random factor, person. Time is considered a *fixed* effect because the experimenter chose the five time points for the purposes of this study. Were one to replicate the study, one would use the same time points for assessment (because time is fixed) but would use a different random sample of participants (because person is random). The final ANOVA model is known as a *mixed model* because it contains a combination of fixed and random factors.

Calculation of the Test Statistic

To determine whether the means for each level of time differ, I use an omnibus (or overall) F test to evaluate the null hypothesis that the population means do not differ across the five time points. The reader may recall from other statistics texts that an *F* test is a ratio of variances. Variances are measures of the differences among scores, and researchers are interested in explaining those differences. For example, one might ask why all the accuracy scores in the method of loci data set are not the same. One possibility is that there are differences among the accuracy scores because there are differences among the average scores for each of the time points. Thus, I determine whether the variance explained by the time factor is significantly greater than zero. This is accomplished by comparing the variance that is due to the effect with the variance that is due to error. In the case of a repeated measures ANOVA, the *F* test for time effect is based on the following ratio of variances: time effect /(Person × Time interaction).

This ratio makes intuitive sense if one considers what it means to have a significant amount of variance that is due to the Person × Time interaction. The interaction specifies that the differences among the five time points may vary depending on which person is being observed. For example, Person 1's scores appear to increase over time, whereas Person 2's scores seem to decrease (see Table 10.1). Thus, the differences between the time points are different for Persons 1 and 2. In other words, the Person × Time interaction assesses the inconsistency of the time effect across people. Ideally, the conclusion about the effect of time should be the same for all people under study. Therefore, if the amount of variance that is due to the Person × Time interaction is large relative to the variance that is due to the time effect, one can conclude that there is too much variation from person to person to make any meaningful inference about the differences between the group means for the five time points. This is not the case for these data because the test for the time effect gives the following equation: $F(4,396) = 28.77$, $p < .0001$. That is, one would reject the null hypothesis of no differences in group means across time.

As alluded to earlier, the error term for the repeated measures ANOVA *F* test, the Person × Time interaction, does not include variation that is due to individual differences in the overall mean score. In Table 10.1, these individual overall means are shown in the right-most column. Note that there is a great deal of variability across people in these means (e.g., Person 3's mean is about twice as large as Person 99's mean). In a standard between-subjects design, these individual differences would contribute to the error variance, making the denominator of the *F* ratio larger, thus making it more difficult to obtain a significant effect. In the mixed-model ANOVA, the variation that is due to differences between people's mean scores is captured in the random person effect rather than counting it as error. This is the reason that within-subjects designs are generally more powerful than between-subjects designs.

A Note About Effect Size

The significance test tells whether the variance in the dependent variable explained by the independent variable is significantly different from zero. An equally important question, however, concerns how much variance is explained. In statistical parlance, this is an example of an *effect size.* There are two types of effect sizes: raw (or unstandardized) and standardized.

Raw (or Unstandardized) Effect Size

In examining the differences between the means shown at the bottom of Table 10.1, one might express the effect size as the largest difference between any of the means. In this case, it would be the difference between Time 4 (47.30) and Baseline (34.06), which is 13.24. This is the *raw or unstandardized* effect because it is expressed in the original metric of the dependent variable: the memory accuracy scores. These scores are percentages, so the effect detected corresponds to a difference of 13 percentile points on the measure. Similarly, it could be said that the effect is equal to a 100 × (13.24/34.06) = 39% increase in memory scores from baseline. Raw effect sizes are useful when the dependent variable is measured on such well-known metrics as percentile points, T scores, z scores, and inches.

When the dependent variable metric is not easily interpreted or when one is interested in comparing results from multiple studies, one generally wants an effect size that is expressed in the same metric, no matter what the conditions. In this case, a standardized effect size is required.

Standardized Effect Size

The most frequently reported measure of effect size for repeated measures analyses is *eta squared* (η^2; Maxwell & Delaney, 1990), which indexes the proportion of variance explained by a variable (think of it as similar to an R^2). For the method of loci data, η^2 for the effect of time was .23, which means that 23% of the variability in memory scores was related to the variability in the time of the measurements. There are other measures of effect size available (see Maxwell & Delaney, 1990; Rosenthal & Rosnow, 1985; and Rosenthal & Rubin, 1982), but I focus on η^2 in this chapter.

How big should η^2 be for one to be impressed by the effect? One way to approach the problem is to consider how large the obtained effect is relative to other effects found in the social sciences. Cohen (1977) offered a classification scheme for measures of the explained proportion of variance that considers .01 as small, .09 as medium, and .25 or greater as large. Most social sciences studies obtain small to medium effects using this scheme. In the method of loci study, the obtained effect of .23 would probably be characterized as a relatively large effect using Cohen's scheme.

What does an effect size of, for example, .10 really mean? One way to answer this is with a *binomial effect size display* (BESD; Rosenthal, Rosnow, & Rubin, 2000; Rosenthal & Rubin, 1982), which expresses the effect size in terms of the change in success rate (e.g., survival rate, improvement rate) attributable to an experimental treatment in a hypothetical treatment outcome study. For example, if η^2 were .10, this would correspond to a treatment group survival rate of 66% and a control group survival rate of 34%, which is a whopping difference. Calculation of the success rate for the treatment and control groups using η^2 is straightforward (see Rosenthal & Rubin, 1982). First, take the square root of η^2 and call it r. The experimental success rate is then $100 \times (.50 + r/2)$, and the control group success rate is $100 \times (.50 - r/2)$. So the success rates for the two groups on the basis of the obtained $\eta^2 = .23$ would be calculated as follows:

$$r = \sqrt{.23} = .50$$

$$\text{experimental success rate} = 100 \times (.50 + .50/2) = 75\%$$

$$\text{control success rate} = 100 \times (.50 - .50/2) = 25\%.$$

Thus, the effect of time of measurement is equivalent to an improvement in success rate of $75 - 25 = 50\%$.

Recasting Cohen's (1977) guidelines in terms of BESD figures, a small effect ($\eta^2 = .01$) corresponds to a 10% difference in mortality rates between treatment and control, a medium effect ($\eta^2 = .09$) corresponds to a 30% difference, and a large effect ($\eta^2 = .25$) corresponds to a 50% difference. Although researchers have been slow to use the BESD when reporting their results, the critical consumer should be aware of this enlightening way of characterizing the size of an experimental effect.

Follow-Up Tests

Having rejected the null hypothesis, the next step is to determine which means are significantly different from one another. As in the traditional ANOVA, there are two categories of follow-up tests: (a) post hoc comparisons between means and (b) a priori contrasts.

Post Hoc Tests

If the researcher had no specific a priori expectation regarding the differences between the individual means, then a post hoc analysis would be conducted to compare all pairs of means. In the method of loci example, the five time points yield $k(k - 1)/2 = 5(5 - 1)/2 = 10$ unique comparisons among means, where k is the number of levels of the W-S variable. Performing multiple tests of statistical significance increases the chance of a Type I error (sometimes called "alpha inflation"). Accordingly, there are several post hoc procedures that control the overall Type I error. A discussion of the merits and limitations of the various procedures is too complicated for the purposes of this chapter (see Maxwell & Delaney, 1990, for a thorough review of post hoc procedures).

For illustrative purposes, I present the results of all pairwise comparisons using paired samples (or dependent) t tests in Table 10.2. To maintain an overall Type I error rate of $p < .05$, each t test was evaluated

Table 10.2

Post Hoc Pairwise Comparisons of Means for the Method of Loci Example

Time points compared	Mean difference	Standard error of the mean difference	*t* value (99 *df*)
Baseline vs. T1	−2.91	0.72	4.04
Baseline vs. T2	−6.42	1.23	5.27
Baseline vs. T3	−9.42	1.78	5.28
Baseline vs. T4	−13.24	2.39	5.54
T1 vs. T2	−3.51	0.71	4.92
T1 vs. T3	−6.51	1.24	5.24
T1 vs. T4	−10.33	1.82	5.69
T2 vs. T3	−3.00	0.73	4.12
T2 vs. T4	−6.82	1.27	5.37
T3 vs. T4	−3.82	0.75	5.07

Note. All *t*s significant by Bonferroni-adjusted $p < .05$.

using the Bonferroni-adjusted critical p equal to the overall desired alpha (.05), divided by the number of comparisons (10) = .005. For these data, every t test produced a p value less than .005, so all are considered statistically significant by the Bonferroni criterion.

A Priori Contrasts

Recall that the first hypothesis for the method of loci example stated that the memory aid should improve older adults' memories over time. This is known as an *a priori hypothesis* because the researcher predicted a particular pattern of means prior to conducting the analysis. Note that as it is stated, the hypothesis is ambiguous. It predicts that the mean accuracy scores are increasing over time, but it does not specify exactly how they are increasing (e.g., in a straight line, a curved line, or in a few abrupt "jumps"). This is typical of most published hypotheses regarding scores over time. Technically, when the patterns of means are expected to increase in an unspecified way over time, one says that the means are a *monotonically increasing* function of time. To test a priori hypotheses, researchers use planned contrasts. The reader can consult other treatments of ANOVA for an in-depth discussion of contrasts (see especially Rosenthal & Rosnow, 1985; and Rosenthal et al., 2000). For the purposes of this chapter, I focus on two commonly used contrasts for repeated measures data: repeated and polynomial.

A *repeated contrast* compares each level of the W-S variable with the adjacent level. For the method of loci example, this would entail an examination of the differences between the baseline and T1, T1 and T2, T2 and T3, and T3 and T4. Note that the maximum number of contrasts one can examine at once is $k - 1$, where k is the number of levels of the W-S variable. Because the repeated contrasts consist of a limited set of pairwise comparisons, the relevant contrasts are all found in Table 10.2.

A *polynomial contrast*, sometimes called a "trend analysis," is conducted to determine the form of the relationship among the levels of the W-S variable. For a W-S variable with k levels, $k - 1$ contrasts are constructed. The first represents a linear (straight-line) relationship among the levels, the second represents a quadratic relationship (U shape or a parabola: W-S variable^2), the third a cubic relationship (S shape: W-S variable^3), and so forth. Note that the shape of a polynomial curve can be determined by looking at the power to which the variable is raised. That number minus 1 tells how many "bends" there are in the curve. Thus, time^2 is represented by a curve with $2 - 1 = 1$ bend (a U shape), and $time^3$ is represented by a curve with $3 - 1 = 2$ curves (an S shape).

In the method of loci data, there are $k - 1 = 4$ polynomial contrasts for time. This means that it is possible to test for a linear effect of time, a quadratic effect ($time^2$), a cubic effect ($time^3$), and a quartic effect ($time^4$). A separate F test is used for each polynomial to determine whether each particular trend (linear, quadratic, cubic) explains a statistically significant amount of variance in the dependent variable. For the method of loci data, only the first polynomial term corresponding to the linear effect of time is significant, $F(1, 99) = 30.96$, $p < .001$, $\eta^2 = .24$. Thus, one can conclude that the older adults' memory scores did improve over time and that the form of this improvement was a linear (i.e., straight-line) increase over time.

Assumptions

Repeated measures ANOVA has the same basic assumptions as a standard ANOVA with a few twists. *Independence of observations* requires that one participant's scores should not be influenced by another participant's scores. For example, in a study examining the effect of an educational intervention using groups of 8–10 students, the performance of a particular child in the group may be influenced by the other children in that group. Therefore, the observations are not independent from one another. This assumption cannot be violated or else the F tests will not provide accurate estimates of the Type I error rate (i.e., the p value).

Normality requires that the distribution of the dependent variable in the population be normal. In the case of repeated measures ANOVA, this means that the values of the dependent variable at each level of the W-S variable are normally distributed in the population. When a B-S variable such as treatment group is involved, the dependent variable must be normally distributed in the population for each group as well. Generally speaking, the repeated measures ANOVA is considered robust to violations of the normality assumption (Maxwell & Delaney, 1990; Stevens, 1996). This means that the Type I and Type II error rates for the F test are significantly distorted only when the distribution of the data is an extreme deviation from normal.

There is one consideration related to normality that receives little attention, however. The ANOVA model assumes that the mean is an adequate representation of the population. When the dependent variable is skewed to the right or left, the mean becomes an inaccurate "summary" of the data. In that case, the researcher might not want to compare the raw means. Instead, the researcher might opt to transform the data to make the distribution more normal (Emerson & Stoto, 1983), or he or she might choose a nonparametric analysis that tests for differences in the medians or the overall distributions (see Siegel, 1956).

Homogeneity of variances is required in a standard ANOVA, so that the estimate of the error variance is accurate. The same basic idea applies to repeated measures ANOVA, but the presence of repeated measures makes it more complicated. Technically, the assumption is known as *sphericity* and describes a form of the relationship between scores at all levels of the W-S variable. An explanation of sphericity is beyond the limits of this chapter (see Hertzog, 1994), but something can be said of a special case of sphericity known as compound symmetry. As Maxwell and Delaney (1990) noted, in all but a few cases, compound symmetry must be met in one-way repeated measures ANOVA. Repeated measures data are *compound symmetric* when (a) the variances of the measures at each level of the repeated factor are equal, and (b) the covariances, and hence correlations, between the measures at each level of the repeated factor are also equal. For the method of loci study, this would mean that the variances of the accuracy scores at baseline and Weeks 1–4 must be equal and that the correlations between peoples' scores at any two time points must be equal. When there are only two levels of the W-S variable, the sphericity assumption is met by default if the variances are equal because there is only one covariance term.

In a factorial repeated-measures design that includes a B-S variable such as group, an additional requirement is that the variances and covariances of the repeated measures must be the same for each group. This assumption, known as *homogeneity of covariance matrices*, can be tested using Box's (1950) M test. If the dependent variable is not normally distributed, however, Box's M tends to indicate that the covariance matrices are different, when in fact they are not (Stevens, 1996). The combination of the sphericity and homogeneity of covariance matrices assumptions is known as *multisample* sphericity (Stevens, 1996).

In practice, it is difficult to satisfy the sphericity assumption when there are more than two levels of the W-S variable. Particularly when data are collected over time for the same person, it is odd to think that the correlation between Time 1 and Time 2 is the same as the correlation between Time 1 and Time 5. (In the method of loci example, the correlations between Time 1 and each successive time point are .87, .66, .39, and .20, respectively.) When the sphericity assumption is not met, the F test is biased toward showing significant differences when there really are none. For this reason, researchers often assess the degree of sphericity with a value called "epsilon" (ε).[2] An ε of 1.0 indicates that sphericity is met, and a value of $\varepsilon = 1/(k - 1)$ indicates the worst possible violation of sphericity, where k is the number of levels for the W-S variable. There are two ways of calculating ε: the Greenhouse–Geisser (1959) estimator and the Huynh–Feldt (1976) estimator. The former is thought to underestimate ε, whereas the latter overestimates it (Maxwell & Delaney, 1990). In the example here, the two estimates of ε are very close: Greenhouse–Geisser = .290, Huynh–Feldt = .291. These values suggest the method of loci data depart significantly from sphericity, given that the worst possible ε for these data is $1/(5 - 1) = .25$.

Authors of research articles in professional journals seldom report their values of ε, much less indicate which of the two methods was used to calculate it. In cases where ε is reported, it is plain to see how far the data depart from the sphericity assumption. Recall that the smallest possible value ε can take in any research study with k levels of the W-S variable is $1/(k - 1)$, indicating the farthest departure from sphericity. Thus, the worst possible ε for a study with $k = 3$ levels is .50, whereas the worst possible ε for a study with $k = 6$ is .20. So what is a "good enough" value of ε to indicate the sphericity assumption is not unduly violated? There seems to be a consensus among researchers that $\varepsilon > .90$ is fairly safe, but there is little agreement as to how low it can go. Different recommendations can be found in Hertzog and Rovine (1985), Maxwell and Delaney (1990), and Stevens (1996).

Adjusted F Tests

Two options are available when sphericity appears to be violated. One is to use repeated measures MANOVA, which is covered later in this chapter. In cases where the violation of sphericity is not too extreme (ε is between .75 and .90; Hertzog & Rovine, 1985), an adjusted F test can be used within the mixed-model framework. The adjusted tests are obtained by multiplying the numerator and denominator degrees of freedom (df) by ε, making it necessary to obtain a larger F value to reach

[2]The Mauchly test of sphericity is sometimes used by researchers, but most methodologists do not recommend this test because it is highly influenced by how far the data depart from multivariate normality (Kesselman, Rogan, Mendoza, & Breen, 1980).

significance. (Recall that when sphericity is not met, the *F* value is biased toward being larger than it actually is.) Either the Greenhouse–Geisser or the Huynh–Feldt estimate of ε can be used, resulting in a more conservative or more liberal adjustment, respectively. (Stevens, 1996, recommended taking the average of the two estimates of ε and using it to correct the df.) In the method of loci example, the df for the *F* test (4, 396) would be adjusted to (1.16, 114.80) using the Greenhouse–Geisser ε and (1.16, 115.31) using the Huynh–Feldt ε. In both cases, the obtained *F* of 28.77 is highly significant, $p < .0001$.

Sphericity and Contrasts

As with the overall *F* test for the W-S effect, post hoc and a priori contrast analyses are vulnerable to violations of the sphericity assumption. In fact, the validity of the *p* values obtained from a priori and post hoc contrast analyses is more dependent on the assumption of sphericity than the overall *F* test. The topic of selecting the appropriate error term for such contrasts is beyond the confines of this chapter, but excellent discussions can be found in Maxwell and Delaney (1990) and Rosenthal and Rosnow (1985). As a critical consumer of research, the most important thing to know is that when sphericity is not met, these statistical tests may be biased toward showing significant results that are not in fact significant. It is the data analyst's responsibility to explore this possibility and take the necessary steps to correct any biases that may exist.

A Factorial Example

Until now my discussion has been limited to the case where there is only one independent variable or a so-called one-way repeated measures ANOVA. Recall that the original method of loci example included a B-S variable that specified whether the study participant was an artist or nonartist. Including this variable in the analyses now makes the design a 2 (group) × 5 (time) mixed-model ANOVA. (A study with one B-S and one W-S variable is sometimes referred to as a "split-plot design," which is a holdover from ANOVA's early beginnings in agronomy; "plot" refers to a plot of land.) The effects to analyze are as follows: (a) the B-S effect of group addresses whether the mean scores for the two groups differ, regardless of time; (b) the W-S effect of time addresses whether the scores vary as a function of time, regardless of group; and finally, (c) the interaction between group and time addresses whether the effect of group (or time) depends on the level of time (or group). I review these in greater detail below.

Between-Subjects Analysis

The main effect for group is tested after collapsing across the different levels of time. For the method of loci data, this involves (a) taking each person's average performance score across the five testing occasions and (b) comparing the means of these mean scores for each of the two groups. (Recall that these means are the same as those found in the right-most column of Table 10.1.) In this case, the mean for the artist group was 35.88 (SD = 13.74), and the mean for the nonartist group was 45.04 (SD = 13.67). The ANOVA comparing these two means produced $F(1, 98) = 11.18$, $p < .005$, $\eta^2 = .10$. Hence, performance scores averaged across time differed between the two groups, with the nonartist group having a larger mean across time than the artist group. This effect is not in itself very interesting because what is truly sought is whether the people in the study improved their memories and whether this improvement was different for the artist versus nonartist older adults. These latter two points are addressed by the within-subjects analyses.

Within-Subjects Analysis

The W-S effects include the main effect for time and the Group × Time interaction. The analysis of the W-S effects begins with an assessment of multisample sphericity. Recall that this assumption requires that sphericity holds for both groups and that the covariance matrix of the levels of the dependent variable must be the same for each group. The first condition is assessed using ε, which gives values of .28 and .30 for the nonartist and artist groups, respectively. This indicates a gross violation of the sphericity assumption in both groups. The second condition is tested using Box's M, which shows that the covariance matrices of the dependent variable do not differ between the artists and nonartists [$M = 16.29$, $F(15, 38{,}669) = 1.026$, $p = .423$]. I proceed with the analysis for the sake of illustration using the Geisser–Greenhouse correction to offset the bias introduced by the failure to meet the multisample sphericity assumption.

When this correction is made to the df for the time and Group × Time F tests (both of which have unadjusted df of 4 and 392), the resulting df are 1.16 and 114.15. Using the Geisser–Greenhouse corrected F test, there was a significant main effect for time, $F(1.16, 114.15) = 29.85$, $p < .001$, $\eta^2 = .23$. Taken by itself, this effect is interpreted as indicating that the mean percentage correct on the memory test (collapsing across the two between-subjects groups) was not the same for all five testing occasions. An examination of Figure 10.1 suggests that the means increased over time. Additionally, there was a significant Group × Time interaction $F(1.16, 114.15) = 4.70$, $p < .05$, $\eta^2 = .05$. That is, the differences in mean performance across time depended on whether the group was the artists or the nonartists. Visual inspection of Figure 10.1 would lead one to expect that the nonartist group's means increased to a greater extent than the artist group's means. A more detailed exploration of this pattern requires polynomial contrasts to determine the form of the effects just reviewed (e.g., straight line, U shaped, S shaped).

Polynomial Contrasts

With five levels of the W-S variable time, I construct 5 − 1 = 4 polynomial contrasts corresponding to the linear, quadratic, cubic, and quartic effects of time. There was a significant Group × Time interaction for the linear component of time, $F(1, 98) = 4.96$, $p < .05$, $\eta^2 = .02$, but no significant effects for the other polynomial terms. This significant contrast indicates that (a) the memory accuracy scores change over time in a linear fashion and (b) the *slopes* (i.e., steepness) of the lines for the two groups of participants are significantly different from one another. This is an important finding because it is consistent with the researcher's second hypothesis: Scores for the artist group increased more rapidly than scores for the nonartist group, presumably because the artists are more adept at using the method of loci.

I now pause to summarize the main points about the traditional mixed-model ANOVA approach to repeated measures data. For data sets with repeated measures, the standard ANOVA is inappropriate because the independence of observations assumption is violated and the model disregards the systematic individual differences in the dependent variable. These problems are remedied using the mixed-model ANOVA in which person is considered a random effect and the W-S variable is considered a fixed effect. This analysis generates more appropriate F tests. The size of the effects can be interpreted as raw effect sizes or by using a standardized measure of effect size such as eta squared or BESD. The mixed-model ANOVA assumes independence of observations, normality, and sphericity (or multisample sphericity in a B-S design). Adjusted F tests are available when the sphericity assumption has been violated. Finally, I reviewed post hoc and a priori follow-up tests of significance, including repeated and polynomial contrasts. These follow-up tests of significance clarify the exact nature of the relationships found by the omnibus F tests.

The method of loci example serves to demonstrate the chief features of a mixed-model ANOVA approach to analyzing repeated measures data. The extreme violation of the sphericity assumption, however, indicates that this model may not be the most appropriate for these data. As mentioned earlier, another option for analyzing these data is the multivariate analysis of variance (MANOVA).

Repeated Measures Multivariate Analysis of Variance

In principle, there is nothing different between a repeated measures MANOVA and a standard MANOVA. For that reason, the reader is encouraged to consult other treatments of MANOVA for an introduction to this procedure (e.g., Stevens, 1996; Weinfurt, 1995). Here I review briefly how the MANOVA can be used to analyze repeated measures data.

First, it is helpful to know that the only difference between the ANOVA and MANOVA approaches to repeated measures data is in the handling of the W-S effects. The analysis of the B-S effects is exactly the same. In other words, the analysis of the group effect in the method of loci study is identical for the ANOVA and MANOVA approaches. The time and Group × Time interaction effects, however, are analyzed differently. It is also helpful to know that when the W-S variable has only two levels (e.g., Dose 1 and Dose 2 in a drug trial), the two procedures produce identical results for all effects.

Calculation of Test Statistics

A MANOVA is used to examine the effects of one or more independent variables on two or more dependent variables. Specifically, a MANOVA tests the null hypothesis that a collection of means on the dependent variables (called a "vector of means") is equal for all levels of the independent variables. To fit repeated measures data into the MANOVA framework, it is necessary to construct a new set of dependent variables, which, for these purposes, I call *D*s. One or more D variables is created to represent each effect of interest in the repeated measures analysis. For instance, in the method of loci example, I need a set of D variables to represent the effects of time, group, and the Time × Group interaction. The best way to describe D variables is with an example, so I proceed with a method of loci analysis.

First, I need to test the omnibus B-S effect of group. As stated earlier, this test is equivalent to the ANOVA approach. That is, the D variable used to test the group effect is simply each person's mean accuracy score across time periods (the right-most column in Table 10.1). Thus, the test of the group effect is a one-way ANOVA, giving $F(1, 98) = 11.18$, $p < .005$, $\eta^2 = .10$. This indicates that averaging over all the time points, the artists ($M = 35.88$, $SD = 13.74$) had a lower mean performance than the nonartists ($M = 45.04$, $SD = 13.67$).

Next, I test the omnibus W-S effect of time. For any given W-S or interaction effect with k levels, $k - 1$ D variables are created. There are several ways to construct the D variables, but I use a fairly straightforward way equivalent to the repeated contrasts, as described in the ANOVA section. That is, $k - 1 = 5 - 1 = 4$ D variables for the time effect are calculated for each person:

$$D_1 = \text{Time } 1 - \text{Time } 0$$

$$D_2 = \text{Time } 2 - \text{Time } 1$$

$$D_3 = \text{Time } 3 - \text{Time } 2$$

$$D_4 = \text{Time } 4 - \text{Time } 3.$$

The test of the W-S effect of time is equivalent to the test that $D_1 = D_2 = D_3 = D_4 = 0$ in the population. That is, there is no difference in the memory scores for each time point. For the W-S effect, an exact F value is calculated, giving $F(4, 95) = 8.36$, $p < .00001$, $\eta^2 = .25$. This effect indicates that regardless of whether the participants are artists or nonartists, the memory scores vary according to time.

Testing the Time × Group interaction is a little less obvious. The effect of time has already been incorporated into the creation of the new dependent variables D_1 through D_4. That is, these new variables measure the extent of change in scores between levels of time. To determine whether this effect of time varies according to group, I simply include group as an independent variable that is used to predict the vector of D_1 through D_4. When the B-S variable has only two levels, like the group variable, then an exact multivariate F test can be conducted. In this case, the interaction is significant, $F(4, 95) = 2.60$, $p < .05$, $\eta^2 = .10$. When the B-S variable has more than two levels, however, one of four multivariate test statistics is used instead: Pillai's trace, Wilks's lambda, Hotelling's trace, and Roy's largest root, the most popular of which is Wilks's Lambda.[3] These values are then converted to approximate F values to derive a significance level for the test. Often a researcher reports only the approximate F test but does not include the actual multivariate test statistic value.

Follow-Up Tests

To better understand the nature of the effects found with the omnibus tests, one can conduct any number of a priori or post hoc comparisons, as in the ANOVA approach. MANOVA follow-up tests are discussed in detail in Weinfurt (1995), so I present only some general considerations here. The construction of D variables makes it very easy to test contrasts in the context of the main analysis. Recall that the way the D variables for the time effect is defined is exactly like the way a repeated contrast is defined. In other words, a contrast analysis is built in to the MANOVA. Separate F tests could be conducted for each of the D variables to determine which contrast was significant and to determine whether Group related to some but not other contrasts. Alternatively, the four D variables could be defined using a polynomial contrast, such that the four new variables would represent the linear, quadratic, cubic, and quartic components of time.

As always, when multiple tests of significance are conducted, whether a priori or post hoc, Type I error should be controlled. For most applications, the Bonferroni procedure ensures that the overall Type I error is kept at the desired level (Maxwell & Delaney, 1990). To review, this approach involves computing a critical p value equal to α/C, where α is the desired overall Type I error (e.g., .05) and C is the number of comparisons (i.e., number of tests of statistical significance). In some cases, however, the investigator wants to conduct a *complex comparison,* such as comparing the average scores for two levels of the W-S variable with the third level. For example, in a drug trial, patients might receive each of three doses (e.g., low, medium, high) of a drug over time. The investigator may be interested in whether the response to the high dose is significantly greater than the average of the low and medium doses. When complex comparisons such as these are conducted, researchers sometimes use the Roy–Bose multivariate extension of the Scheffé test (Maxwell & Delaney, 1990).

[3] Wilks's lambda has a handy property: When subtracted from 1, Wilks's lambda equals η^2. This is helpful when researchers do not present effect sizes but do give the value of Wilks's lambda.

Assumptions

A thorough discussion of the assumptions underlying MANOVA and how they are tested can be found by Weinfurt (1995). Briefly, MANOVA assumes that the variables under consideration have a multivariate normal distribution, that the covariance matrices of the *D* variables are the same for all levels of the B-S variables, and that the observations are independent from one another. MANOVA is generally considered sensitive against violations of the first assumption, somewhat sensitive against violations of the second, and extremely sensitive to the last assumption (Weinfurt, 1995). (Box's *M* can be used to test the equality of the covariance matrices of the *D* variables across groups. For the method of loci data, $M = 16.29$, $p = .423$, which indicates the assumption is not violated.) The important point about the MANOVA approach to repeated measures data, however, is that it does not assume sphericity, as does the ANOVA approach. When sphericity is violated, MANOVA is a more valid and statistically powerful procedure. (Statistical *power* is the ability to detect statistically significant effects that exist in nature.) Note, though, that when sphericity is met (e.g., $\varepsilon = 1.00$), the MANOVA approach is actually less powerful than the ANOVA approach.

As a summary of the MANOVA approach to repeated measures data, the MANOVA creates a new set of variables (*D*s) for each effect of interest (e.g., W-S, B-S, and the interaction). These are then treated as multiple dependent variables that the MANOVA is designed to analyze. The important difference between the MANOVA and the mixed-model ANOVA approach is that the MANOVA does not assume sphericity, whereas the ANOVA does. Given that sphericity is difficult to meet in many cases, the MANOVA approach provides a more viable way of analyzing repeated measures data.

Analysis of Pretest–Posttest Designs

The pretest–posttest design is a frequently encountered repeated measures design in psychology that requires special attention. In such a design, measurements are taken before and after an intervention of some sort. Often, two or more groups of people will participate (e.g., a treatment group and a control group), and the researcher is interested in determining whether one group's posttest scores changed from the pretest significantly more than another group's posttest scores. A critical consumer of research should be aware that an investigator's choice of analysis for pretest–posttest designs can make a big difference in the final results. In the worst case, the analysis the investigator conducts may not even answer the question of interest. Thus, it is important to understand the options for analyzing these data. Researchers in the behavioral sciences tend to use one of three statistical analyses in this case (Huck & McLean, 1975; Jennings, 1988). The first approach is to conduct a repeated measures ANOVA with time as the W-S variable and group as the B-S variable. The second approach would be to use an analysis of covariance (ANCOVA) using the pretest scores as covariates. Less frequently used would be an ANOVA on the gain scores. (The *gain score* for a person is simply the posttest score–the pretest score.) Which analysis is the most appropriate?

The first approach, the repeated measures ANOVA, is probably the most frequently used, but it is the least parsimonious. Typically, a researcher who conducts a 2 (time) × 2 (group) repeated measures ANOVA will report the main effects for time, group, and the Group × Time interaction. Actually, the effect of interest in the pretest–posttest design is expressed by the Group × Time interaction, and the other two main effects are irrelevant (Huck & McLean, 1975).[4] Once a significant interaction is found, however, the analyses of simple effects or interaction contrasts are necessary to characterize the nature of the interaction. Thus, multiple steps are required to answer the simple question of which group changed more. Technically, the F value for the Group × Time interaction is equivalent to the F value generated by an ANOVA on the gain scores, although the latter has fewer degrees of freedom in the denominator.[5] Because performing a one-way ANOVA on gain scores is more efficient, this procedure is always better than the repeated measures ANOVA.

The question then becomes when to use an ANOVA versus an ANCOVA on gain scores. Assuming that the correlation between pretest and posttest scores is less than 1.0 (which is almost always the case), the chief determinant of which approach to use is the type of study design.

[4]Sometimes a researcher wants to know, additionally, whether scores on some measure increased from pretest to posttest for all people on average, regardless of group membership. Hence, the researcher would examine the main effect of time for evidence of an overall time effect.

[5]In fact, for a randomized design (i.e., people are randomly assigned to treatment groups) with no mean differences between the groups on the pretest scores and a pretest–posttest correlation of 1.0, the ANOVA, ANCOVA, and gain score analysis are for all practical purposes equivalent (Maxwell & Delaney, 1990). There is almost never a pretest–posttest correlation of 1.0 in behavioral science research, however.

For a randomized design, where there is random assignment to treatment groups and the pretest group means can be assumed to be the same, the ANCOVA provides a more powerful analysis of the group effect (for a technical discussion, see Huck & McLean, 1975, pp. 516–517, and Maxwell & Delaney, 1990, pp. 392–393).

When the study is a quasi-experimental design (Campbell & Stanley, 1963), however, the situation is more complex. In a quasi-experimental study, people are not randomly assigned to groups, but rather the groups are often pre-existing or "intact." Examples of intact grouping variables include gender, race, and age group (young or old). The crucial point to understand is that the ANCOVA and the gain-score analysis both answer different questions when analyzing *intact groups.*

In the ANCOVA, the question of interest is the following: If the groups were equivalent on the pretest, would there be a significant difference between the groups on the posttest? To use a well-known example, consider a study in which male and female university students were weighed before and after their freshman year of school (Lord, 1967). In this case, an ANCOVA using weights of students before entering school as the covariate would test whether the mean weight of men and women differed at the end of the school year, once the end-of-year mean weights had been statistically adjusted. Maxwell and Delaney (1990) described the nature of this adjustment and the results it yielded:

> Here we could phrase the question, For subpopulations of males and females having identical pretest weights—say, at the grand mean weight for the two sexes combined—would a difference in mean weight at the end of the year be expected? For example, if the mean weight of females is 120 pounds and that of males is 160, the grand mean, assuming equal numbers of males and females, is 140 pounds. Males weighing 140 are lighter than the average male and some regression back toward the mean of the whole population of males would be expected over the course of the year. Females weighing 140 similarly would be unusually far above the mean weight for females and would be expected as a group to show some regression downward over the academic year. (p. 394)

In this case, an ANCOVA on the adjusted posttest means would show a statistically significant difference, leading the unaware reader to believe that men and women underwent different weight changes over the year. In fact, for this example (Lord, 1967), the mean weight for men remained the same over the course of the year, as did the mean weight for women. Not only was there no differential change in mean weight between men and women, but there was no mean change at all. These misleading results arise because the question answered by the ANCOVA is the wrong question. Comparing group posttest scores as if the groups had the same mean pretest score is totally useless when the groups do not have the same mean scores on the pretest.[6]

An ANOVA on the gain scores would answer a very different and simple question: Did men and women undergo differing amounts of mean weight change over the school year? The answer to this is found by calculating a gain score for each person (i.e., end-of-year weight–beginning-of-year weight) and performing a one-way ANOVA on the mean gain scores for men and women. (Of course, this amounts to an independent samples *t* test because there are only two groups, but the ANOVA terminology is retained for the sake of generalizability.) When this is done, the analysis correctly shows that the gain scores for men and women are zero and thus do not differ between groups.

Now, if there are no group differences on the pretest mean, then the ANCOVA would be statistically more powerful than the gain-score analysis, as in the randomized designs discussed earlier. This is because the ANCOVA is always more powerful than the gain-score analysis when there are no pretest group differences, whether as a result of randomization to treatment or simply because the intact groups do not differ on the pretest for some reason. At this point, all of these details might seem bewildering, so I include a decision tree (Figure 10.2) to help the reader organize this material.

Analysis of Longitudinal Data Using Hierarchical Linear Models

Research involving two or more assessments of the same dependent measure over time is known as *longitudinal*. The analysis of longitudinal data is notoriously difficult in the social sciences. The good news is that statistical methods have been evolving over the past 2 decades to remedy this situation (see Collins & Horn, 1991). The bad news is that there are many new techniques, and it is often difficult for the uninitiated to keep them straight. Even more troublesome is the fact that researchers performing analyses for longitudinal research designs have fallen prey

[6]The issue of statistical control when intact groups are involved has received much attention. Cohen and Cohen (1983) provided a wonderful example of the essential problem. Imagine comparing the mean altitude of the Himalayan and Catskills mountain ranges, controlling for air pressure. Because air pressure is a direct function of altitude, the mean altitudes adjusted for air pressure would be equivalent—hardly an interesting result.

Figure 10.2

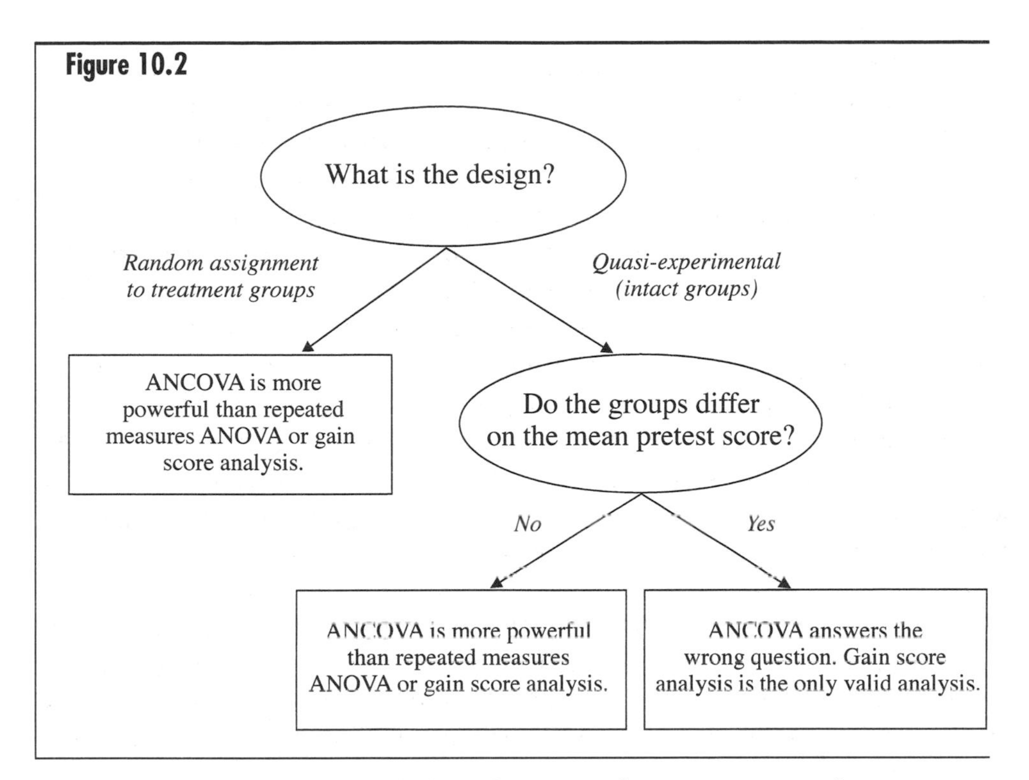

A decision tree for deciding which analysis to use for pretest–posttest data, assuming a correlation between pretest and posttest is less than 1.0.

to several misconceptions, many of which were discussed in a little-read but important chapter by Rogosa (1988).[7] My goal of this portion of the chapter is to present an analysis that offers a more viable alternative to traditional ANOVA- or MANOVA-based analyses, known as HLM.

Philosophical Rationale

Recall that the research questions motivating the method of loci training study concerned the method's effectiveness for older people, visual artist and nonartist people in particular. To address these questions using the traditional ANOVA or MANOVA approach, I had to rephrase the questions. Specifically, any reference to the behavior of people is translated as a reference to the behavior of group means. This is most explicit in the ANOVA approach to repeated measures analysis, where the data to analyze are just those means plotted in Figure 10.2. The

[7]Rogosa's 1988 chapter was reprinted with supplemental material in Rogosa (1995). This chapter is a must read for any serious researcher interested in studying phenomena over time.

underlying assumption (consistent with the sphericity assumption) is that all members of a group (e.g., visual artist older people) can be characterized by the same profile of mean performance across time; any individual deviations from that profile are attributed to error. Thus, although the original research question might have been framed in terms of how particular people perform, the unit of statistical analysis is the group mean profile. Is this a problem?

Although it may not be apparent to many practicing researchers, the above situation is a problem, and one that has been couched in terms of the debate over the distinction between general and aggregate propositions (Bakan, 1967; Danziger, 1990; Lamiell, 1998; Porter, 1986; Yarnold, Feinglass, Martin, & McCarthy, 1999). That which is true in general is true for all individuals in a population, whereas something that is true in aggregate is true only as a characteristic of a group (e.g., an average value). One of the most pertinent restatements of this general and aggregate distinction is Estes's (1956) article in which he argued that except under stringent and infrequently realized conditions, inferences about individual learning curves cannot be made on the basis of learning curves for a particular group. The same argument could be made for inferences concerning curves of the sort displayed in Figure 10.1. On the basis of mean growth curves for the artist and nonartist groups, one would like to infer something about the rate of increase in the performances of artist and nonartist people. Unfortunately, there is no sound basis for such an inference. The mean profiles for each group support inferences about the two groups on average, but they do not support inferences about the performance of any of the persons in general.[8]

If the goal of the research was to explain how the people changed across time, then the appropriate level of analysis should be individual growth curves (Rogosa, 1988). This is the philosophy behind a family of procedures for analyzing repeated measures that is used with greater frequency in clinical psychology (Tate & Hokanson, 1993; Willett, Ayoub, & Robinson, 1991), personality psychology (Moneta & Csikszentmihalyi, 1996), developmental psychology (Reese, 1995; van den Boom & Hoeksma, 1994), and clinical neuropsychology (Francis,

[8]When all individuals share the exact same growth curve, the assumption of compound symmetry required in univariate repeated measures ANOVA is met. In other words, the only case in which an analysis of an aggregated profile of scores across time can yield information about individual's growth curves is when the individual curves are all the same. For more technical treatments of the fallacies inherent in drawing conclusions about individuals from aggregated data, see Bakan (1967), Estes (1956), Lamiell (1998), and Rogosa (1988, 1995).

Fletcher, Stuebing, Davidson, & Thompson, 1991). The approach is known alternatively as "individual growth curve analysis" (Rogosa, Brandt, & Zimowski, 1982; Shock, 1951), "random effects modeling" (Laird & Ware, 1982), "covariance components modeling" (Dempster, Rubin, & Tsutakawa, 1981), "multilevel modeling," and "hierarchical linear modeling" (Bryk & Raudenbush, 1987, 1992).

For the purposes of this chapter, I refer to this general class of models as *hierarchical linear models* (HLMs).[9] At present, the available techniques differ mostly in the way they estimate parameters of the models (which is discussed shortly), but the hierarchical form of the models is essentially the same. The models are known as hierarchical because there are two stages or levels required to describe the data. Each of these is reviewed below. (Please note that this section is intended only as a conceptual introduction to this family of analyses. See Suggestions for Further Reading at the end of this chapter for more detailed treatments of HLMs.)

Overview of the Model

Although an HLM analysis is computationally sophisticated, its rationale is simple. To understand how people change over time and how that change might be related to other variables (such as treatment group), two things need to be done: The first is to find a way of quantifying how much each person changed, and the second is to use this quantity as the dependent variable in a new analysis. This new analysis uses a variable (or variables), such as treatment group, gender, and age to predict individual differences in change. (See also Yarnold et al., 1999, for a conceptually similar method that uses single-case analysis of longitudinal data as an initial preprocessing step.) The most simple HLM analysis pursues these objectives using an overall model with two levels.

Level I Model

The first stage involves specifying the model for the individual time paths. Consider the data points for the five measurement occasions for

[9]Confusion is sometimes created because the older sense of *hierarchical* refers to how the predictors are entered into a multiple regression equation (Cohen & Cohen, 1983). When they are entered hierarchically (or *sequentially*), a predictor is allowed to explain variance in the dependent variable that could not be explained by the predictors preceding it. In other words, if the predictors are entered A, B, and C, then A is allowed to explain all available variance in Y; B is allowed to explain variance in Y left over after A has been entered; and C is allowed to explain residual variance in Y remaining after A and B have been entered. This is in contrast to *simultaneous* entry, where the effect of each predictor is estimated while simultaneously controlling for all other predictors in the model.

two of the adults from the memory study (Person 6 and Person 20), displayed in Figure 10.3A. How might one describe the paths of the points, or the *growth trajectories*, for these two people? One needs to posit some parsimonious model that describes what one believes to be each person's true growth trajectory.[10] This *Level I model*, or within-subject model, is expressed as the equation of a line that best fits an individual time path. It could be a straight line, a U-shaped curve, an S-shaped curve, or another type of line. In Figure 10.3A, clearly a straight line would be appropriate for both people under study. The best-fitting lines are added in Figure 10.3B.[11]

Recall that the equation of a straight line is expressed as $Y = a + bX$, where a is the *intercept* of the Y axis when X is zero and b is the slope that describes how much Y changes for every one unit increase in X. For our example, the Level I equation describing straight-line growth in Y as a function of time is similar, except that an additional term is added to indicate that there is error in the prediction of Y (i.e., all of the data points do not lie exactly on the lines in Figure 10.3B). So the two key parameters of the model describing each person's growth curve in this case are the intercept (the level of Y at Time 0) and the slope (the rate of change in Y for every one unit increase in time). Initially, the parameters of the Level I models are estimated using *ordinary least squares* (OLS), which is exactly the same as a simple regression analysis. The Level I models for the two people displayed in Figure 10.3A are shown next to their respective growth curves in Figure 10.3B. For example, Person 20's model states that the person begins with a baseline performance of 16% and increases 12.8 percentile points for every 2-week testing period.

Figure 10.3B illustrates the fact that at least for these data, not everyone has the same value of Y at Time 0, indicating that the intercept is different for some of the people's growth models. Additionally, some people's lines are steeper than others, reflecting the fact that not everyone's slope is the same. Hence, a growth model must be custom made for each person's data.

As a summary, the Level I model for each person consists of an

---

[10]The assumption that each person's observed data are the result of some underlying "true" growth model + random error is based on classical test theory. The reader should be aware that there are alternative models that divide the observed score into (a) a stable trait component, (b) a state component, and (c) an error of measurement (Nesselroade, 1991a, 1991b). There is seldom a single, agreed-on way of doing things in data analysis.

[11]It is possible to fit curvilinear models of individual growth, and these would simply be polynomial regression equations. See Francis et al. (1991) for an example of a polynomial within-subject model.

intercept term (*a*) indicating his or her performance at baseline (time = 0) and a slope (*b*) indicating the increase (or decrease) in memory performance associated with an increase of one time period (i.e., 1 week).

Level II Model

At the between-subjects level, or *Level II model*, the goal is to predict the values of each individual's intercept and slope. The primary question of interest for the method of loci study concerns whether differences in the amount of change in memory performance can be explained by group membership (artist or nonartist). In other words, one must determine the proportion of variance in individual slopes (*b*) that is accounted for by group membership. The first step, then, is to determine how much total variation there is in the slope parameters. Toward this end, one would construct an unconditional model (Bryk & Raudenbush, 1992). The model is *unconditional* because no attempt is made to explain variance in the individual Level I model parameters. This model consists of two equations predicting the individual intercepts (*a*) and slopes (*b*), respectively. In each case, the individual parameters (*a* or *b*) are expressed as random deviations from a population mean. Thus, one gets

$$a = \beta_{00} + \text{random deviation}$$

$$b = \beta_{10} + \text{random deviation},$$

where β_{00} is the estimated population average intercept (i.e., the average baseline status) and β_{10} is the estimated population average slope (i.e., the average change). The *random deviation* terms indicate that there is variation around these population averages because not everyone has the same intercept and slope. Table 10.3 displays the values for these parameters for the method of loci data.

As shown in Table 10.3, the model estimated a mean intercept (or baseline value) of 33.86, but individuals deviated from that average intercept such that the variance of these random deviations was 211.88. Likewise, the people varied about the estimated mean slope of 3.30, as reflected by the variance of 34.23. Two things need to be said about these numbers.

First, I said earlier that one feature of HLM analysis is that each person is allowed to have his or her own intercept and slope describing memory scores over time. What if there are not substantial differences between people's intercepts (i.e., everyone starts out with the same base-

Figure 10.3A

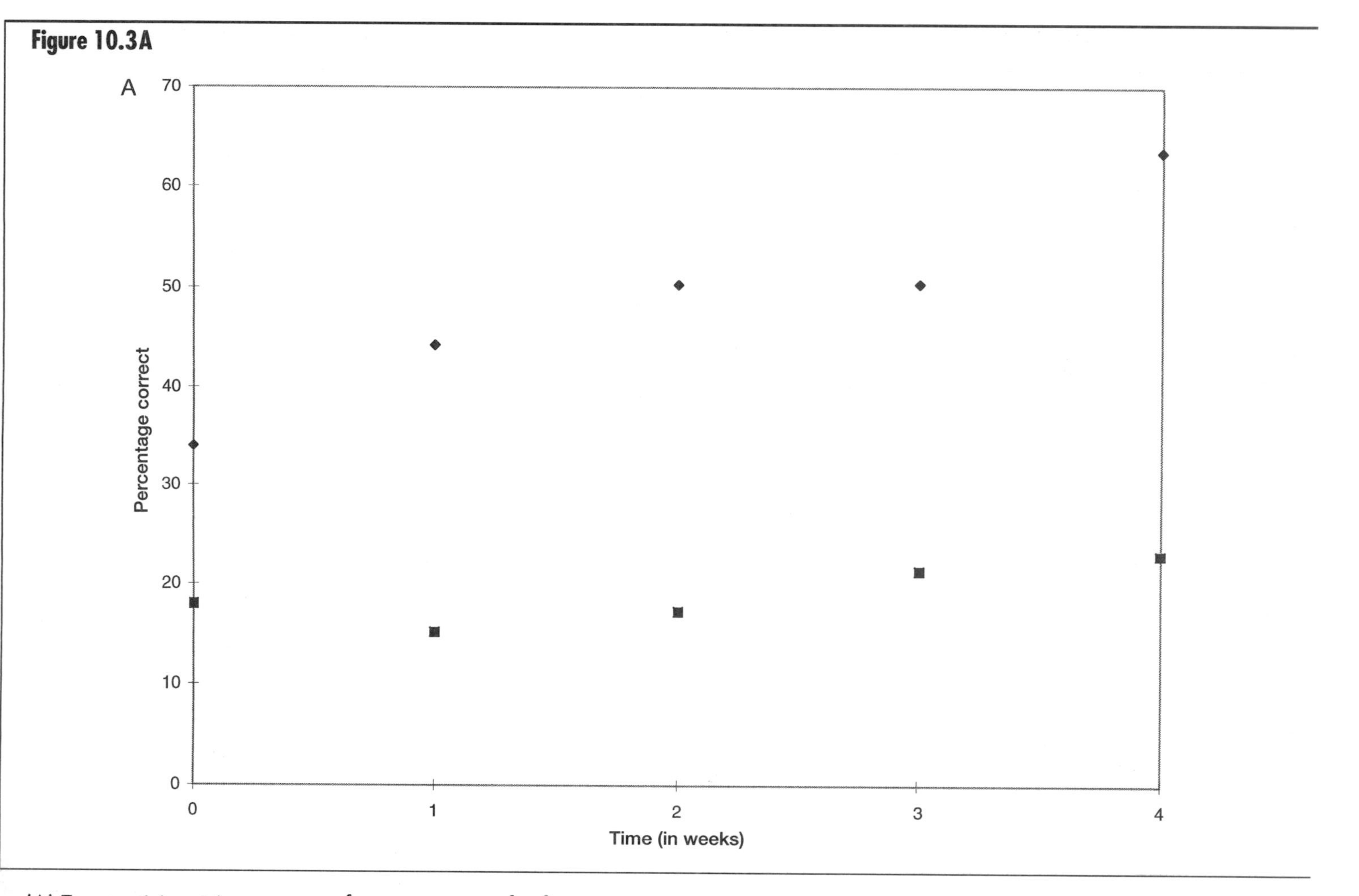

(A) Two participants' memory performance scores for five measurement occasions.

Figure 10.3B

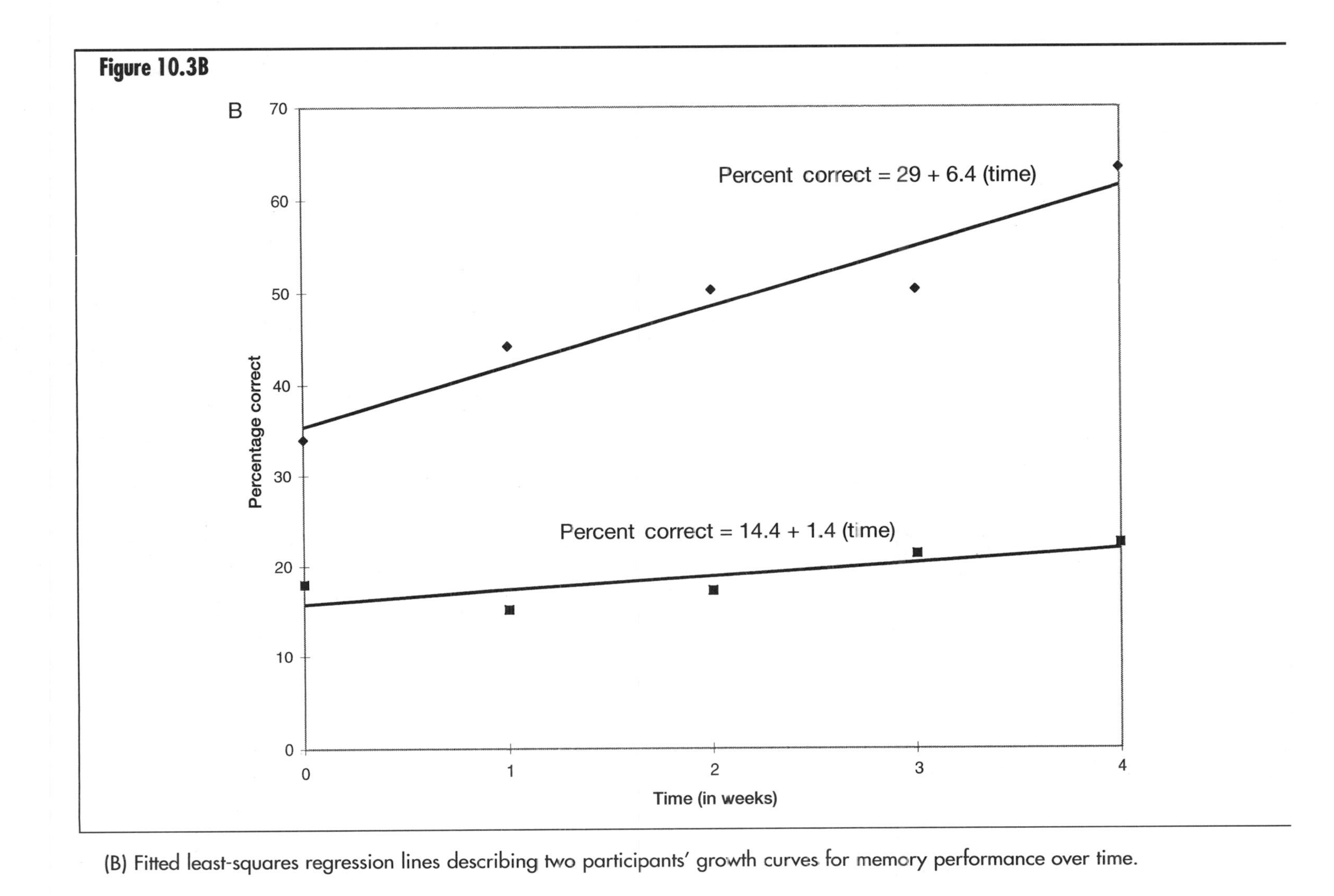

(B) Fitted least-squares regression lines describing two participants' growth curves for memory performance over time.

Table 10.3

Unconditional Model Results for the Method of Loci Data

Level I parameters and their predictors	Value
Intercepts (*a*)	
β_{00}	33.86**
Variance of random deviations	211.88**
Slopes for time (*b*)	
β_{10}	3.30**
Variance of random deviations	34.24**

Note. β = fixed-effect coefficient. **Coefficient or variance term differs from zero, $p < .001$.

line memory score)? Similarly, what if people do not differ in their rates of change (i.e., everyone has essentially the same slope)? Consider the case of the intercept first. One can test whether it makes sense to allow everyone to have their own intercept value by using a *maximum likelihood chi-square test* (Bryk & Raudenbush, 1992). This statistic tests the null hypothesis that the intercepts are homogeneous for all the people in the study. Stated otherwise, the statistic tests whether the variance in the intercepts is equal to zero. In this case, the chi square with 99 *df* is 3,925, $p < .001$, making it doubtful that the variance in intercepts equals zero. One can use the same test to examine the variance in slopes, yielding a chi square (df = 99) of 3,808, $p < .001$. So on the basis of these maximum-likelihood chi-square tests, one can conclude that it is appropriate to allow each person to have his or her own intercept and slope values.[12] (Note that the statistical significance of these variance terms is indicated in Table 10.3 with asterisks. Additionally, the coefficients β_{00} and β_{10} are tested to see if they differ from zero using a procedure discussed later. The significance levels of these tests are also reported in the table.)

The second point one should understand is that the variance of the random deviation terms can be useful for describing the variations in the Level I model parameters. Because one is primarily interested in rates of change (and, one would hope, improvement) in memory scores, one

[12]The maximum-likelihood chi-square tests for the random effects are very popular, but there have been publications criticizing their use (e.g., Stram & Lee, 1994). Those readers who are interested in performing their own HLM analyses are advised to follow this issue in the literature.

uses the slope parameter as an example. One would want to get a sense of the variability in slopes in the sample. From basic statistics courses, I know that if the individual slope values (*b*) are approximately normally distributed, then 68% of the individual slopes fall between +1 and −1 standard deviations (*SD*s) of the estimated average slope. To calculate the *SD* of the slopes, I take the square root of the variance, $\sqrt{34.24} = 5.85$. The slope for a person 1 *SD* lower than the average is $(3.30 - 5.85) = -2.55$. This slope is a weekly rate of change in memory performance score, so a person who is −1 *SD* experienced a change in memory score of −10.2 points out of 100 (4 weeks after baseline × −2.55 points/week = −10.2). Similarly, for a person falling 1 *SD* above the mean slope, the weekly rate of change is 9.15, resulting in a net increase of 36.6 points ($4 \times 9.15 = 36.6$). Thus, although the average person experienced a change of 13.2 points ($4 \times 3.30 = 13.2$), about 68% of the people experienced changes between −10.2 and +36.6 points throughout the duration of the study. This underscores the importance given to individual change in the HLM approach to longitudinal data.

Now that I concluded that there is considerable variation in the rates of memory performance change (*b*), I want to determine whether knowing a person's treatment group membership helps to explain some of the variation in the intercepts and the slopes. Accordingly, I construct a *conditional model*, which predicts the intercept and slope terms using group (a dummy variable coded 0 = nonartist, 1 = artist) as a predictor variable:

$$a = \beta_{00} + \beta_{01}(\text{group}) + \text{random deviation}$$

$$b = \beta_{10} + \beta_{11}(\text{group}) + \text{random deviation.}$$

These equations can be said to constitute the conditional model because it is stating that the intercept and slope parameters are partly determined by, or are conditional on, group membership (group). In contrast, the unconditional model described earlier stated simply that the intercepts and slopes are determined by random deviations from a population mean value and hence are not conditional on anything.

Table 10.4 displays the results of the conditional model, along with the results of the unconditional model for the sake of comparison. Recall that the objective of the conditional model is to use group membership to help explain variations in individual intercepts (*a*) and slopes (*b*). Because I am primarily interested in whether group had an effect on rate of change in memory scores, I begin with the numbers in the bottom half of Table 10.4. The total variation in rates of change are

Table 10.4

Unconditional and Conditional Model Results From the Method of Loci Data

Level I parameters and their predictors	Unconditional model	Conditional model (adding group as a Level II predictor)
Intercepts (*a*)		
β_{00}	33.86**	40.70**
β_{01} (group)		−13.96**
Variance of random deviations	211.88**	164.37**
Slopes for time (*b*)		
β_{10}	3.30**	2.15*
β_{11} (group)		2.34*
Variance of random deviations	34.24**	33.20**

Note. β = fixed-effect coefficient. β_{10} and β_{11} are not estimated in the unconditional model, so no values are shown in the table. *Coefficient or variance term differs from zero, $p < .05$. **Coefficient or variance term differs from zero, $p < .001$.

expressed as the variance of the random deviations in the unconditional model. This so-called unconditional variance is 34.24. In the conditional model predicting *b*, there is another random deviation term; it too has a variance. This conditional variance of 33.20 reflects the variations in rates of chance that could not be explained by treatment group membership. Note that the unexplained variance in slopes decreases by 1.04 after adding group as a predictor of individual slopes. The proportion of variance in slopes explained by group membership is unconditional variance − conditional variance/unconditional variance or (34.24 − 33.20)/34.24 = .03. Thus, group membership is useful for explaining .03 × 100 = 3% of the variations in rate of change.[13] Is this enough to be statistically significant?

The statistical significance of a group's effectiveness in predicting variations in slope is obtained by examining the parameter estimate describing the effect of group on the rates of change, β_{11}. This parameter is estimated using the generalized least squares method. In this example, $\beta_{11} = 2.34$, with a standard error of 1.17. The test statistic used is the familiar *t* test for evaluating the null hypothesis that a regression parameter equals zero (i.e., the parameter estimate divided by its stan-

[13]This proportion of variance explained expresses the same information as η^2. The Group × Time interaction η^2 was .04 using the MANOVA approach.

dard error). In this case, $t = 2.01$, $p < .045$. Following convention, the significance of this parameter is indicated in Table 10.4 with an asterisk. This indicates that part of the reason that participants in the method of loci study varied in their memory performance over time is that participants also varied with respect to whether they were artists.

How much did the artists and nonartists differ in their average slopes? Because group was dummy coded (nonartist = 0, artist = 1), the coefficient β_{11} is the difference in mean slopes between the artists and nonartists. This value is 2.34, meaning that the average change in memory scores over the 4 weeks of the study was 9.36 points ($4 \times 2.34 = 9.36$) higher for the artists than the nonartists. To obtain the estimated slope for each group, I calculate $\beta_{10} + \beta_{11}$ (group) for each group. For the nonartists (group = 0), the estimated mean slope is simply β_{10} or 2.15. For the artists, the estimated mean slope is $2.15 + 2.34 = 4.49$. Thus, the HLM results agree with the mixed-model ANOVA and MANOVA results that also demonstrate a significant effect for group membership.

Prediction of Individual Growth Curves

In some cases, a researcher might be interested in knowing how much each person in the study changed. For instance, in the method of loci study, the investigator might want to compare participant characteristics of the "slow" learners with "quick" learners on the basis of their rates of change over the 4 weeks. One of the more advanced features of the HLM approach to repeated measures data is the ability to generate improved estimates of individual change over time. Recall that each person's data were initially described by the equation of a line derived from a simple regression estimation (OLS) in the Level I model. As one can imagine, it is easier to fit a line to some people's data than to others. For those people whose data are "noisy" or even incomplete (i.e., missing data at various points), my estimates of their intercepts (a) and change (b) will not be as accurate. In fact, they could be wildly different from the rest of the sample's estimates. To adjust for this, the researchers uses the HLM approach to generate more refined predictions of people's within-subjects model parameters a and b using a method known as *empirical Bayes estimation* (Bryk & Raudenbush, 1992). How is this done?

Note that when one looks at the Level I and Level II models, one sees there are two ways to get estimates of a and b. On the one hand,

I have the estimates of *a* and *b* calculated for each person by fitting a line to their plot of data over time. These are the OLS estimates. On the other hand, I have the estimates of *a* and *b* from the two Level II equations that predict individuals' *a* and *b* values using other variables (e.g., Group). Thus, there are two possible sources of information about the "true" values of each person's *a* and *b*. Empirical Bayes methods calculate an estimate of these true values by using a weighted average of the two estimates provided by the OLS and the Level II equations, respectively. If the person's data points fit well with his or her OLS regression line, then more weight is given to that estimate. If the person's data do not fit well with the OLS line, however, then more weight is given to the *a* and *b* values predicted on the basis of that person's group membership using the Level II model equations. So the empirical Bayes procedure uses all the available information in the data set to arrive at the most reliable estimate of that person's growth curve parameters.

To make this adjustment more concrete, consider two participants from the method of loci study. Person 53 was a nonartist whose OLS-estimated slope was −6.19. The empirical Bayes procedure examines the fit of Person 53's line to determine how trustworthy the OLS value is as an estimate of his true slope. In this case, the OLS estimate is pretty good but not perfect. Thus, Person 53's individual slope of −6.19 is slightly "shrunk" back toward the nonartist group's mean slope of 2.15, resulting in an empirical Bayes estimate of −5.92 for Person 53's slope. The same shrinkage is observed for Person 70, an artist with an OLS-estimated slope of 16.11. Her slope was shrunk back toward the artist group's mean of 4.49, resulting in an empirical Bayes estimated slope of 15.62. This illustrates the manner in which the HLM approach allows better estimation of individual change.

Assumptions

HLM is a newer method for analyzing data; for that reason, there is still much work to be done regarding the assumptions underlying the models. Most of the models assume that all the variables are normally distributed in the population and that the errors of prediction for the Level I model have the same variance for all people under study. For other assumptions, see Bryk and Raudenbush (1992).

Conclusion

In this chapter, I reviewed three approaches to analyzing repeated measures data. In the ANOVA framework, repeated measures are handled by treating person as a random effect. In effect, this allows the researcher to remove variations due to individual differences, which makes it easier to test hypotheses about the means across the levels of the W-S variable. The ANOVA, however, is heavily dependent on the validity of the sphericity assumption. The MANOVA approach does not require the assumption of sphericity and offers a more powerful and flexible way of analyzing the data. A newer approach to longitudinal repeated measures designs is HLM analysis, which focuses more on individual growth curves over time. Along the way, I also reviewed the various approaches to analyzing pretest-posttest designs, for which Figure 10.1 provides a succinct summary.

Repeated measures designs offer a statistically powerful way of addressing theoretical hypotheses and can result in compelling evidence for or against hypotheses of interest. For this reason, a study of analytic procedures for repeated measures data will be richly repaid.

Suggestions for Further Reading

This chapter was intended for the critical consumer of research rather than the person who actually conducts the analyses. The choice of how to analyze a repeated measures data set is a difficult one because of the many options. Fortunately, there are some excellent references available that will help anyone wishing to understand how to choose and perform a repeated measures analysis. A comprehensive introduction to all of the methods presented in this chapter can be found in Bijleveld and van der Kamp (1998). For a thorough treatment of repeated measures ANOVA and MANOVA, see Maxwell and Delaney (1990). Stevens (1996) provided a discussion of repeated measures ANOVA and MANOVA with numerous examples of statistical software package output. Hertzog (1994) wrote an excellent review and critique of the traditional ANOVA–MANOVA approach as it is applied to developmental research. Rosenthal and Rosnow (1985) and Rosenthal et al. (2000) presented concise, readable guides to constructing linear contrasts in repeated measures ANOVA.

Nesselroade (1991a, 1991b) presented a brief introduction to the structure of longitudinal databases and the conceptual questions addressed by examining different facets of a longitudinal database. A "must-read" review of the problems inherent in many longitudinal modeling procedures is found by Rogosa (1995), and alternative strategies that are based on individual growth curve analysis are described by Rogosa, Brandt, and Zimowski (1982); Rogosa and Willett (1985); and Willett, Ayoub, and Robinson (1991). Bryk and Raudenbush (1987) provided a somewhat technical (but necessarily so) introduction to hierarchical linear models, and these same authors wrote one of the best introductory books on the subject (Bryk & Raudenbush, 1992). Applications of hierarchical linear models can be found by Francis et al. (1991) and Raudenbush and Chan (1993).

Glossary

A PRIORI CONTRAST Any contrast between means or combinations of means that was prespecified by the researcher prior to conducting the analysis (also know as a "planned contrast").

BETWEEN-SUBJECTS (B-S) VARIABLE A variable on which each person is measured at only one level of the variable (e.g., treatment group is a B-S variable because each person is assigned to either the treatment or control group).

BINOMIAL EFFECT SIZE DISPLAY (BESD) A way of understanding the magnitude of the effect of some independent variable on a dependent variable by converting the effect into survival rates associated with medical treatment and control conditions in a hypothetical treatment outcome study (Rosenthal & Rubin, 1982).

COMPLEX COMPARISONS In repeated measures ANOVA or MANOVA, a comparison between combinations of levels of the independent variables (e.g., comparing the mean response of the control group to the average of the mean responses of two different treatment groups. Such a complex comparison would test whether any treatment was better than the control).

COMPOUND SYMMETRY A specific case of *sphericity* that requires that (a) the variances of the measures at each level of the *W-S variable* are equal and (b) the covariances between the measures at level of the W-S variable are also equal.

CONDITIONAL MODEL In HLM, a model in which individual differences in the *Level I model* parameters are explained by another predictor variables in the *Level II model.* That is, the Level I parameter values are conditional on the values of the predictor variables.

CONDITIONAL VARIANCE In HLM, the variability in the *Level I model* parameters that is explained by the *Level II model.*

EFFECT SIZE An indication of how important is the magnitude of some effect. This is different from statistical significance, which tells only whether an effect is present.

EMPIRICAL BAYES ESTIMATION In HLM, estimates of each person's *Level I model* parameters are derived from a weighted average of the *OLS*-estimated values and the values are predicted using the *Level II model.*

EPSILON (ε) A measure of the degree of *sphericity* in repeated measures ANOVA. It ranges from 1 (*perfect sphericity*) to $1/k - 1$ (*worst violation of sphericity*), where k is the number of levels for the *W-S variable.*

ETA SQUARED (η^2) A measure of effect size that indexes the proportion of variance in the dependent variable as explained by the independent variable.

F STATISTIC A test statistic based on the ratio of variance that is due to the effect and to the variance that is due to error. It is used in both ANOVA and MANOVA approaches to repeated measures analyses.

FIXED EFFECT An independent variable whose levels are specified in advance by the researcher. Anyone wanting to replicate a study that used a fixed effect would have to use the same levels of the independent variable as in the original study (e.g., dosage level of an experimental drug is fixed; the dosage levels were not selected at random from a population of dosages). See *random effects.*

GAIN SCORE For studies with measurements taken at two time points, the gain score is the difference between the second and first measurement (i.e., how much was "gained" from the first to second measurement).

GENERALIZED LEAST SQUARES In HLM, the method for estimating the values of the *Level II model* parameters.

GREENHOUSE–GEISSER ESTIMATOR A way of adjusting the degrees of freedom for the *F* test in repeated measures ANOVA. The degrees of freedom are multiplied by an estimate of *epsilon* (ε). This correction is thought to be more conservative than a similar procedure, the *Huynh–Feldt estimator.*

HIERARCHICAL LINEAR MODELS (HLM) An approach to longitudinal repeated measures data that begins by modeling individual growth curves (*Level I model*). Then, variations in people's growth curve parameters are modeled using other predictors (*Level II model*).

HOMOGENEITY OF VARIANCES A statistical assumption of ANOVA that requires the variance of the dependent variable to be the same for each level of the *B-S variables.*

HOMOGENEITY OF COVARIANCES In the case of repeated measures ANOVA or MANOVA, this statistical assumption specifies that the matrix of variances and covariances among the repeated measures (or the *D* variables for MANOVA) is the same for each level of the *B-S variables.*

HOTELLING'S TRACE A multivariate test statistic used in MANOVA, which is converted to an *F* statistic to obtain a significance level.

HUYNH–FELDT ESTIMATOR Another method of adjusting the degrees of freedom in the *F* test in repeated measures ANOVA. This method is thought to be more liberal than the *Greenhouse–Geisser estimator.*

INDEPENDENCE OF OBSERVATIONS A statistical assumption that denotes that each observation or data point is not influenced by another observation or data point.

INTACT GROUPS Groups that exist outside of the research context, over which the experimenter has no control (e.g., gender, race, religion).

INTERCEPT (*a*) The point where a line intersects with the Y axis. In other words, the value of the dependent variable when the independent variable is equal to zero.

LONGITUDINAL A type of *repeated measures* study design in which measurements are taken of the same dependent variable over time.

LEVEL I MODEL In HLM, the linear model that describes each person's data over time. In the simplest case, it is comprised of an *intercept* (*a*) and a *slope* (*b*).

LEVEL II MODEL In HLM, the higher order model that uses independent variables to predict variations in people's *Level I model* parameters (e.g., *a* and *b*).

MAXIMUM LIKELIHOOD CHI SQUARE In HLM, the statistical test used to determine whether the variance of a Level 1 model parameter (e.g., *a*) is equal to zero. A significance test result is taken as evidence that there are substantial variations between people in their values of the Level 1 model parameters.

MIXED MODEL Another name for a repeated measures ANOVA, referring to the fact that the model mixes at least one *random effect* and one *fixed effect*.

MULTISAMPLE SPHERICITY An assumption in repeated measures ANOVA when there is a *W-S variable* and at least one *B-S variable*. The assumption requires that (a) *sphericity* is satisfied for each B-S group individually and (b) the covariance matrix of the levels of the W-S variable is the same for all groups defined by the B-S variable.

NORMALITY A statistical assumption that specifies that the dependent variable is normally distributed in a population.

ORDINARY LEAST SQUARES (OLS) A method for estimating the parameters of a linear regression equation.

PILLAI'S TRACE A multivariate test statistic used in MANOVA, which is converted to an *F* statistic to obtain a significance level.

PLANNED CONTRAST See *a priori contrast*.

POLYNOMIAL CONTRAST A set of contrasts that model polynomial components of the *W-S variable*. For a W-S variable with *k* levels, there are $k - 1$ polynomial contrasts. The first is a linear effect, the second is a quadratic effect (variable^2), the third is a cubic effect (variable^3), and so forth.

RANDOM EFFECT The effect of an independent variable whose levels are selected at random from a population of levels. In repeated measures ANOVA, person is a random effect because the particular people (who are levels of the variable person) are selected at random from a population of people.

REPEATED CONTRAST A way of constructing contrasts that compares each level of the *W-S variable* to its adjacent level.

REPEATED MEASURES DESIGN A research design in which each person is measured at two or more levels of the independent variable. Alternatively, a design in which two or more assessments of the dependent variable are made.

ROY'S LARGEST ROOT A multivariate test statistic used in MANOVA, which is converted to an *F* statistic to obtain a significance level.

SLOPE (*b*) A parameter of a regression equation that indicates the amount of change in the dependent variable associated with a one-unit increase in the independent variable.

SPHERICITY An assumption in repeated measures ANOVA that specifies how the covariance matrix of the levels of the *W-S variable* should be structured. The degree of sphericity is assessed by *epsilon* (ε), and violations of sphericity can lead to serious biases in ANOVA test statistics.

SPLIT-PLOT DESIGN A repeated measures design with one *W-S variable* and one *B-S variable.*

TREND ANALYSIS See *polynomial contrast.*

UNCONDITIONAL MODEL In HLM, a model in which the values of the *Level I model* parameters are seen as varying randomly about an overall mean value. In other words, no predictor variables are used to model between-person variations in Level I parameters.

UNCONDITIONAL VARIANCE In HLM, the amount of variation between people in the values of the parameters of the *Level I models.* It indexes the amount of differences between people that could potentially be explained by other predictor variables.

WILKS'S LAMBDA The most popular multivariate test statistic used in MANOVA, which is converted to an *F* statistic to obtain a significance level. Subtracting lambda from one gives *eta squared* or the proportion of variance explained by the independent variable.

WITHIN-SUBJECTS (W-S) VARIABLE A variable on which each person is measured at two or more levels of the variable.

References

Agresti, A. (1990). *Categorical data analysis.* New York: Wiley.

Bakan, D. (1967). *On method: Toward a reconstruction of psychological investigation.* San Francisco: Jossey-Bass.

Beck, A. T., & Steer, R. A. (1987). *Beck Depression Inventory manual.* San Antonio, TX: The Psychological Corporation.

Bijleveld, C., & van der Kamp, L. (1998). *Longitudinal data analysis.* London: Sage.

Bosman, E. A., & Charness, N. (1996). Age-related differences in skilled performance and skill acquisition. In F. Blanchard-Fields & T. M. Hess (Eds.), *Perspectives on cognitive change in adulthood and aging* (pp. 428–453). New York: McGraw-Hill.

Box, G. E. P. (1950). Problems in the analysis of growth and wear curves. *Biometrics, 6,* 362–389.

Bryk, A. S., & Raudenbush, S. W. (1987). Application of hierarchical linear models to assessing change. *Psychological Bulletin, 101,* 147–158.

Bryk, A. S., & Raudenbush, S. W. (1992). *Hierarchical linear models: Applications and data analysis methods.* Newbury Park, CA: Sage.

Campbell, D. T., & Stanley, J. C. (1963). *Experimental and quasi-experimental designs for research.* Chicago: Rand McNally.

Cohen, J. (1977). *Statistical power analysis for the behavior sciences.* Hillsdale, NJ: Erlbaum.

Cohen, J., & Cohen, P. (1983). *Applied multiple regression/correlation analysis for the behavioral sciences.* Hillsdale, NJ: Erlbaum.

Collins, L. M., & Horn, J. L. (Eds.). (1991). *Best methods for the analysis of change: Recent advances, unanswered questions, future directions.* Washington, DC: American Psychological Association.

Danziger, K. (1990). *Constructing the subject.* Cambridge, England: Cambridge University Press.

Dempster, A. P., Rubin, D. B., & Tsutakawa, R. K. (1981). Estimation in covariance components models. *Journal of the American Statistical Association, 76,* 341–353.

Emerson, J. D., & Stoto, M. A. (1983). Transforming data. In D. C. Hoaglin, F. Mosteller, & J. W. Tukey (Eds.), *Understanding robust and exploratory data analysis* (pp. 97–128). New York: Wiley.

Estes, W. K. (1956). The problem of inference from curves based on group data. *Psychological Bulletin, 53,* 134–140.

Francis, D. J., Fletcher, J. M., Stuebing, K. K., Davidson, K. C., & Thompson, N. M. (1991). Analysis of change: Modeling individual growth. *Journal of Consulting and Clinical Psychology, 59,* 27–37.

Greenhouse, S. W., & Geisser, S. (1959). On methods in the analysis of profile data. *Pyschometrika, 24,* 95–112.

Hertzog, C. (1994). Repeated measures analysis in developmental research: What our ANOVA text didn't tell us. In S. H. Cohen & H. W. Reese (Eds.), *Life-span developmental psychology* (pp. 187–222). Hillsdale, NJ: Erlbaum.

Hertzog, C., & Rovine, M. (1985). Repeated measures analysis in developmental research: Selected issues. *Child Development, 56,* 787–809.

Huck, S. W., & McLean, R. A. (1975). Using a repeated measures ANOVA to analyze the data from a pretest-posttest design: A potentially confusing task. *Psychological Bulletin, 82,* 511–518.

Huynh, H., & Feldt, L. S. (1976). Estimation of the Box correction for degrees of freedom from sample data in the randomized block and split plot designs. *Journal of Educational Statistics, 1,* 69–82.

Jennings, E. (1988). Models for pretest–posttest data: Repeated measures ANOVA revisited. *Journal of Educational Statistics, 13*(3), 273–280.

Kesselman, H. J., Rogan, J. C., Medoza, J. L., & Breen, L. L. (1980). Testing the validity conditions of repeated measures *F* tests. *Psychological Bulletin, 87,* 479–481.

Lamiell, J. T. (1998). "Nomothetic" and "idiographic": Contrasting Windelband's understanding with contemporary usage. *Theory and Psychology, 8,* 23–38.

Laird, N. M., & Ware, J. H. (1982). Random-effects models for longitudinal data. *Biometrika, 65*(1), 581–590.

Lindenberger, U., Kliegl, R., & Baltes, P. B. (1992). Professional expertise does not eliminate age differences in cognitive plasticity of a mnemonic skill. *Psychology and Aging, 7,* 585–593.

Lord, F. M. (1967). A paradox in the interpretation of group comparisons. *Psychological Bulletin, 68,* 304–305.

Maxwell, S. E., & Delaney, H. D. (1990). *Designing experiments and analyzing data.* Pacific Grove, CA: Brooks/Cole.

Moneta, G., & Csikszentmihalyi, M. (1996). The effect of perceived challenges and skills on the quality of subjective experience. *Journal of Personality, 64*(2), 275–310.

Nesselroade, J. R. (1991a). Interindividual differences in intraindividual change. In L. M. Collins & J. L. Horn (Eds.), *Best methods for the analysis of change* (pp. 92–105). Washington, DC: American Psychological Association.

Nesselroade, J. R. (1991b). The warp and woof of the developmental fabric. In R. M. Downs, L. S. Liben, & D. S. Palermo (Eds.), *Visions of asthetics, the environment & development: The legacy of Joachim F. Wohlwill* (pp. 213–240). Hillsdale, NJ: Erlbaum.

Porter, T. M. (1986). *The rise of statistical thinking: 1820–1900.* Princeton, NJ: Princeton University Press.

Raudenbush, S. W., & Chan, W. (1993). Application of a hierarchical linear model to the study of adolescent deviance in an overlapping cohort design. *Journal of Consulting and Clinical Psychology, 61,* 941–951.

Reese, E. (1995). Predicting children's literacy from mother–child conversations. *Cognitive Development, 10,* 381–405.

Rogosa, D. (1988). Myths about longitudinal research. In K. W. Schaie, R. T. Campbell, W. Meredith, & S. C. Rawlings (Eds.), *Methodological issues in aging research* (pp. 171–209). New York: Springer.

Rogosa, D. (1995). Myths and methods: "Myths about longitudinal research" plus supplemental questions. In J. M. Gottman (Ed.), *The analysis of change* (pp. 3–66). Mahwah, NJ: Erlbaum.

Rogosa, D., Brandt, D., & Zimowski, M. (1982). A growth curve approach to the measurement of change. *Psychological Bulletin, 92*(3), 726–748.

Rogosa, D. R., & Willett, J. B. (1985). Understanding correlates of change by modeling individual differences in growth. *Pyschometrika, 50*(2), 203–228.

Rosenthal, R., & Rosnow, R. L. (1985). *Contrast analysis: Focused comparisons in the analysis of variance.* Cambridge, England: Cambridge University Press.

Rosenthal, R., Rosnow, R. L., & Rubin, D. B. (2000). *Contrasts and effect sizes in behavioral research.* Cambridge, England: Cambridge University Press.

Rosenthal, R., & Rubin, D. B. (1982). A simple, general purpose display of magnitude of experimental effects. *Journal of Educational Psychology, 74,* 166–169.

Shock, N. W. (1951). Growth curves. In S. S. Stevens (Ed.), *Handbook of experimental psychology* (pp. 330–346). New York: Wiley.

Siegel, S. (1956). *Nonparametric statistics for the behavioral sciences.* New York: McGraw-Hill.

Stevens, J. P. (1996). *Applied multivariate statistics for the social sciences* (3rd ed.). Hillsdale, NJ: Erlbaum.

Stram, D. O., & Lee, J. W. (1994). Variance components testing in the longitudinal mixed-effect model. *Biometrics, 50,* 1171–1177.

Tate, R. L., & Hokanson, J. E. (1993). Analyzing individual status and change with hierarchical linear models: Illustration with depression in college students. *Journal of Personality, 61*(2), 181–206.

van den Boom, D., & Hoeksma, J. B. (1994). The effect of infant irritability on mother–infant interaction: A growth-curve analysis. *Developmental Psychology, 30,* 581–590.

Weinfurt, K. P. (1995). Multivariate analysis of variance. In L. G. Grimm & P. R. Yarnold (Eds.), *Reading and understanding multivariate statistics* (pp. 245–276). Washington, DC: American Psychological Association.

Willett, J. B., Ayoub, C. C., & Robinson, D. (1991). Using growth modeling to examine systematic differences in growth: An example of change in the functioning of families at risk of maladaptive parenting, child abuse, or neglect. *Journal of Consulting and Clinical Psychology, 59,* 38–47.

Yarnold, P. R., Feinglass, J., Martin, G. J., & McCarthy, W. J. (1999). Comparing three pre-processing strategies for longitudinal patients: An example in functional outcomes research. *Evaluation and the Health Professions, 22,* 254–277.

11

Survival Analysis

Raymond E. Wright

Fifteen smokers complete smoking-cessation therapy. How long will they abstain? A group of students begins graduate studies. Who will complete his or her doctorate first? A young couple gets married. Which psychosocial and economic factors influence how long they will wait to have their first child?

Each of these questions focuses on the time interval between two events: between smoking cessation and relapse, between the start and the completion of a doctoral program, or between marriage and the birth of a first child. These questions are typically investigated by following a group of individuals over a period of weeks, months, or years to see how long it takes for the event to occur. Although it might seem reasonable to estimate duration by computing the average time between events or by using linear regression to model the time interval, duration data often have special characteristics that preclude application of these popular statistical techniques. The most serious problem occurs if smokers relapse after the study is over, if students leave graduate school to pursue other interests, or if the researcher loses contact with couples before they have their first child. In each case, the event times are unknown to the investigator, that is, they are "censored." The most common methods for dealing with censored data can lead to biased duration estimates or can ignore important information.

Survival analysis (Kleinbaum, 1996; Lee, 1992) is a family of statistical methods specifically designed for analysis of duration data. When used properly, survival analysis can handle censored data without bias. Moreover, although survival analysis has roots in medicine and engineering, it is appropriate for studying a variety of time-related outcomes, including marriage, goal attainment, achievement, family breakdown, and employee turnover. The simplest application of survival analysis involves estimating the amount of time to the occurrence of an event (e.g., relapse, completion of a degree, marriage) for a group of individuals. Like analysis of variance (ANOVA) and linear regression analysis, however, survival analysis can also be used to compare durations for two or more groups and to build multivariable models that explain variation in duration. Survival analysis can also be more informative than other techniques because it gives insight into the relationship between time and the outcome of interest. That is, it enables the researcher to determine not just whether an outcome is likely to occur but whether it will occur early or late and whether the chances of event occurrence increase gradually or sharply over time.

In this chapter, I give an overview of survival analysis. I focus primarily on nonparametric methods for analysis of continuous duration data. These methods are popular in the survival analysis literature, and knowledge of these techniques is useful for understanding related methods, such as parametric and discrete-time survival analysis. First, I discuss essential survival analysis concepts and show how survival analysis can be used to estimate duration for a sample of participants. Next, I describe the basic assumptions of survival analysis. Then, I show how to compute survival estimates and how survival methods can be used to compare duration for two or more groups. Next, I show how survival regression analysis can be used to build univariable and multivariable models of duration. Finally, I summarize special extensions and adaptations of these techniques.

Survival Analysis Concepts

Survival analysis focuses on the time interval between two events or *survival time*. The start of the interval varies, depending on the purpose of the study. In studies of clinical or educational effectiveness, it is often marked by completion of an intervention (as in the smoking-cessation example) or assignment to a treatment or control group (as in a study to determine whether a new treatment prevents the recurrence of depression). The interval ends when the event of interest occurs. Although the event is sometimes called a "terminal event," it need not have negative implications; for example, the event might be a job promotion or medical cure. The time interval during which participants are followed is called an *observation* period or *follow-up* period; it, too, varies from study to study. For example, in a study of time to complete a cognitive task, the observation period might be a few minutes long; in studies of marriage duration, couples might be followed for years or even decades. Studies that track individuals over a period of time are called *longitudinal follow-up* studies.

In many longitudinal studies, however, the observation period ends before all individuals experience the event. Survival times for these individuals are unknown. Although the researcher knows that the individual was event free at the end of the study, the actual survival time must be longer. Survival times that are known only to exceed a value are called *censored survival times*. Censored data can also occur because a person never experiences the event in his or her lifetime (e.g., because he or she never marries), because he or she withdraws from participation in a study (e.g., because of illness or lack of interest), or because the investigator loses contact with the participant (e.g., because he or she moves without providing new contact information). This last situation represents a kind of censoring known as *loss to follow-up*.

Actually, statisticians differentiate between two kinds of censoring: right censoring and left censoring. Each example discussed so far is an instance of *right censoring* because the end of the interval is unknown. Censoring can also occur at the start of the interval or *left censoring*. For example, left censoring could occur in a study of the duration of a depressive episode if a patient were depressed at the start of the study, but the investigator did not know when the episode began (e.g., because complete psychiatric records were unavailable). In other words, the start of the interval precedes the beginning of the observation period. The distinction between left and right censoring is important because most survival analysis techniques (including the methods described in this chapter) are not designed to handle left censoring. Left censoring is much more difficult to deal with, so it should be avoided if possible (Luke, 1993; Willett & Singer, 1991). (Because discussions of survival analysis often disregard left censoring, statisticians often use the term *censoring* to refer to right censoring. I follow this convention throughout this chapter.)

Common Methods for Dealing With Censoring

How can duration be estimated when some event times are censored? Most intuitive methods are, in fact, problematic. For example, suppose that 24 clients completed smoking-cessation therapy during a 3-year period. To determine how long clients abstain from smoking after treatment, one might consider computing the average time between completion of the intervention and resumption of smoking. If some people are still cigarette free at the end of the 3-year period, however, one would need to decide how to handle the censored durations before analyzing the data. One approach would be to ignore the censored durations and compute average duration using only the data for people who were known to have relapsed. This approach would tend to underestimate abstinence duration, however, because the abstinence durations for people whose data are censored might be much longer than those for people who relapsed during the study period. Another approach, treating the censored cases as if a relapse had occurred at the time of censoring, could also lead to a serious underestimation of duration if people whose data were censored remained abstinent beyond the time of censoring.

A third strategy for handling censored data would be to ignore abstinence times and instead focus on whether a relapse occurred. The analysis would determine the proportion of clients who relapsed during the 3-year period. How could one deal with people who complete treatment near the end of the observation period? Because they have less opportunity for relapse to occur, including them in the analysis would tend to bias the overall proportion of events downward. Ignoring abstinence times leads to other problems as well. First, it makes it difficult to compare the results of the study with those from other studies that had a longer or shorter observation period. It could also mean losing insight into one's data, including the ability to determine when the risk of relapse is greatest.

One-Sample Survival Analysis

An alternative approach, *survival analysis*, can be used to estimate abstinence duration without bias when censored data are present. In survival analysis, duration is estimated by computing a *survival function*, which estimates the probability that a client will survive (i.e., be cigarette free) past a specified time. Figure 11.1 is a survival function for hypothetical smoking-cessation clients. The plot shows that the proportion of clients who were abstinent through the first 400 days was about .92. For 600 days, the proportion was about .39. Thus, clients who completed the intervention had a 92% probability of remaining abstinent for at

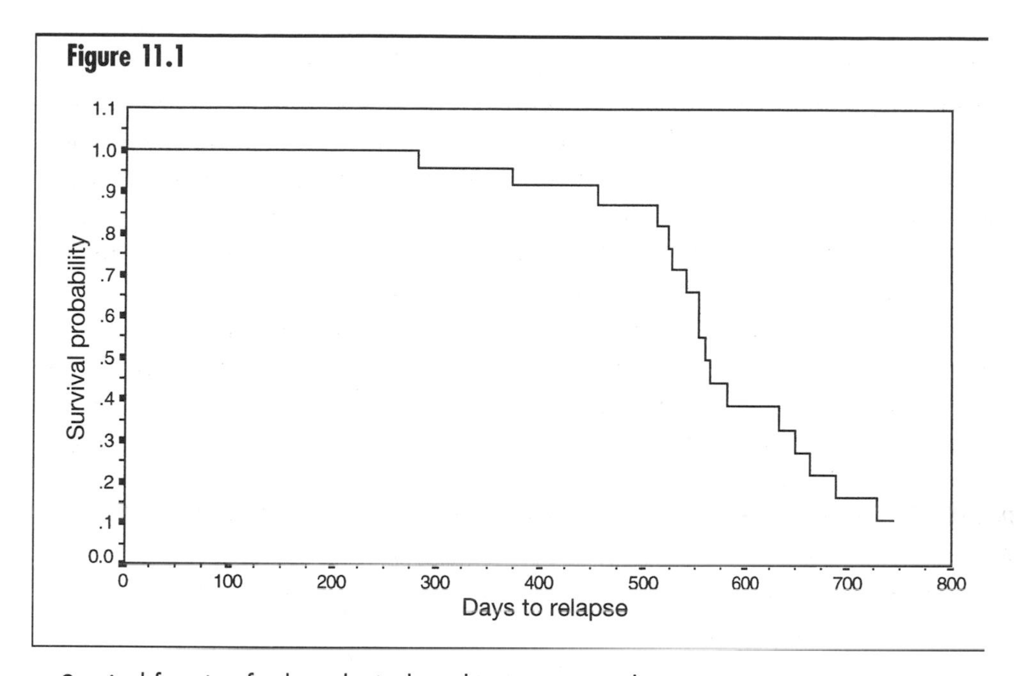

Survival function for hypothetical smoking-cessation clients.

least 400 days and a 39% chance of being cigarette free for 600 days or more. Moreover, because the probability of relapse and the probability of abstinence must add up to 100%, the plot can also be used to determine the probability of relapse within a specific time period. For example, the probability of relapse during the first 400 days after treatment is 8% (100%–92%), and the probability of relapse within 600 days is 61% (100%–39%). In addition, the steepness of the survival function reveals when the probability of relapse is unusually low or high. For example, the plot is relatively flat through the first 500 days after program completion, which indicates that the likelihood of relapse was low during this period. (In fact, the plot shows that the probability of relapse within the first 500 days is about .13.) The plot is steepest between about 500 and 600 days after completion of therapy. Thus, clients were at greatest risk of relapse during that period.

The plot can also be used to estimate abstinence time for a typical (i.e., randomly chosen) client. In survival analysis, the central tendency is generally estimated using *median survival time*; which can be estimated from the survival plot by linear interpolation. To estimate the median, first draw a horizontal line from the .5 point on the y axis to the survival function. Then, starting at the intersection of the horizontal line and the survival function, draw a vertical line down to the time axis. For the smoking cessation data, the median is about 560 days, which means that clients in the sample typically remained abstinent through the first 560 days after completion of the intervention. The median is only a point estimate of duration, however, so it is often useful to estimate a likely range of durations for a typical client. For example, half of the clients' durations fall within the interquartile range (IQR). The upper and lower boundaries for the IQR are estimated by interpolating from the .25 and .75 points on the y axis. Thus, abstinence durations tended to range between 525 and 660 days for clients who completed the smoking-cessation program.

There are a few important features to notice about the survival function. First, it starts at one and tends to drop toward zero as time passes. The value of the survival function is one at Time 0 because all individuals were cigarette free when they completed the intervention. Moreover, survival functions cannot rise over time because the sample proportion of events cannot decrease. In other words, a client who relapses may be able to quit again, but the original relapse cannot be undone. Also the survival function ends at 745 days, a result of the method used to compute the survival function, the Kaplan–Meier method (Kaplan & Meier, 1958). Survival functions computed using the Kaplan–Meier method end at the largest observed duration (in this example, the longest duration was 745 days). This implies that the probability of surviving for, say, 800 or 1,000 days, is not known, except that it must be less than or equal to the probability of abstaining for 745 days (about .11). There is an important exception to this rule, however. If the survival function drops to zero, the estimated probability of surviving to that time or longer is zero.

Assumptions of Survival Analysis

Several conditions must be met for a survival analysis to be valid. Here I discuss general assumptions common to all methods used in this chapter. Later, I describe additional assumptions that are specific to survival regression analysis.

The first assumption is that the events recorded for individuals in the sample are *statistically independent* (Kalbfleisch & Prentice, 1980); this assumption is violated if any individual is represented more than once in the data set. If the events are not independent, hypothesis tests may be invalid.

Second, the event must represent a change from one of two mutually exclusive and collectively exhaustive states to another (e.g., from life to death, marriage to divorce, or abstinence to relapse). In other words, an individual cannot be in more than one state at a time, and every case must be a member of one of the two states under analysis. For example, life and death are mutually exclusive and exhaustive categories because one cannot be alive and dead simultaneously and because everyone is either alive or dead. In contrast, the categories "baseball fan" and "football fan" are neither mutually exclusive (because one can be a fan of both baseball and football) nor exhaustive (because one can be a fan of another sport or a fan of no sport at all). In the smoking-cessation study, a client is either smoke free or abstinent at any time (collectively exhaustive), yet he or she cannot be both smoke free and abstinent at the same time (mutually exclusive).

Third, all individuals are event free when they enter the study. If some of the clients in the smoking-cessation example did not become cigarette free on completion of the intervention, they would have been excluded from the survival analysis. Otherwise, if they were included, the survival curve could not legitimately start at one.

Fourth, although the methods assume no particular theoretical distribution of survival times, they are intended primarily for use with continuous data. That is, the survival times should be measured precisely, instead of being recorded in large intervals such as months or years. In practice, methods designed for continuous-time survival model data are often used for data recorded in small intervals (e.g., days). Methods specifically designed for data that are grouped into intervals are known as *discrete-time survival models* (Singer & Willett, 1993; Willett & Singer, 1993). These techniques are summarized at the end of this chapter.

The final assumption is the *independent-censoring assumption*; this condition requires that censoring is unrelated to the probability of event occurrence. The independent-censoring assumption is met in studies in which censoring is a purely random process or if censoring occurs because the study ends before a participant experiences the event (Kalbfleisch & Prentice, 1980). It is violated if a participant leaves or is excluded from a study because he or she is at unusually high or low risk of event occurrence, for example, if he or she withdraws from a smoking-cessation study because relapse is imminent. Independent censoring is needed to get unbiased estimates of survival time (Luke, 1993), so studies in which nonindependent censoring is suspected should be interpreted with caution.

Computing Survival Estimates

Now that I have explained essential concepts and assumptions that underlie survival analysis, I discuss how a survival function is actually computed.

Kaplan–Meier Survival Function

Computation of the Kaplan–Meier survival function is straightforward. In fact, for small samples, it can be done with a hand calculator. To explain, I explore how the Kaplan–Meier survival function was computed for the smoking-cessation data. Data for the study are given in Table 11.1. For each client, two essential pieces of information are recorded. The first is the known abstinence time for each client, in days. The second is the client's status on the last day on which he or she was observed. Event status is recorded as "event" for people who relapsed during the observation period and "censored" for patients who were event free.

Table 11.2 is a Kaplan–Meier survival table computed from the relapse data. Notice that in the first two columns, the durations from Table 11.1 are sorted from shortest to longest and that for any particular time censored cases are listed after relapses. Ordering the data is the first step in computing a Kaplan–Meier estimate. Next, for each duration, the cumulative number of events (i.e., relapses) and the number of clients who remained at risk (i.e., under observation) are both recorded. The column "Cum. events" gives the cumulative number of events. For example, one event occurred at 283 days, and a total of three events occurred by Day 457. The column labeled "*N* at risk" shows the number of clients who remained under observation after each survival time. Notice that each time an event occurs or a duration is censored, the number of clients at risk decreases by one. For example, after the first event (at Day 283), 23 of the 24 clients in the sample remained at risk, and after the second event (which occurred on Day 373), 22 clients remained at risk. At Day 399, the number at risk drops to 21 because one individual is censored at that time.

Next, I compute the conditional survival probability for each event time (this is an intermediate step; the conditional probabilities are not the values shown in the survival plot). The conditional probability is the probability of surviving past a particular time for individuals who were at risk for relapse. For example, among the 23 clients who re-

Table 11.1

Survival Times for Hypothetical Smoking-Cessation Clients

Client	Day	Status
1	457	Event
2	565	Event
3	461	Censored
4	581	Event
5	541	Event
6	283	Event
7	729	Event
8	553	Event
9	559	Event
10	417	Censored
11	527	Censored
12	481	Censored
13	527	Event
14	399	Censored
15	745	Censored
16	523	Event
17	373	Event
18	689	Event
19	665	Event
20	651	Event
21	635	Event
22	553	Event
23	745	Censored
24	513	Event

mained in the study after the first event occurred (on Day 283), 22 survived past the next time an event occurred (at Day 373). Thus, the conditional probability of surviving past Day 373 is .9565 (22 / 23). This value is shown in the third column of Table 11.2. Similarly, among the 12 clients who were at risk after Day 541, 10 clients survived through Day 553. Therefore, the conditional probability of surviving past Day 553 is .8333 (10/12).

Now I am ready to compute the probabilities that are plotted in Figure 11.1. These values are known as cumulative survival probabilities or simply as survival probabilities. For the first event time (283 days), the *cumulative probability* is equal to the conditional survival probability (.9583). This is the value shown in Table 11.2 in the column labeled "Cum. prob." For subsequent event times, survival probabilities are ob-

Table 11.2

Kaplan–Meier Survival Table for Hypothetical Smoking-Cessation Clients

Day	Status	Cond. prob.	Cum. prob.	*SE*	LCL	UCL	Cum. events	*N* at risk
283	Event	.9583	.9583	.0408	0.8767	1.000	1	23
373	Event	.9565	.9167	.0564	0.8039	1.000	2	22
399	Censored						2	21
417	Censored						2	20
457	Event	.9500	.8708	.0698	0.7312	1.000	3	19
461	Censored						3	18
481	Censored						3	17
513	Event	.9412	.8196	.0824	0.6548	0.9844	4	16
523	Event	.9375	.7684	.0918	0.5848	0.9520	5	15
527	Event	.9333	.7172	.0989	0.5194	0.9150	6	14
527	Censored						6	13
541	Event	.9231	.6620	.1056	0.4508	0.8732	7	12
553	Event	[a]					8	11
553	Event	.8333	.5517	.1132	0.3253	0.7781	9	10
559	Event	.9000	.4965	.1145	0.2675	0.7255	10	9
565	Event	.8889	.4413	.1143	0.2127	0.6699	11	8
581	Event	.8750	.3862	.1126	0.1610	0.6114	12	7
635	Event	.8571	.3310	.1092	0.1126	0.5494	13	6
651	Event	.8333	.2758	.1040	0.0678	0.4838	14	5
665	Event	.8000	.2207	.0967	0.0273	0.4141	15	4
689	Event	.7500	.1655	.0869	0.0000	0.3393	16	3
729	Event	.6667	.1103	.0734	0.0000	0.2571	17	2
745	Censored						17	1
745	Censored						17	0

Note. Cond. prob. = conditional probability; Cum. prob. = cumulative probability; *SE* = standard error; LCL = lower confidence limit; UCL = upper confidence limit; Cum. events = cumulative number of events. [a]See data on the next row.

tained by multiplying conditional probabilities for all event times up to and including the event time of interest. For example, the cumulative probability of surviving past 373 days is .9167, which is computed as the conditional probability for 283 days (.9583) multiplied by the conditional probability for 373 days (.9565). For 457 days, the survival probability is equal to .9583 × .9565 × .9500 or .8708. Equivalently, survival probabilities can be obtained by multiplying the conditional probability for the event time of interest and the cumulative probability for the preceding event time. For example, the survival probability for 457 days (.8708) could be computed as .9500 × .9167.

Although survival probabilities are easy to compute, they are only point estimates of the population value for a particular event time. Therefore, it is often important to compute 95% confidence intervals (CIs) for each sample probability that estimate the likely range of values for the population survival probability. CIs are computed using estimated *standard errors* (SEs), which measure the expected variability in a statistic from sample to sample. (SEs are difficult to calculate by hand but are available in many statistical software packages.) For the smoking cessation data, standard errors are shown in Table 11.2 in the column labeled "SE." For example, the standard error for the probability of surviving past Day 541 is .1056. The approximate lower limit for a CI is computed as the sample value − two × its standard error. The approximate upper limit is computed as the sample probability + two × the standard error (Lee, 1992). (For a more precise estimate of the lower and upper bounds, use 1.96 × SE.) Thus, for 541 days, the lower confidence limit is .6620 − 2 (.1056) or .4508, and the upper confidence limit is .6620 + 2 (.1056), or .8732. Therefore, it is estimated that in the population, the cumulative survival probability past 541 days is between .4508 and .8732. The 95% CI for a probability has the same interpretation as it has for a mean: If 95% CIs were computed for an infinite number of random samples from a population, 95% of the intervals would contain the population value. Lower and upper confidence limits are shown in Table 11.2 for each relapse time in the columns labeled "LCL" and "UCL," respectively.[1]

There are a few things to notice about the Kaplan–Meier table. First, survival probabilities are not computed for every time point but

---

[1]Although probabilities must fall between zero and one, the confidence interval formula sometimes gives values that are beyond this range. One way to deal with this is to truncate confidence limits that exceed one or fall below zero. The values shown as 0 or 1 in Table 11.2 are truncated values.

only when an event occurs. This is why the Kaplan–Meier function has a staircase appearance. Moreover, censored cases are used only in computations that precede the time of censoring. For example, the clients whose times were censored at 399 and 417 days were included in the risk set for survival probabilities associated with earlier times (283 and 373 days) but were excluded when computing the conditional probability of surviving beyond the time of censoring. Thus, the Kaplan–Meier method uses what is known about censored data (i.e., that an individual survived at least to a particular time) without forcing readers to guess when the event actually took place. It is this treatment of censored data that allows for unbiased survival estimates when censoring is present.

Two-Sample Survival Analysis

In the previous example, a single survival function for the sample was obtained. There is often reason to suspect that survival differs for individuals with different characteristics, however. For example, abstinence from smoking might be affected by the strength of a smoker's addiction or his or her motivation to quit. Group differences in survival experience can be determined by comparing Kaplan–Meier survival functions for each group. For example, one might compare survival functions for smokers who are rated highly or only moderately addicted or survival functions for smokers whose motivation to quit is classified as high or low. Formal procedures are available to test the hypothesis that two or more survival curves are equivalent.

To illustrate comparison of survival functions, I compare the outcomes for two groups of patients who received medical rehabilitation, an interdisciplinary field of medicine that is designed to increase the capacities of people with disabilities to function independently. For the purposes of treatment planning and patient and family counseling, it is important for clinicians to be able to predict how long it will take a patient to successfully complete treatment. I use the Kaplan–Meier method to determine whether severely impaired patients progress more slowly than less impaired patients. Specifically, I compare the treatment completion times for brain-injured patients whose cognitive status at admission is classified as high or low on the basis of assessments by a neuropsychologist and a speech and language pathologist. Data for the analysis were obtained for 390 adults who were discharged from an inpatient rehabilitation facility during a 2.5-year period. Treatment completion times were censored for 13.6% of patients, who withdrew from the program for various reasons such as personal or financial issues or transfer to another facility.

Survival functions for high- (bottom plot) and low-status (top plot) patients are given in Figure 11.2. The plot shows the probability of remaining in the program (i.e., not completing treatment) as a function of treatment days. If one interpolates from the y axis, it can be seen that median time to completion was lower for high-status patients (about 24 days) than for low-status patients (about 41 days). For high-status patients, the IQR was from about 16–32 days. For low-status patients, it was about 26–56 days. Moreover, across time points, low-status patients were less likely to have completed rehabilitation. For example, interpolating from the time (x) axis, one can see that at 20 days, about 90% of low-status patients remained in the program, whereas about 60% of high-status patients were still undergoing treatment. Similarly, at 50 days, about 30% of low-status patients were undergoing treatment, whereas only about 5% of high-status patients remained in the program. The group survival functions appear different, but are these group differences statistically significant, or do they simply represent chance differences due to sampling variability? To determine whether the group

Figure 11.2

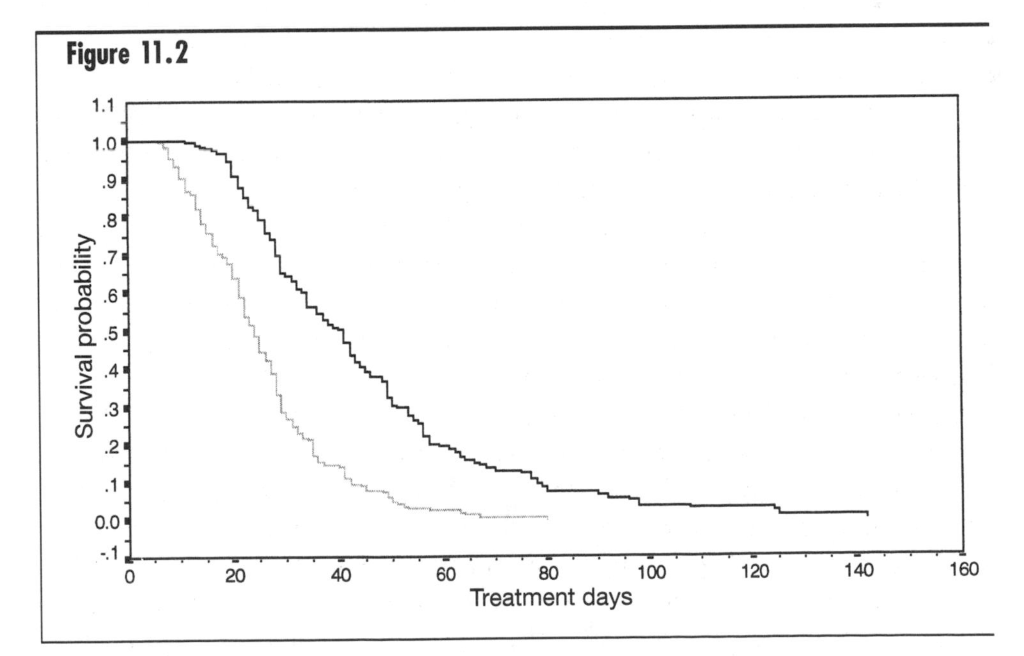

Kaplan–Meier survival functions by cognitive status. The dark line represents low-status patients; the light line represents high-status patients.

survival functions could have been obtained from samples from the same population, hypothesis tests are performed. As for *t* and *F* tests, hypothesis testing in survival analysis involves reasoning by contradiction. I first assume that the population survival functions for the two groups are identical (this assumption is the null hypothesis). Hypothesis tests tell whether there is sufficient evidence in the sample data to reject the null hypothesis and, therefore, to accept the alternative hypothesis that the samples are from different populations. This problem is analogous to the use of *t* tests or ANOVA to compare group means. The difference is that the goal here is to compare survival functions instead of averages.

One way to test whether population survival functions are identical is to use the *log-rank* (*Mantel–Cox*) statistic (Cox, 1972; Mantel, 1966). Specifically, a large value of the log-rank statistic means that the survival functions were probably obtained from two populations. The criterion for "large" is the probability (p) associated with the log-rank statistic. The null hypothesis is rejected if the observed log-rank statistic occurs in 5% or fewer of random samples of two groups from the same population. (The cutoff probability for the hypothesis test, .05, is known as "alpha.") For two-group comparisons, the probability for the log-rank statistic is obtained from a chi-square distribution with 1 degree of freedom (df). The log-rank statistic for the rehabilitation example is 84.92, with $p < .00005$, which indicates that the two samples are probably not from the same population. Therefore, it is concluded that patients with low cognitive status typically take longer to complete the program compared to high-status patients. Note that the test does not reveal why the survival functions differed. What seems likely in this case, however, is that cognitively impaired persons have lower attention spans and need more careful instructions from the treatment staff. Thus, their treatment proceeds more slowly. Nonetheless, other factors that were not considered may also have a role in determining treatment completion time.

A few additional comments are needed about the log-rank test. First, the log-rank test is a *whole-pattern* test; *t* compares entire survival functions rather than focusing on differences at a particular time point. Moreover, the test is most sensitive to group differences when one group has a consistently higher or lower probability of experiencing an event across time points. Other tests may be more appropriate in other situations. For example, the *Breslow test* (i.e., the generalized Wilcoxon test) is more sensitive to group differences at early time points (Breslow,

1970). The *Tarone–Ware test* (Tarone & Ware, 1977) is more sensitive if the survival functions do not differ by a constant factor (e.g., if they diverge over time or if they intersect). These tests differ from the log-rank test in the weight that they give to different parts of a survival function. For the rehabilitation data, however, all tests are highly significant.

Multisample Survival Analysis

With only slight adjustments, survival functions for three or more groups can be compared. For example, Hart, Kropp, and Hare (1988) used the log-rank test to determine whether knowledge about a prisoner's personality characteristics could help predict his behavior following release on parole or mandatory supervision. Data for the study were obtained for 231 white male prisoners released from a federal prison during an 8-year period. On the basis of a psychological assessment completed during their imprisonment, each subject was classified into one of three groups: high, medium, or low psychopathy status. Time to the first release violation was recorded in days. Some violation times were censored, however, because no violation occurred during the period of supervision (i.e., the prisoner successfully completed terms of his supervised release) or because the study ended before a violation occurred. The authors used the Kaplan–Meier method to estimate time to the first violation for each group. They hypothesized that highly psychopathic prisoners would be likely to violate the conditions of their release.

Figure 11.3 illustrates the sample survival functions for each group. The plot gives the percentage of prisoners who were "violation free" as a function of time after release. Aside from the early crossing of the survival functions, the differences between the groups were fairly consistent. For all but the earliest times, low-status prisoners had the lowest risk of violation. Moreover, consistent with the researchers' predictions, highly psychopathic prisoners were most likely to violate terms of their release, although consistent differences between high- and medium-status prisoners were not evident until about 250–300 days after release.

Approximate group medians, which can be obtained from the plot, were 230 days for high-status prisoners and nearly double (460 days) for medium-status prisoners. The median could not be computed for low-status prisoners, however, because their survival function did not

Figure 11.3

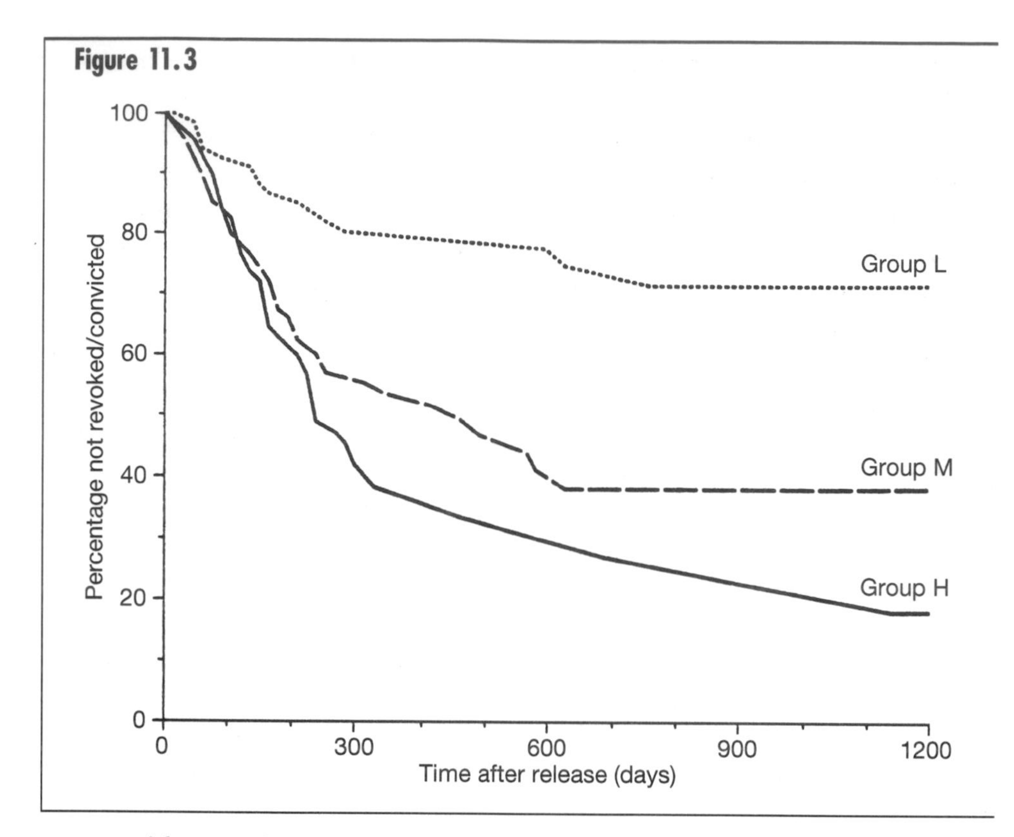

Survival functions by psychopathy status. From "Performance of Male Psychopaths Following Conditional Release From Prison," by S. Hart, P. Kropp, and R. Hare, 1988, *Journal of Consulting and Clinical Psychology, 56,* p. 230. Copyright 1988 by the American Psychological Association. Reprinted with permission.

drop to the .5 point on the y axis. Thus, to compare all three groups, the authors computed the probability of being violation free for two time points of interest: 1 year and 3 years. The probability of being violation free for 1 year was .80 for low-status prisoners, .54 for medium-status prisoners, and .38 for high-status prisoners. For 3 years, the probabilities were .71, .38, and .18 for low-, medium-, and high-status prisoners, respectively. The authors also used the log-rank statistic to test the equivalence of survival functions for the low-, medium-, and high-psychopathy groups. (To use the log-rank test for more than two groups, the dfs must be adjusted. The rule is that the dfs are set to the number of groups minus one. Thus, for the present example, there were 2 dfs.) The value of the log-rank statistic was 19.68, with an associated probability less than .0001. Thus, the hypothesis that group survival functions were identical was rejected.

On the basis of their analysis, Hart, Kropp, and Hare (1988) concluded that the psychopathy assessment made a significant contribution to the prediction of postrelease outcomes. In particular, they suggested that the assessment method may ultimately be useful to judges who need to decide whether a prisoner should be granted conditional release. A few notes about the analysis are in order, however. First, the log-rank statistic used in the prisoner behavior study, like an omnibus F test in ANOVA, indicated only that the survival functions differed. It did not say where the specific group differences were. It cannot be concluded, for example, that all groups differed from each other, or that any two specific pairs of groups (e.g., the medium- and high-status groups) differed. Pairwise log-rank tests are available in many software packages, however, to identify specific differences between groups. (Note that the distinction between omnibus and pairwise test is relevant only for studies involving three or more groups. For two-group comparisons, only pairwise tests are performed.)

Finally, Hart et al. (1988) noted that further research would be necessary to separate the effects of psychopathy and other factors that are associated with prisoners' behavior following release, (e.g., a long history of criminal offenses). Controlling for these confounding factors could be achieved using stratification. In a stratified Kaplan–Meier analysis, group comparisons are made within levels of a categorical confounding factor. For example, to adjust for the effects of being a repeat offender, the psychopathy groups would be compared both among repeat offenders and among individuals without a history of criminal behavior. A discussion of stratification was given by Kleinbaum (1996). A more flexible way to adjust for confounding factors is to build a survival regression model. This approach is discussed in the next section.

Survival Regression Analysis: Cox Regression

Each of the analyses described above involved categorical predictor variables, such as high- or low-cognitive skills or high-, medium-, or low-psychopathy status. What if you want to predict survival using a continuous variable such as age? Or, what if you want to know how each of several predictor variables relates to survival? You might suspect, for example, that a more accurate prediction of abstinence could be obtained by taking into account how many cigarettes a client smoked per day, how long he or she had been a smoker, and the number of years of experience of his or her therapist. Unfortunately, use of the Kaplan–Meier method is limited to analyses that involve a single categorical predictor. Another technique, *Cox regression analysis*, is used to build multivariable models that relate one or more continuous or categorical variables to survival (Cox, 1972; Cox & Oakes, 1984). Cox regression shares many features with linear regression. Both are used to estimate the strength of association between a set of independent variables and a dependent variable. Both yield regression coefficients and predicted values, allow for the inclusion of interactions between variables, and adjust for confounding factors. Whereas linear regression is used to predict scores on a continuous dependent variable such as height or weight, Cox regression is used to predict the rate of event occurrence. Moreover, only Cox regression can handle censored data appropriately.

The Hazard

A key concept in Cox regression analysis is the *hazard*, which is defined as the immediate potential or "risk" of event occurrence (Cox & Oakes, 1984). The hazard tells whether an event is likely to occur for someone who has survived to a particular time, for example, whether an individual who is abstinent at 6 months is likely to relapse in the near future. (Note that "hazard" is neutral in survival analysis; the events it refers to may have positive or negative implications.) The hazard also tells the rate of event occurrence at a particular instant in time (Kalbfleisch & Prentice, 1980).[2] A large hazard means that events occur at a fast rate; a small hazard indicates that events occur at a slower rate. The hazard can be plotted over time to obtain a *hazard function*, which is a mathematical relationship that describes changes in the risk of event occurrence over time. Unlike the survival function, however, the hazard can increase, decrease, or fluctuate over time. For example, the hazard function for human mortality is U-shaped because the death rate is highest among infants and elderly people and lowest during young adulthood. Furthermore, whereas the survival probability ranges from 0 to 1, the continuous-time hazard has no upper limit.

In Cox regression analysis, the hazard is treated as the dependent variable. The hazard estimate for an individual is computed on the basis of three factors. The first factor is the *baseline hazard function*, which

[2]The hazard is a rate only for continuous-time survival models; for discrete-time models, the hazard is a probability.

estimates the overall risk of event occurrence as a function of time. The baseline hazard can increase, decrease, or fluctuate over time, but like the constant term in linear regression, it is always the same for all persons in the sample. The second factor is a set of coefficients estimated from the model. The *model coefficients* describe the relationship between each predictor variable and the rate of event occurrence (e.g., whether a variable is positively or negatively associated with the hazard). The third factor is the individual's values on the predictor variables. In Cox regression analysis, the estimated hazard for a particular individual can be thought of as an adjustment to the overall risk for the combined sample. The model coefficients and the individual's predictor values determine how much the hazard estimate for a particular individual differs from the baseline hazard.

Types of Models

Statisticians distinguish between two kinds of Cox regression models (Kleinbaum, 1996). I first consider the simpler model, the standard Cox regression model. Later, I discuss a generalization of the standard model, known as the extended Cox regression model. The distinction between the types of Cox models is important because they differ both in flexibility and in terms of an important underlying assumption about the relationship between time and the rate of event occurrence. In particular, the standard model requires "proportional hazards," whereas the extended model does not. This assumption is discussed below.

To illustrate the application of the standard Cox regression model, I present a reanalysis of the rehabilitation treatment completion data. This time Cox regression is used to predict treatment completion time on the basis of cognitive status. (Although this example uses a single categorical variable to predict hazard, it illustrates essential Cox regression concepts are extended to situations involving continuous variables below.) One can include cognitive status in the analysis if it is first converted into a dummy-coded variable. To do so, one would treat one category (low status) as a baseline, or reference group, and the other category (high status) as our comparison group. By convention, the reference group is arbitrarily coded 0, and the comparison group is coded 1. Note that the procedure for dummy coding a dichotomous predictor is identical for linear regression and Cox regression.

As in linear regression, a Cox model yields a regression coefficient for the predictor variable (b). In linear regression, the model coefficient has a straightforward interpretation. The coefficient for the predictor variable estimates the change in the dependent variable associated with a one-unit increase in the predictor. To interpret a Cox regression coefficient, the concept of a hazard ratio (*HR*) must be understood. For a dichotomous predictor, the hazard ratio estimates a relative hazard (or "relative risk") that compares the rate of event occurrence for the comparison and reference groups. A hazard ratio is computed by using the model coefficient as the exponent of e, the base of the natural logarithms. For the rehabilitation data, the model coefficient is 1.02. Thus, the hazard ratio equals $e^{1.02}$ or 2.76. Therefore, the hazard for high-status patients (the comparison group) was 2.76 times greater than the hazard for low-status patients (the reference group). In other words, high-status patients completed rehabilitation 2.76 times more quickly than did low-status patients. Another way to interpret the hazard ratio is to calculate the percentage difference in hazard, which is computed by subtracting one from the hazard ratio and multiplying the result by 100% (Blossfeld, Hamerle, & Mayer, 1989). For the rehabilitation data, the percentage difference is $[(2.76 - 1) \times 100\%]$, or 176%. This value indicates that the rate of treatment completion was 176% greater for high-status patients than for low-status patients.

The raw Cox regression coefficient (b) represents the natural logarithm of the hazard ratio, which is harder to interpret than a hazard ratio. The raw coefficient does have a useful function, however: a positive coefficient means that the rate of event occurrence is higher for the comparison group than for the reference group, a negative coefficient indicates that the rate is smaller for the comparison group, and a value of zero means that the rate is identical for both groups. In other words, the hazard ratio is one. If the coding scheme had been reversed, that is, if low-status patients were defined as the comparison group and high-status patients were the reference group, then the coefficient for cognitive status would have been negative. Thus, regardless of which group is designated as the reference group, the same conclusion would be reached about the relationship between cognitive status and the rate of treatment completion.

Although the discussion thus far has focused on the hazard function, survival plots are nonetheless available from a Cox regression analysis. Cox survival functions for the rehabilitation data are shown in Figure 11.4. The Cox survival functions are interpreted in the same way as Kaplan–Meier survival functions. Cox survival functions, however, are

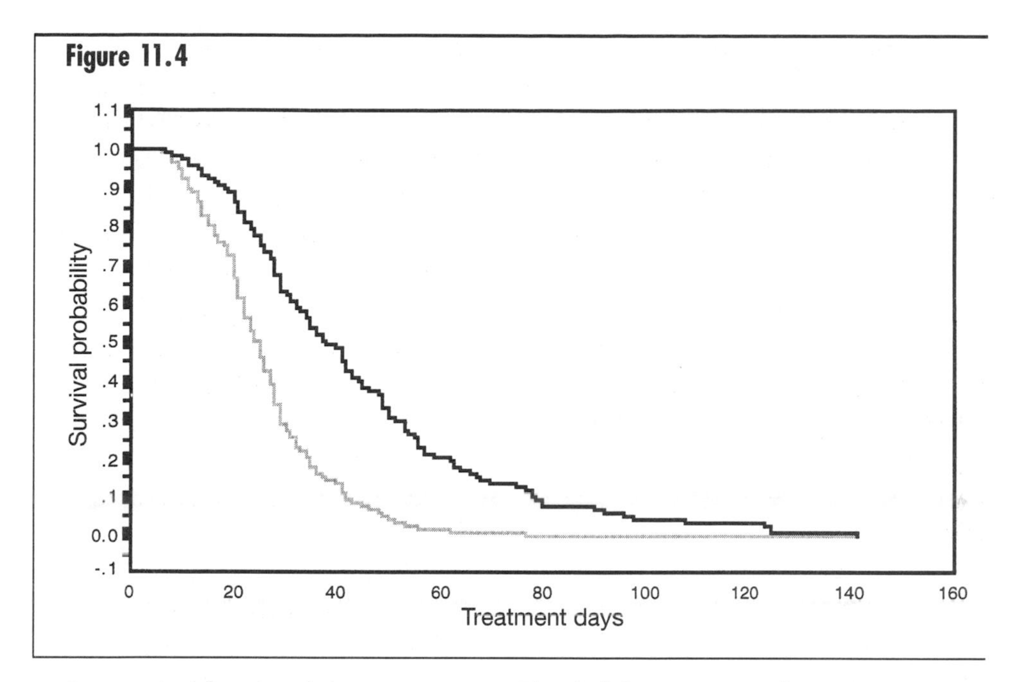

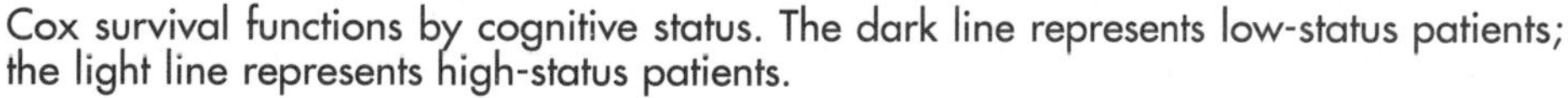
Cox survival functions by cognitive status. The dark line represents low-status patients; the light line represents high-status patients.

calculated differently. Specifically, they are computed as exponents of the *baseline survival function*, which is the survival function for the reference category (Kleinbaum, 1996). For example, it can be seen from the plot that at 20 days the baseline survival probability (the survival probability for the low-status group) is about .90. The 20-day survival probability for high-status patients is computed as the baseline probability raised to the power of the hazard ratio. Thus, the 20-day survival probability for high-status patients is approximately $.90^{(2.76)}$, or .75. This is the value shown in the plot. Similarly, the baseline value for 40 days is roughly .50. Therefore, the 40-day predicted survival probability for high-status patients is $.50^{(2.76)}$, or .15. (In general, a positive Cox regression coefficient means that for a particular time point, the survival probability is larger for the reference group; negative coefficients mean that survival probability is higher for the comparison group.) For the rehabilitation data, the Cox survival plot is very similar to that obtained from our Kaplan–Meier analysis. Cox regression and Kaplan–Meier methods do not necessarily produce similar survival functions, though. This is an important issue that I return to later.

Estimation of Coefficients

In linear regression analysis, the model coefficients minimize the sum of the squared differences between the actual values and predicted values of the dependent variable. That is, the coefficients are chosen according to the least-squares criterion. In Cox regression, the *maximum likelihood* (*ML*) criterion is used for selecting parameter estimates. The coefficient values that are chosen yield the highest probability (likelihood) of obtaining the actual event times observed among the sample of individuals (Allison, 1984). Thus, the Cox regression coefficients are known as *maximum likelihood parameter estimates.* (The computational method used to select Cox regression coefficients is actually a special kind of maximum likelihood estimation known as the *partial likelihood* [*PL]* method [Cox, 1975]. The *PL* method is specifically designed for use with censored data.)

Instead of reporting the actual sample likelihood, researchers usually report one of two related measures: the log likelihood (*LL*) or the deviance ($-2LL$). The *LL* is generally negative. The deviance, equal to the $LL \times -2$, is generally positive. The *LL* and the deviance are zero when the likelihood is one. Coefficients that maximize the likelihood also maximize *LL* and minimize deviance (McCullagh & Nelder, 1989).

Hypothesis Testing

In the rehabilitation example, the nonzero model coefficient means that a relationship exists between cognitive status and the time to treatment completion for individuals in the sample. It is not known, however, whether this association generalizes to the population from which the sample of rehabilitation patients was drawn. In other words, it is not known whether the predictor coefficient is zero in the population. One way of testing whether the population coefficient is zero is to use a likelihood-ratio (*LR*) statistic. The null hypothesis that the population coefficient is zero is rejected if the observed likelihood-ratio statistic occurs in 5% or less random samples from a population in which the coefficient is zero. For models containing a single predictor, the probability for the likelihood-ratio statistic is obtained from a chi-square distribution with one df. The likelihood-ratio statistic for the rehabilitation sample is 76.12, with $p < .00005$, which indicates that the population coefficient probably differs from zero. Moreover, because the sample coefficient is positive, it is concluded that high-status patients tend to complete treatment faster than low-status patients.

Another popular test used to determine whether the population coefficient differs from zero is the *Z test.* To compute *Z*, the researcher would divide the predictor coefficient by its estimated standard error, a measure of the expected variability in the coefficient from sample to sample. For example, the *Z* statistic for cognitive status is 1.02/.12, or 8.5. The probability of obtaining a *Z* value of 8.5 is less than .0005, so the hypothesis that the population coefficient is zero can be rejected. (The probability associated with *Z* is obtained from the normal distribution. The probability is less than .05 for *Z* values with absolute values of 1.96 or greater.) Thus, as with the likelihood-ratio test, the result of the *Z* test indicates that a positive association exists between cognitive status and the rate of treatment completion. Both *Z* and the likelihood ratio are commonly used statistics that assume large samples, and they typically give parallel findings when used on the same set of data. The likelihood ratio should be used, however, if the two tests give different results, because it has better statistical properties (Kleinbaum, 1996).

Interval Estimation for Coefficients and Hazard Ratios

The Cox regression coefficient for cognitive status is only a point estimate of the population coefficient. A 95% CI uses sample data to estimate the likely range of values for the population coefficient. The approximate lower limit for the interval is computed as the value of the coefficient minus two times the standard error of the coefficient. The approximate upper limit is computed as the value of the coefficient plus two times the standard error. (This rule gives a rough estimate. For a more precise estimate of the lower and upper bounds, use 1.96 the SE.) For example, for cognitive status, the lower limit is 1.02 − 2 (0.12) or 0.78, and the upper bound is 1.02 + 2 (0.12) or 1.26. Thus, it is estimated that the population coefficient is between .78 and 1.26. The 95% CI has the same interpretation as in linear regression: If 95% CIs were computed for an infinite number of random samples from a population, 95% of the intervals would contain the population value. Furthermore, when the 95% CI includes 0, the coefficient is nonsignificant at the 5% alpha level.

A 95% CI can also be computed for the hazard ratio if *e* is raised to the power of the upper and lower confidence bounds for the model coefficient. In the present example, the lower bound for the hazard ratio is $e^{(.78)}$ or 2.18, and the upper bound is $e^{(1.26)}$ or 3.53. CIs can be used for hypothesis testing here, as was the case for Cox regression coefficients. If the 95% CI contains the number *1*, the hazard ratio is nonsignificant at the 5% level of significance. This would mean that the time to complete rehabilitation does not reliably differ between high- and low-status patients.

Assumptions

In addition to the general assumptions described in the introduction, two conditions must be met for a Cox regression analysis to be valid. First, to test hypotheses involving Cox regression coefficients, large samples are required (Blossfeld et al., 1989; Cox, 1975). This is because the standard errors for maximum likelihood coefficients have a normal distribution only when very large samples are used. In other words, maximum likelihood coefficients have an "asymptotically" normal distribution. Therefore, hypothesis tests performed on small samples should be regarded with caution.

The second assumption is that the hazards for the reference and comparison group are proportional over all time points; this is known as the *proportional-hazards assumption* (Lee, 1992). It is violated when one group has a higher risk of event occurrence at early time points but a lower risk at later time points. For example, nonproportional hazards have been noted in medical studies that compare invasive procedures (e.g., surgery) with noninvasive procedures for treatment of cancer (Kleinbaum, 1996). Although surgery yields better prospects for long-term survival, it increases short-term mortality risk due to medical complications that result from the operation. The most severe violations of the proportional-hazards assumption occur when hazard or survival functions intersect. Nonetheless, violations may occur even when survival functions do not cross (Kleinbaum, 1996).

The proportional-hazards assumption is implicit in applications of the standard Cox regression model. The reason is that the hazard ratio estimated by the model does not vary over time. For example, for the rehabilitation analysis, the model yielded a single hazard ratio (2.76) that applied to all time points. Thus, if the relative hazard actually varied over time (as in the surgery example), using the standard Cox model would have been misleading. An informal way to test the proportionality assumption is to compare the survival plots obtained from the Kaplan–Meier method with the Cox regression analysis (Kleinbaum, 1996). If the plots from the two methods are similar, the assumption is considered to be met. Otherwise, if the plots differ substantially, it can be concluded that the assumption is violated. (It has already been verified that the Cox and Kaplan–Meier survival functions were similar for the rehabilitation data, so there was no reason to suspect a violation of the proportional-hazards assumption.)

Another informal way to test the proportional-hazards assumption is to compare transformed Kaplan–Meier survival functions for the groups (Kleinbaum, 1996). The Kaplan–Meier functions are plotted after they are transformed using the *log-minus-log* transformation (the name comes from the fact that the natural logarithm is applied twice to the survival function. First, the natural logarithm of the each survival probability is computed. Then the negative of each logged value is taken. Finally, the resulting values are logged again, yielding the log-minus-log values.) If the functions are nonparallel, there is reason to suspect nonproportionality. Several formal tests of the proportional-hazards assumption are also available (Kleinbaum, 1996). One approach, which is described below, involves using a generalization of the Cox regression model: the extended Cox regression model. An advantage of this approach is that it can be used to model survival when the proportional-hazards assumption is violated. (For univariable situations, the Kaplan–Meier method can also be used to describe the survival experience when the proportional-hazards assumptions is violated.)

Presentation of Results

Results from a study that uses Cox regression are typically presented in a format that is similar to that of linear regression analyses. Generally, an overall (or omnibus) goodness-of-fit measure is given: the *LL*, the deviance ($-2LL$), or the likelihood ratio (*LR*) for the model. Next, model coefficients are presented with standard errors, test statistics (*Z* or likelihood ratios), probabilities, and CIs. (Often the hazard ratio [e^b] is reported because it is easier to interpret than model coefficients.) Most studies also present plots of survival or hazard functions and discuss whether the proportional hazards assumption was met. Presentation of results varies widely across studies, however. Some studies may report only a few of the statistics described above.

Multivariable Cox Regression Analysis

The earlier analysis of the rehabilitation data found that patients with poor cognitive skills took longer to finish treatment compared to patients with greater cognitive ability; however, other variables that might be related to treatment completion time were not considered. For example, treatment might be faster for psychologically well-adjusted patients, who may be more motivated to complete treatment. As in linear regression analysis, a multivariable Cox regression model can be formulated to determine the relationship between a set of predictors (as a whole) and the hazard of event occurrence, as well as the relationship between each predictor and the hazard, while statistically adjusting for the other predictors in the model. The Cox regression model can contain continuous or categorical predictors.

To illustrate multivariable Cox regression analysis, I again use the rehabilitation data, but I add three other variables to the model. Predictors of treatment time include a continuous variable (age in years) and three dichotomous predictors: sex, cognitive status, and psychosocial status at admission. The status measures were coded low or high. Psychosocial status classification was based on assessment by a clinical psychologist and a social worker. Thus, for the status variables I create two dummy variables for analysis that are arbitrarily coded 0 for low status (the reference group) and 1 for high status (the comparison group). For sex, I create a dummy variable for analysis that is coded 0 for men (the reference group) and 1 for women (the comparison group). I examine whether treatment completion time is associated with cognitive status when controlling for sex, age, and psychosocial status. This analysis helps one to determine whether a relationship exists between treatment time and cognitive status that is independent of the other variables in the model.

Here it is important to recall the prior discussion concerning the choice of the likelihood-ratio statistic or Z to evaluate the probability of the obtained results. For multivariable problems, only the likelihood-ratio statistic can be used to test the overall model (i.e., the system of coefficients for all predictor variables), which is analogous to using an omnibus F test in linear regression. Either test statistic can be used to evaluate individual coefficients in a multivariable model. Recall that the Z test is used to compare a model coefficient with its standard error; for an individual coefficient, the likelihood ratio is analogous to partial F in multiple regression analysis.

Although I am interested primarily in the coefficient for cognitive status, I can use an omnibus likelihood-ratio test to determine whether any predictor coefficient differs from zero. The dfs for the omnibus likelihood-ratio test are 4. (For multivariable analysis, dfs are equal to the number of predictor variables in the model.) For the rehabilitation study, the likelihood-ratio statistic (107.15) has a probability less than .00005, which means that at least one of the population coefficients differs from zero (the omnibus test does not tell which of the coefficients differs from zero).

Coefficients and *Z* statistics for the multivariable model are shown in Table 11.3. The *Z* statistic for cognitive status is positive and has an associated probability of less than 0.05, so it can be concluded that patients with high cognitive status tend to complete treatment faster than do low-status patients, when taking into account age, sex, and psychosocial status. That is, there is a positive relationship between cognitive status and the rate of treatment completion that does not depend on whether a patient is old or young, on whether a patient is male or female, or on the patient's level of psychosocial adjustment. The coefficient associated with cognitive status is .707. Thus, the model estimates that, when controlling for age, sex, and psychosocial status, the rate of treatment completion was 2.03 [$e^{(.707)}$] times greater for high-status patients than for low-status patients. Furthermore, the rate of treatment completion is 103% faster for high-status patients than for low-status patients [(2.03 − 1) × 100%].

Based on the *Z* statistics in Table 11.3, it can also be concluded that age is positively related to the rate of treatment completion. Thus, older patients tend to have shorter treatment times in comparison with younger patients, when controlling for the other variables in the model. Raising *e* to the power of the coefficient for a continuous variable gives the hazard ratio associated with a one-unit increase in the predictor, when statistically adjusting for the other predictors in the model. Thus, the hazard ratio for age (1.01) indicates that for each 1-year increase in a patient's age there is a small (1%) increase in the rate of treatment

Table 11.3

Statistics for Multivariable Cox Regression Model

					95% CI (e^b)	
Variable	**b**	***SE***	***Z***	e^b	**Lower**	**Upper**
Cognition	.707	.141	5.01*	2.030	1.540	2.670
Psychosocial	.562	.141	3.99*	1.750	1.330	2.310
Age	.012	.003	4.00*	1.010	1.010	1.020
Sex	−.184	.121	−1.52	0.832	0.657	1.050

Note. CI = confidence interval. *$p < .05$.

completion [($e^{(.01)}$ − 1) × 100%]. A hazard ratio can also be computed for a different size increase in a continuous predictor, for example, to compare 50-year-olds with 60-year-olds. To do so, multiply the Cox regression coefficient by the size of the increase, then raise *e* to the power of the product (Kleinbaum, 1996). For a 10-year increase in age, the estimated hazard ratio is $e^{(10 \times .01)} = e^{(.10)}$ or 1.105. Thus, the model estimates that the rate of treatment completion is 10.5% greater [($e^{(.01 \times 10)}$ × 1) × 100%] for 60-year-olds than for 50-year-olds. This percentage estimate applies to any 10-year difference in age, so it can also be used to compare 30-year-olds with 40-year-olds or 70-year-olds with 80-year-olds.

Patients with high psychosocial status also tend to complete treatment more quickly. There is insufficient evidence, however, to reject the hypothesis that the population coefficient is zero for sex. Nevertheless, one may still examine the coefficient to uncover any hints of a sex trend in this particular sample. The negative coefficient for sex means that males (the reference category) tended to complete treatment faster than females (the comparison group), when taking into account the effects of the other predictors in the model. If the coding scheme had been reversed, the coefficient for sex would have been positive, but the interpretation would be the same.

In our univariable analysis of the rehabilitation data, the survival functions compared high- and low-status patients without adjustment for other explanatory variables. A multivariable Cox regression analysis allows the researcher to obtain survival functions for each predictor that adjust for the effects of other predictors in the model; these are known as *adjusted survival functions* (Kleinbaum, 1996). The adjusted survival functions for men (bottom line) and women (top line) are shown in Figure 11.5. Consistent with the result of the *Z* test for sex, the plot shows that treatment times were very similar for males and females when the effects of the other predictors were taken into account. (Treatment time tended to be slightly lower for males, however, which is reflected in the negative sign of the sex coefficient.) Adjusted survival functions could also be plotted for the other predictors in the model. Note, however, that the Cox model estimates an adjusted survival function for each value of a variable. Thus, if a variable has more than a few values, plots of survival functions can become dense and difficult to read. Therefore, for continuous variables such as age, researchers often plot only selected values (e.g., quartiles).

Figure 11.5

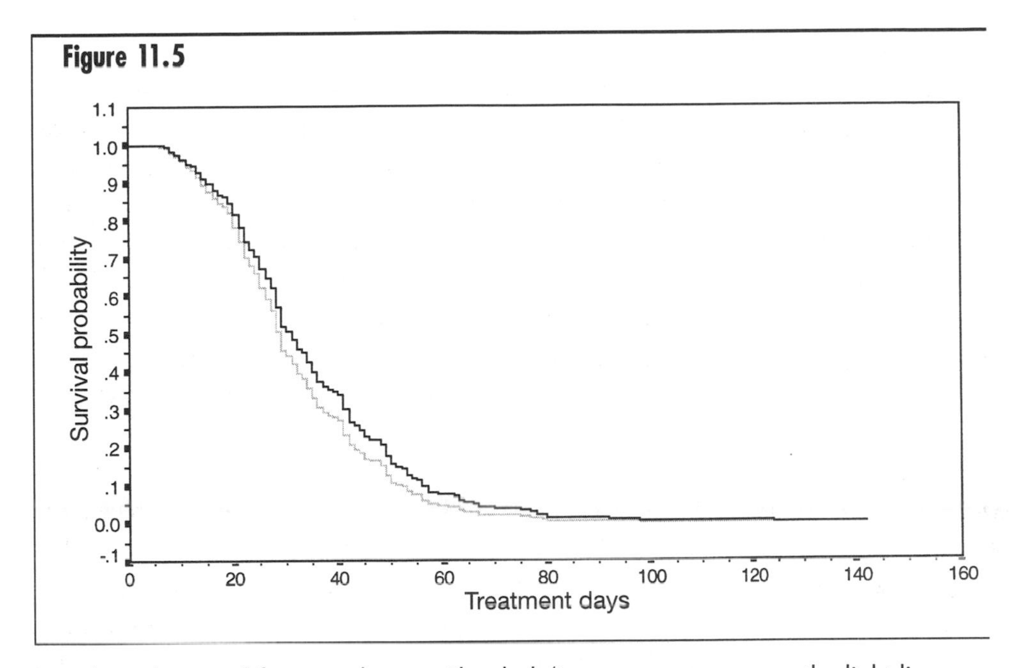

Adjusted survival functions by sex. The dark line represents women; the light line represents men.

Formal Tests of the Proportional-Hazards Assumption: Extended Cox Regression Model

As for the univariable Cox analysis, the proportional-hazards assumption for the multivariable model must be tested. Recall that for the earlier analysis, which used a single categorical predictor, the proportional-hazards assumption required that the hazard ratio for the high- and low-cognitive status groups remained constant over time. For multivariable models, this assumption extends to each predictor in the model. Moreover, for continuous predictors, the hazard ratio must be constant for any two values of the predictor. Thus, for the rehabilitation data, hazard ratios must not vary over time for males and females, for people with high- and low- cognitive or psychosocial status, or for older and younger patients. An extension of the Cox regression model can be used to formally test the proportional-hazards assumption. This approach can be used for univariable and multivariable survival analysis.

The proportional-hazards assumption requires that the effect of a predictor variable does not change over time. In other words, there is no interaction between time and the predictor (Kleinbaum, 1996). The hypothesis that hazards are proportional for each predictor in the model can therefore be tested if a model is developed that includes an interaction between each predictor and time. (A Cox model that includes Time × Predictor interactions is an instance of a special type of Cox model known as an *extended Cox regression model.*) If a statistically significant interaction is found, it can be concluded that the proportional-hazards assumption is violated for the associated predictor. Otherwise, if all interactions are nonsignificant, it is concluded that the proportional-hazards assumption is met for variables in the model. (An omnibus likelihood-ratio test is also available that assesses several predictors simultaneously. For details regarding this approach, see Kleinbaum, 1996.)

For example, reconsider the multivariable rehabilitation analysis. To test the proportional hazards assumption, an eight-variable extended Cox regression model was fit to the data. The predictor variables included sex, age, cognitive status, psychosocial status, and Time × Predictor interactions involving each predictor. Because none of the Z statistics for the interaction terms were statistically significant, I conclude that there is no evidence of nonproportionality. Thus, the use of the standard Cox regression model (i.e., the model without interactions) is appropriate for the four-variable rehabilitation analysis. Suppose, however, that I add another variable, medical status—a dichotomous variable coded 0 for low status and 1 for high status—to the model. When an extended Cox model is estimated that includes medical status and the Medical Status × Time interaction, Medical Status × Time interaction is statistically significant ($Z = -2.71$, $p < .01$). Thus, for patients with high- and low-medical status, hazards are not proportional over time.

Because the proportional-hazards assumption is violated for the medical status variable, I conclude that a single hazard ratio cannot accurately characterize the ratio of treatment completion for patients with high- and low-medical status. Fortunately, the extended model can be used not only to detect nonproportionality, but also to describe it. To use the extended model for this purpose, examine both the main effect term for medical status and the Medical Status × Time interaction term. The term for medical status was positive and statistically significant ($b = 1.665$; $Z = 5.12$; $p < .001$), which suggests that high-status patients complete treatment more quickly, on average, compared with low-status patients. Because our analysis indicated that the hazard ratio varies over time, though, I must also take into account the time-by-status coefficient, which reveals how the hazard ratio for high- and low-status changes, or how much it increases or decreases over time. The sign of the interaction term was negative ($-.0250$), which means that the hazard ratio for high- and low-status patients tended to shrink as time elapsed. (In general, a positive Time × Predictor interaction implies that the hazard ratio increases with time; a negative coefficient implies a decreasing trend over time; Cox & Oakes, 1984.) In other words, the hazard ratio was greater among patients with short rehabilitation stays than among patients with longer stays.

The extended model can be used to summarize the rehabilitation data because it does not assume proportional hazards; the Time × Predictor interactions allow hazard ratios to vary over time. This is an important difference between the extended model and the standard Cox model, which assumes proportionality. Further examples of the use of the extended Cox model to test the proportional-hazards assumption and model nonproportionality are provided by Kleinbaum (1996); Stablein, Carter, and Novak (1981); and Stablein, Carter, and Wampler (1980).

Multivariable Cox Regression Analysis: A Research Application

I now examine how the Cox regression model was used in a research study to determine the relationship between a set of predictors and the time to occurrence of an event. Greenhouse and colleagues (Greenhouse, Kupfer, Frank, Jarrett, & Rejman, 1987) used Cox regression to estimate the time to stabilization for 98 patients who received somatic treatment for clinical depression. Predictors of time to stabilization included 12 biological and clinical measures collected before treatment began: sex, age, age of first depressive episode, number of previous episodes, duration of current episode, Hamilton rating score (a measure of the severity of depression), four sleep variables (sleep latency, sleep efficiency, REM density, and REM latency), and two neuroendocrine measures (Cortisol nadir and Cortisol latency). All variables were continuous measures except sex. Patient stabilization status was recorded at regular clinical follow-up visits during a 3-year period. Stabilization times were censored for 22 patients who withdrew from the study.

Greenhouse et al. developed a multivariable model to determine which variables were associated with stabilization time when controlling for a patient's age and clinical severity. When adjusting for the control variables, two predictors were positively associated with time to stabilization: number of previous episodes and Cortisol nadir. Table 11.4 gives model statistics for the control variables and the statistically significant

Table 11.4

Model Statistics for the Depression Study

Variable	*b*	*SE*	*Z*	e^b	95% CI (e^b)
Age	.0212	.011	1.93*	1.02	1.00–1.04
Hamilton score	−.0459	.024	−1.91*	0.96	0.91–1.00
Number of episodes	.0306	.016	1.91**	1.03	1.00–1.06
Cortisol nadir	.2300	.092	0.50**	1.26	1.05–1.51

Note. e^b = hazard ratio; CI = confidence interval. *Z* statistics and CIs are computed from coefficients and standard errors reported by Greenhouse et al. Due to rounding of decimals, the *Z* statistic for number of episodes is less than the 1.96 cutoff value needed for statistical significance. The CI includes one for the same reason. From "Analysis of Time to Stabilization in the Treatment of Depression: Biological and Clinical Correlates," by J. B. Greenhouse, D. J. Kupfer, E. Frank, D. B. Jarrett, & K. A. Rejman, 1987, *Journal of Affective Disorders, 13,* p. 263. Copyright 1987 by Elsevier Science, Amsterdam, The Netherlands. Reprinted with permission. $^*p < .10$. $^{**}p < .05$.

predictors. When controlling for age and severity of depression, patients with high Cortisol nadir values or many earlier episodes tended to respond to treatment most quickly. For example, the hazard ratio for Cortisol nadir estimates that rate increased 1.26 × [$e^{(.230)}$] for each one-unit increase in Cortisol nadir. Furthermore, the hazard ratio for number of previous episodes indicates that the rate of stabilization increased 1.03 × [$e^{(.0306)}$] for each previous depressive episode. Although Greenhouse et al. found that the control variables (age and Hamilton score) were significantly associated with time to stabilization in a univariable analysis, the effects for these variables only approached significance in the multivariable analysis ($p < .10$). Model coefficients for these variables indicate that there was a trend in the sample for older patients and patients with lower Hamilton scores (i.e., less-severely depressed patients) to stabilize most quickly.

CIs for hazard ratios are also shown in Table 11.4. Recall that to compute a 95% CI for a hazard ratio, first compute a 95% CI for the Cox regression coefficient. For example, for Cortisol nadir, the lower confidence limit is .23 − 2 (.092) or .046 and the upper limit is .23 + 2 (.092) or .414. Raising *e* to the power of the upper and lower limits gives a 95% CI for the hazard ratio: 1.05 to 1.51. Thus, with 95% confidence, it is inferred that in the population, the rate of stabilization increases between 5% and 51% for each one-unit increase in Cortisol nadir at baseline. No fitted survival plots were provided by Greenhouse et al. This is common practice, however, when the main focus of the study is to identify statistically significant predictors of event occurrence rather than to predict outcomes for individual patients.

Greenhouse et al. argued that their study was a first step in the search for patient characteristics that predict recovery from clinical depression. The observation that patients with many previous episodes responded more quickly was unexpected, however, and therefore suggests a need for replication. Moreover, the authors suggested that using a fuller set of psychological baseline measures may help sort out the relative roles of biological and psychosocial factors that influence the rate of patients' responses to somatic treatment of depression. Finally, it is worth noting that the results may have been dependent, in part, to the definition of stabilization. Patients were considered to have stabilized if they maintained a Hamilton score of 7 or lower for three or more consecutive weeks. The results may have been different if the authors used another criterion (e.g., a smaller or larger Hamilton score or a longer or shorter minimum time requirement). The definition of the outcome variable is an important consideration when any survival analysis is reviewed.

Special Topics in Survival Analysis

Techniques for analysis of survival data encompass a broad set of methods. This section outlines some adaptations and analogues of the methods described in this chapter that are appropriate for special data analysis situations.

Iterative Multivariable Cox Regression

In applied research, it is often desirable to identify a parsimonious subset of predictors from a larger set. Iterative Cox regression analysis is a popular technique that is used to try to find a useful subset of variables; that is, a subset that includes only statistically significant predictors and that results in good fit to the data. The most common forms of iterative model-building strategies include forward, backward, and stepwise Cox regression (Blossfeld et al., 1989; SPSS, 1994). These are analogous to model-building procedures used in multiple linear regression.

In forward selection, variables are tested, one at a time, for entry into the model. The first variable added is the variable that has the smallest probability associated with the likelihood ratio statistics among the statistically significant predictors. Other variables are then added to the model if their likelihood ratios are also significant when taking into account the variables already in the model. Model building stops when all variables have been entered, or when the likelihood ratio is nonsignificant for all variables that have not been entered.

In backward Cox regression, the model starts with all predictors, whether or not they are statistically significant. Variables are tested, one at a time, for removal from the model. The first variable that is removed is the variable whose likelihood ratio statistic has the largest probability greater than alpha. The procedure continues to remove variables from the model until the model contains only variables that are statistically significant.

Stepwise Cox regression is a combination of forward and backward model building. Each variable is tested for entry into the model. Whenever a predictor is entered into the model, other variables in the model are tested for removal. The model building process continues until no more variables can be entered or removed.

In many statistical software programs, other criteria can be used for variable selection, such as the probability associated with the Wald statistic (e.g., SPSS, 1994). Iterative procedures, however, involve many tests of statistical hypotheses (i.e., tests of individual coefficients) and therefore dramatically increase the Type 1 error rate for the overall study. In such analysis, cross-validation samples are highly recommended (Thompson, 1995).

Time-Dependent Covariates

In most social science research, values of the predictor variables are recorded at the beginning of the study and subsequent changes in a person's status are disregarded. In other words, the predictors are treated as *time-independent covariates* (or time-constant covariates). This approach is most sensible for variables such as year of birth and race (which do not change over time) and sex (which changes only in rare cases), but in some research situations the predictor variables vary substantially over time. For example, in a study of employee turnover, potential predictors that are likely to change over time include educational status (i.e., highest level of education attained) and attitudes about one's work environment. Variables whose values can change over time are known as *time-dependent covariates* (Cox & Oakes, 1984). Other examples of time-dependent covariates include health status and job market conditions.

Time-dependent covariates are important in statistical analysis because an individual's risk for event occurrence is often more strongly influenced by his or her current status than by the status at the beginning of a study. For example, an employee who has recently attained a higher educational level may be more apt to search for a higher paying, more challenging position than he or she was before finishing the degree. Thus, if time-varying values are available to the researcher, it may be possible to improve model fit by taking into account the effects of time-dependent covariates on survival. Time-dependent predictors can be incorporated in survival analysis using the extended Cox regression model (Kleinbaum, 1996). (I have already used the extended Cox model to examine the effects of Time × Predictor interactions, but the model can be used to estimate the effects of any variables whose values change over time. A Time × Predictor interaction is just a special kind of time-dependent covariate.) Examples of time-dependent Cox regression analyses are provided by Hser, Anglin, and Liu (1991), who studied the effects of employment status, narcotics use, marital status, and legal status on the risk for dropping out of treatment among methadone patients, and by Crowley and Hu (1977), who studied mortality among heart transplant patients.

Discrete-Time Survival Analysis

The methods used in this chapter are designed for continuous duration data, or survival times that are measured precisely. Discrete-time survival methods are available for analysis of duration data that are grouped into intervals (e.g., months, years, or semesters; Allison, 1982). Discrete time methods are analogous to those described in this chapter. The discrete-time analogue of the Kaplan–Meier method is the *life table* or *actuarial table* used by demographers and actuaries (Lee, 1992). An adaptation of logistic regression (Wright, 1995) can be used to build multivariable survival regression models in situations involving discrete data (Abbott, 1985; Singer & Willett, 1993; Willett & Singer, 1991). Despite the availability of discrete-time methods, however, continuous-time survival analysis is commonly used when survival time is recorded in small intervals. In general, this poses no problem if the size of the time interval is small compared to the rate of event occurrence (Luke, 1993; Peterson, 1991).

Parametric Survival Analysis

The Kaplan–Meier and Cox regression methods require no assumptions about the distribution of survival times. In this sense, they are regarded as *nonparametric* techniques.[3] This is a convenient fact because the researcher often has no reason for choosing one theoretical distribution over another. When the correct distribution of event times can be specified in advance, however, *parametric* survival methods often can be more informative than nonparametric techniques (Luke, 1993). Use of parametric survival methods involves a choice between one of several theoretical survival time distributions. The *exponential distribution* is used when the hazard is assumed to be constant over time. The *Weibull distribution*, a generalization of the exponential distribution, is used when the hazard is thought to increase or decrease over time. Other commonly used distributions include the log-normal, gamma, and Gompertz distributions. Nonetheless, nonparametric methods can be useful even when the ultimate goal is a parametric analysis. For example, nonparametric methods can often help suggest an appropriate survival time distribution (Lee, 1992).

Survival Analysis for Multiple and Repeated Events

The methods described in this chapter are designed primarily for analysis of time to a single event. In some situations, however, it is important to distinguish between two or more events of interest. For example, a researcher who studies job transitions may want to avoid lumping together all types of job changes in a single analysis. In particular, he or she may wish to differentiate job changes due to firing, retirement, and promotion. In studies of disease-related mortality, it is common to distinguish between deaths due to disease, accidents, and suicide. *Competing-risk* survival analysis methods are for situations in which there is more than one outcome of interest (Cox & Oakes, 1984). Special adaptations are also available for situations in which some individuals experience the event more than once, a common occurrence in longitudinal studies of job changes, marriage, and divorce. These methods are known as *repeated-event* (or repeated-spell) survival methods. For an introduction to survival analysis for multiple and repeated events, see Allison (1984) and Willett and Singer (1995).

---

[3]The Cox regression model actually has both parametric and nonparametric components, so it is also called a *semiparametric* or *partially parametric method* (Luke, 1993; Tuma, 1982).

Conclusion

Survival analysis is a family of statistical methods designed for analysis of the time it takes for an event to occur. It is useful for studying a wide variety of time-related outcomes that are of interest to social scientists, including therapeutic effectiveness, treatment completion, parole behavior, and stabilization of symptoms. When used properly, survival analysis has important advantages over more familiar techniques such as computing a mean or using linear regression to model duration. Two of the most important advantages of survival analysis are that it can handle censored data without bias and that it can give insight about the time course of an outcome. This chapter focused primarily on two popular methods for analysis of continuous duration data: Kaplan–Meier and Cox regression analysis. Both methods are nonparametric, so they can be used even when the distribution of event times cannot be specified in advance.

The Kaplan–Meier method is used to obtain a survival function for a single sample of individuals or to compare survival functions for two or more groups. Formal procedures are available to test the hypothesis that group survival functions are equivalent. The most popular is the log-rank test, which is analogous to an *F* test in ANOVA. Cox regression analysis is used to build univariable and multivariable models of duration. Cox regression shares many features with linear regression. Both kinds of analysis are used to estimate the strength of association between a set of variables and an outcome measure, and both can incorporate continuous and categorical predictors and interactions. Moreover, hypothesis tests and CIs for model coefficients can be computed for both procedures; however, whereas linear regression is used to predict scores on a continuous dependent variable, Cox regression is used to predict the rate of event occurrence.

Statisticians distinguish between two kinds of Cox regression models. The standard model is simplest. It includes only time-independent predictors, such as year of birth and race. The extended Cox model accommodates both time-independent and time-dependent predictors. The time-dependent predictors can be variables such as blood pressure or educational level, or they can be interactions between time and one or more time-independent predictors. The distinction between the standard and extended models is important because only the standard Cox regression model assumes proportional hazards. Indeed, the extended model is a useful option for modeling nonproportionality when the proportional hazards assumption is violated.

Several extensions and adaptations of these techniques are available for special data analysis situations. For example, iterative model-building programs can be used to find a useful subset of a large set of predictor variables. Discrete-time survival methods can be used to analyze duration data that are grouped into intervals. Parametric survival methods are available for situations in which the distribution of survival times can be specified in advance. Special methods have also been developed for analyses that call for a distinction between two or more kinds of events and for situations in which some individuals experience the event more than once.

It is important to remember, however, that the use of survival analysis requires several important assumptions. In particular, both Kaplan–Meier and Cox regression analysis require independence of observations and independent censoring. Furthermore, for hypothesis tests to be accurate, Cox regression requires large samples. Violation of these assumptions can lead to biased survival estimates or can invalidate hypothesis tests. Thus, before drawing conclusions from a survival analysis, one should consider whether the relevant assumptions have been met.

Suggestions for Further Reading

This chapter gives only an overview of survival analysis. Fortunately, a growing body of literature is available to help the reader to extend his or her knowledge of survival analysis. Excellent introductory treatments are given by Allison (1984), Lee (1992), Kleinbaum (1996), Luke (1993), Peto and colleagues (Peto et al., 1976, 1977), and Singer and Willett (Singer & Willett, 1993; Willett & Singer, 1991, 1993). These introductory treatments require only a basic understanding of statistics. More technical discussions are given by Cox and Oakes (1984), Kalbfleisch and Prentice (1980), Blossfeld et al. (1989), and McCullagh and Nelder (1989), who described survival analysis in the context of the generalized linear model.

Several excellent sources are also available to increase the understanding of special topics in survival analysis. Annotated examples of iterative Cox regression analyses are given in Blossfeld et al. (1989) and the SPSS (1994) *Advanced Statistics User's Guide.* Introductions to discrete-time survival analysis are given by Allison (1982) and Willett and Singer (Singer & Willett, 1993, Willett & Singer, 1991, 1993). For more information on models involving time-dependent covariates, see Blossfeld et al. (1989), Kleinbaum (1996), and Cox and Oakes (1984). Applications of time-dependent Cox regression analysis are provided by Hser, Anglin, and Liu (1991); Crowley and Hu (1977); and Stablein and colleagues (Stablein, Carter, & Novak, 1981; Stablein, Carter, & Wampler, 1980). For an introduction to parametric survival analysis, see Blossfeld et al. (1989), Lee (1992), and Luke (1993). Discussions of the use of survival analysis for studies involving multiple or repeated events are provided by Allison (1984), Cox and Oakes (1984), and Willett and Singer (1995).

Glossary

ADJUSTED SURVIVAL FUNCTION The relationship between a predictor and the *survival probability* when taking into account the effects of other predictors in a multivariable model.

BASELINE HAZARD FUNCTION Estimates the instantaneous risk of event occurrence over time for the sample as a whole.

BRESLOW TEST A *nonparametric* test used to compare survival distributions. This test is more sensitive than the *log-rank test* to group differences at early time points.

CENSORED DATA Durations that are known only to exceed some value because the start or end of the time interval is unknown. Often this means that the *event* being monitored (e.g., heart attack) occurs outside the study observation period.

COMPETEING-RISK MODEL Survival model that distinguishes between two or more *events* (e.g., job change due to firing and job change due to promotion).

CONDITIONAL PROBABILITY The probability that an individual survives beyond a particular time point given that he remained at risk after the previous *event*.

CONTINUOUS-TIME SURVIVAL MODEL Survival model designed for analysis of duration data that are measured precisely. In practice, continuous-time models are often used when times are recorded in small units.

COX REGRESSION MODEL A semiparametric survival model that relates one or more variables to the risk that an *event* will occur at a particular time. There are two kinds of Cox models: the standard model and the extended model.

DEVIANCE ($-2LL$) Computed as −2 times the natural logarithm of the sample likelihood. Generally a positive number, the deviance decreases as the likelihood increases (when the likelihood is 1, the deviance is 0).

DISCRETE-TIME SURVIVAL MODEL A survival model for analysis of duration data that are grouped into intervals (e.g. months, years, or semesters).

DUMMY CODING The process of converting a categorical variable to a numeric variable for statistical analysis. To dummy code a dichotomous variable, one category is defined as a reference group and the other category as defined as a comparison group. The reference group is arbitrarily coded 0 and the comparison group is coded 1.

EXPONENTIAL DISTRIBUTION A theoretical distribution of survival times. The exponential distribution is used in *parametric survival analysis* when the hazard rate is thought to be constant over time.

EXTENDED COX REGRESSION MODEL A generalization of the standard *Cox regression model* that incorporates time-dependent predictors. The extended model can be used to test the *proportional-hazards assumption* and to model *nonproportionality* when the proportional-hazards assumption is violated.

EVENT (TERMINAL EVENT, FAILURE) A change from one of two mutually exclusive and exhaustive states to another (e.g., from marriage to divorce or from abstinence to relapse).

HAZARD A measure of the risk that an *event* will occur for an individual at a particular time. For *continuous-time models,* the hazard is a rate that ranges from zero to infinity. For *discrete-time models,* the hazard is a probability.

HAZARD FUNCTION A mathematical relationship that describes changes in the risk of *event* occurrence over time.

HAZARD RATIO [HR, exp(b)] An effect strength measure computed as the ratio of estimated hazards for individuals with different sets of covariate values. For *proportional-hazards models,* the hazard ratio is assumed to be fixed over time. Also known as *relative risk.*

INDEPENDENT-CENSORING ASSUMPTION An assumption that censoring mechanisms are unrelated to the probability that an individual will experience an event. The independent-censoring assumption is violated if persons withdraw from a study because they are at high or low risk of experiencing the *event.* If the assumption is violated, survival estimates may be biased.

INTERQUARTILE RANGE (IQR) A range that is bounded by the 25th and 75th percentiles of the sample data.

KAPLAN–MEIER METHOD (PRODUCT–LIMIT METHOD) A nonparametric method for estimating survival probabilities at observed event times.

LEFT CENSORING A type of censoring that occurs when the start of a time interval is unknown to the investigator.

LIFE TABLE A nonparametric method for summarizing the survival experience of individuals whose data are grouped into intervals.

LIKELIHOOD The probability of obtaining the observed data given a set of coefficient estimates. The likelihood ranges from 0 to 1.

LIKELIHOOD-RATIO STATISTIC (LR) A chi-square statistic used to test whether one or more survival model coefficients differs from zero. When the likelihood-ratio statistic is large compared with its dfs, the hypothesis that the population coefficients are zero is rejected. A likelihood-ratio test is analogous to an F test in linear regression.

LOG LIKELIHOOD (LL) The natural logarithm of the sample likelihood. Generally a negative number, the LL increases as the likelihood increases (the $LL = 0$ when the likelihood is 1).

LOG-MINUS-LOG SURVIVAL PLOT A graphical tool used to examine whether the *proportional-hazards assumption* is met.

LOG-RANK STATISTIC (MANTEL–COX TEST) A nonparametric statistic that is used to test whether two or more survival distributions are equivalent.

LONGITUDINAL FOLLOW-UP STUDY A study that tracks individuals over time.

LOSS TO FOLLOW-UP A type of censoring that occurs when the investigator loses contact with a study participant (e.g., because he or she moves out of state).

MAXIMUM-LIKELIHOOD METHOD (ML) Criterion used for estimating model parameters in a *Cox regression model.* Coefficient values are chosen that yield the highest probability of obtaining the actual *event* times observed among the sample of individuals. See *partial-likelihood method.*

MEDIAN A nonparametric measure of central tendency that is equal to the 50th percentile.

MODEL COEFFICIENT (*b*) Parameter estimate for a predictor variable.

NONPARAMETRIC SURVIVAL MODEL A model that makes no assumptions about the underlying distribution of *survival times.*

NONPROPORTIONALITY A violation of the *proportional-hazards assumption.*

OBSERVATION PERIOD The time interval during which study participants are followed by the researcher. Also known as the *follow-up period.*

PAIRWISE LOG-RANK TEST A special type of *log-rank test* that compares exactly two groups.

PARAMETRIC SURVIVAL MODEL A survival model that requires the analyst to specify the theoretical distribution of survival times.

PARTIAL-LIKELIHOOD METHOD (PL) Criterion used for estimating model coefficients in a Cox regression analysis. The partial-likelihood method is a variant of the *maximum-likelihood method* that is designed to take into account censoring.

PROPORTIONAL-HAZARDS ASSUMPTION The assumption that the *hazard ratio* is invariant over time. The most severe violations of this assumption occur when group hazard or survival functions intersect.

REPEATED EVENTS *Events* that can occur more than once in an individual's lifetime (e.g., marriage or job promotion).

RIGHT CENSORING A type of censoring that occurs when the end of a time interval is unknown to the investigator.

SEMIPARAMETRIC MODEL A model that has both parametric and nonparametric components. The Cox regression model is a semiparametric survival model.

STANDARD COX REGRESSION MODEL A *Cox regression model* that relates one or more time-independent predictors to survival. The standard model assumes proportional hazards.

STANDARD ERROR (SE) Estimates the variability from sample to sample in a model coefficient or survival estimate. Used to compute *Z statistics* and CIs.

STEPWISE MULTIVARIABLE REGRESSION An iterative procedure for multivariable model building.

STRATIFICATION A method used to adjust survival estimates to control for effects of confounding variables.

SURVIVAL ANALYSIS A set of statistical methods used to analyze the time to occurrence of an *event*. Survival methods are designed to incorporate *censored data* without bias.

SURVIVAL FUNCTION A mathematical relationship that describes the cumulative probability of survival over time.

SURVIVAL PROBABILITY The cumulative probability of surviving (i.e., being event free) past a given time point.

SURVIVAL TIME The time interval between two *events* (e.g., between entry into a study and relapse).

TARONE–WARE TEST A *nonparametric test* used to compare survival distributions. This test is more appropriate than the *log-rank test* when group survival distributions do not differ by a fixed amount.

TERMINAL EVENT The event of interest in a longitudinal study.

TIME-DEPENDENT COVARIATE (TIME-VARYING COVARIATE) An explanatory variable whose values may change over time for an individual. Time-dependent covariates may be "inherent" (e.g., marital status) or "defined" (e.g., an interaction between an explanatory variable and time).

TIME-DEPENDENT COX REGRESSION MODEL See *extended Cox regression model.*

TIME-INDEPENDENT COVARIATE (TIME-CONSTANT COVARIATE) A covariate whose values do not change over time for an individual (e.g., year of birth).

WEIBULL DISTRIBUTION A theoretical distribution for survival time data. Used when the *hazard rate* is thought to increase or decrease over time.

Z TEST Tests that determine whether a model coefficient differs from zero. Large *Z* values (in absolute value) mean that the population coefficient probably differs from zero.

References

Abbott, R. D. (1985). Logistic regression in survival analysis. *American Journal of Epidemiology, 121,* 465–471.

Allison, P. D. (1982). Discrete-time methods for the analysis of event histories. In S. Leinhardt (Ed.), *Sociological methodology* (pp. 61–98). San Francisco: Jossey-Bass.

Allison, P. D. (1984). *Event history analysis.* Beverly Hills, CA: Sage.

Blossfeld, H. P., Hamerle, A., & Mayer, K. U. (1989). *Event history analysis.* Hillsdale, NJ: Erlbaum.

Breslow, N. (1970). A generalize Kruskal–Wallis test for comparing *K* samples subject to unequal patterns of censorship. *Biometrika, 57,* 579–594.

Cox, D. R. (1972). Regression models and life tables (with discussion). *Journal of the Royal Statistical Society, 26B,* 103–110.

Cox, D. R. (1975). Partial likelihood. *Biometrika, 62,* 269–276.

Cox, D. R., & Oakes, D. (1984). *Analysis of survival data.* New York: Chapman & Hall.

Crowley, J., & Hu, M. (1977). Covariance analysis of heart transplant data. *Journal of the American Statistical Association, 72,* 27–36.

Greenhouse, J. B., Kupfer, D. J., Frank, E., Jarrett, D. B., & Rejman, K. A. (1987). Analysis of time to stabilization in the treatment of depression: Biological and clinical correlates. *Journal of Affective Disorders, 13,* 259–266.

Hart, S. D., Kropp, P. R., & Hare, S. D. (1988). Performance of male psychopaths following conditional release from prison. *Journal of Consulting and Clinical Psychology, 56,* 227–232.

Hser, Y., Anglin, M. D., & Liu, Y. (1991). A survival analysis of gender and ethnic differences in responsiveness to methadone maintenance treatment. *International Journal of the Addictions, 25,* 1295–1315.

Kalbfleisch, J. D., & Prentice, R. L. (1980). *The statistical analysis of failure time data.* New York: Wiley.

Kaplan, E. L., & Meier, P. (1958). Nonparametric estimation from incomplete observations. *Journal of the American Statistical Association, 53,* 457–481.

Kleinbaum, D. G. (1996). *Survival analysis.* New York: Springer-Verlag.

Lee, E. (1992). *Statistical methods for survival data analysis.* New York: Wiley.

Luke, D. A., (1993). Charting the process of change: A primer on survival analysis. *American Journal of Community Psychology, 21,* 203–246.

Mantel, N. (1966). Evaluation of survival data and two new rank order statistics arising in its consideration. *Cancer Chemotherapy Reports, 50,* 163–170.

McCullagh, P., & Nelder, J. A. (1989). *Generalized linear models.* New York: Chapman & Hall.

Peterson, T. (1991). Time-aggregation bias in continuous-time hazard-rate models. In P. V. Marsden (Ed.), *Sociological methodology* (pp. 263–290). Washington, DC: American Sociological Association.

Peto, R., Pike, M. C., Armitage, P., Breslow, N. E., Cox, D. R., Howard, S. V., Mantel, N., McPherson, K., Peto, J., & Smith, P. G. (1976). Design and analysis of randomized clinical trials requiring prolonged observation of each patient: I. Introduction and design. *British Journal of Cancer, 34,* 585–612.

Peto, R., Pike, M. C., Armitage, P., Breslow, N. E., Cox, D. R., Howard, S. V., Mantel, N., McPherson, K., Peto, J., & Smith, P. G. (1977). Design and analysis of randomized clinical trials requiring prolonged observation of each patient: II. Analysis and examples. *British Journal of Cancer, 34,* 585–612.

Singer, J. D., & Willett, J. B. (1993). It's about time: Using discrete-time survival analysis to study duration and the timing of events. *Journal of Educational Statistics, 18,* 155–195.

SPSS. (1994). *SPSS advanced statistics.* Chicago: Author.

Stablein, D. M., Carter, W. H., & Novak, J. (1981). Analysis of survival data with nonproportional hazard functions. *Controlled Clinical Trials, 2,* 149–159.

Stablein, D. M., Carter, W. H., & Wampler, G. L. (1980). Survival analysis of drug combinations using a hazards model with time-dependent covariates. *Biometrics, 36,* 537–546.

Tarone, R. E., & Ware, J., (1977). On distribution-free tests for equality of survival distributions. *Biometrika, 64,* 156–160.

Thompson, B. (1995). Stepwise regression and stepwise discriminant analysis need not apply here: A guidelines editorial. *Educational and Psychological Measurement, 55,* 525–534.

Tuma, N. B. (1982). Nonparametric and partially parametric approaches to event history analysis. In S. Leinhardt (Ed.), *Sociological methodology* (pp. 1–60). San Francisco: Jossey-Bass.

Willett, J. B., & Singer, J. D. (1991). How long did it take? Using survival analysis in educational and psychological research. In L. M. Collins & J. L. Horn (Eds.), *Best methods for the analysis of change* (pp. 311–343). Washington, DC: American Psychological Association.

Willett, J. B., & Singer, J. D. (1993). Investigating onset, cessation, relapse, and recovery: Why you should, and how you can, use discrete-time survival analysis to examine event occurrence. *Journal of Consulting and Clinical Psychology, 61,* 952–965.

Willett, J. B., & Singer, J. D. (1995). It's deja vu all over again: Using multiple-spell discrete-time survival analysis. *Journal of Educational and Behavioral Statistics, 20,* 41–67.

Wright, R. E. (1995). Logistic regression. In L. P. Grimm & P. Yarnold (Eds.), *Reading and understanding multivariate statistics* (pp. 217–244). Washington, DC: American Psychological Association.

Index

B

E

G

H

J

K

L

M

N

O

Q

R

T

U

V

W

Y

Z

About the Editors

Laurence G. Grimm received his PhD in clinical psychology from the University of Illinois at Champaign–Urbana. He is currently an associate professor of psychology at the University of Illinois at Chicago, where he serves as the director of clinical training. He has published broadly in the area of clinical psychology; he is the author of *Statistical Applications for the Behavioral Sciences* (Wiley) and the coeditor of *Reading and Understanding Multivariate Statistics* (APA).

Paul R. Yarnold received his PhD in academic social psychology from the University of Illinois at Chicago. He is currently a research professor of medicine at Northwestern University Medical School, Division of General Internal Medicine. He is also an adjunct professor of psychology at the University of Illinois at Chicago. He is a Fellow of the Society of Behavioral Medicine and the American Psychological Association's Divisions of Measurement, Evaluation, and Statistics and of Health Psychology. He serves on the editorial boards of *Perceptual and Motor Skills* and *Educational and Psychological Measurement* and has authored over 120 articles in the areas of medicine, psychology, and statistics. He is the coauthor of the statistical software package Optimal Data Analysis and the coeditor of *Reading and Understanding Multivariate Statistics* (APA).